ST/TCD/5

Department of Technical Co-operation for Development
and Economic Commission for Africa

Natural Resources/Water Series No. 18

GROUND WATER IN NORTH AND WEST AFRICA

UNITED NATIONS
New York, 1988

NOTE

Symbols of United Nations documents are composed of capital letters combined with figures. Mention of such a symbol indicates a reference to a United Nations document.

ST/TCD/5

UNITED NATIONS PUBLICATION

Sales No. E.87.II.A.8

04550

ISBN 92-1-104203-8

CONTENTS

CONTENTS (continued)

Maps and figures

CONTENTS (continued)

FOREWORD

The Economic and Social Council, by resolution 675 (XXV) of 2 May 1958, requested the Secretary-General to take appropriate measures for the establishment, within the Secretariat, of a centre to promote co-ordinated efforts for the development of water resources. It also singled out ground-water problems as one of the priority subjects in the development of a programme of studies. Large-scale Ground-water Development, published in 1960, 1/ was the first study prepared in this field by the Water Resources Development Centre (now the Water Resources Branch of the Division of Natural Resources and Energy, Department of Technical Co-operation for Development).

The Advisory Committee on the Application of Science and Technology to Development, in its World Plan of Action, 2/ gave priority to ground-water exploration and development. In fact, in the course of the First and Second United Nations Development Decades, more than 100 projects assisted by the United Nations Development Programme (UNDP) and other United Nations technical co-operation programmes were entirely or partially devoted to ground-water prospecting, assessment or pilot development. (A list of ground-water projects in the Eastern Mediterranean and Western Asia sponsored by UNDP is contained in the annex to the present report.)

While such operational activities were developing, the need for a comprehensive review of the results of the projects and for a dissemination of relevant information became more evident. As a result, the Economic and Social Council, by resolution 1761 B (LIV) of 18 May 1973, requested the Secretary-General to take the necessary measures, within the budgetary limitations, to improve and strengthen the existing United Nations services for the analysis, evaluation and dissemination of world-wide data on natural resources, including water resources.

With respect to ground water, a first comprehensive review of the African continent was published in 1972 and 1973 under the title Ground Water in Africa 3/ as a synthesis of material available in the records and files of the United Nations. The material of the second volume in this series, Ground Water in the Western Hemisphere, 4/ was drawn from country papers which were prepared by hydrogeologists and by ground-water engineers, specialists of the countries concerned. This was also done for the third volume, entitled Ground Water in the Eastern Mediterranean and Western Asia, 5/ for the

---

1/ United Nations publication, Sales No. 60.II.B.3

2/ United Nations publication, Sales No. E.71.II.A.18.

3/ United Nations publication, Sales No. E.71.II.A.16.

4/ United Nations publication, Sales No. E.76.II.A.5.

5/ United Nations publication, Sales No. E.82.II.A.8.

fourth, entitled Ground Water in the Pacific Region, 6/ for the fifth, entitled, Ground Water in Continental Asia, 7/ and for the present volume, the sixth in the series, which is to be followed by a seventh on ground water in central, eastern and southern Africa and an eighth on ground water in Europe. This will complete the presentation of (a) a necessarily brief but full overview of the world's ground-water resources, (b) the state of knowledge about them and their potential, and (c) information about their exploitation and the problems involved.

The present work indicates the progress made since the publication of the first volume on ground water in Africa. A point to note is the large number of African specialists who have taken part in the drafting of the text. There is now hardly a single African country which does not have among its nationals university graduates or engineers specializing in hydrogeology or ground water.

It is to be hoped that this volume, which deals with a number of arid countries, in particular the "Sahelian" countries affected by long periods of severe drought since 1973, will contribute to the development of ground water which is so vital in this part of the world.

The United Nations wishes to thank for their valuable assistance the governmental organizations and the consultants and experts from Africa and other countries who have collaborated in the preparation of this work, in particular the Department of Water and Energy of Mali, the Office of Water Research and Planning of the Department of Water of the Kingdom of Morocco, the Mataria Desert Research Institute (Arab Republic of Egypt), the National Service for the Installation of Water Points of the Republic of Guinea, and the Office of Geological and Mining Research (Orleans, France), as well as A.M. Abdoul, N.B. Ayibotele, I. Barry, R.M. Blamdandi, A. Cavaco, P. Chaperot, Checkh Becaye Gaye, N.C. D'Almeida, E. De Boer, A. Diallo, M.A. Diallo, S.M. Dossou, J . Dubus, M. Faloci, D. Fernandopulle, R. Friedmen, J.A. Hanidu, M. Haupt, W. Iskander, M.T. Jones, L. Kossakowski, J.C. Lachaud, J. Margat, T. Mba Mpondo, L. Moullard, E. Njié, Saad Ali Sabet, O.M. Salem, M. Simonot, W.G. Strupczewski, D.Z. Sua, P.S. Zahiri, E.H. Zander and H. Zebidi.

The simplified hydrogeological map of Africa appended to this volume was kindly supplied by Mr. J. Marget. He is warmly thanked for that. The Division of Natural Resources of the United Nations Economic Commission for Africa (ECA, Addis Ababa) helped with the collection of information on some countries for this publication, for which ECA is jointly responsible with the United Nations Secretariat in New York.

6/ United Nations Publication, Sales No. E.83.II.A.12.

7/ United Nations Publication, Sales No. E.86.II.A.2.

Explanatory notes

The following symbols have been used in the tables throughout the report:

A dash (-) indicates that data are not available or are not separately reported.

A blank indicates that the item is not applicable.

A full stop (.) is used to indicate decimals.

A slash (/) indicates a crop year or financial year, e.g., 1976-77.

Use of a hyphen (-) between dates representing years, e.g., 1975-1978, signifies the full period involved, including the beginning and end years.

Reference to "dollars" ($) indicates United States dollars.

Details and percentages in tables do not necessarily add to totals because of rounding. Some of the data series are not homogeneous; they have been taken from various reviews and publications; the differences or divergencies may be due to typing errors.

The designations employed and the presentation of the material in this publication do not imply the expression of any opinion whatsoever on the part of the Secretariat of the United Nations concerning the legal status of any country, territory, city or area or of its authorities, or concerning the delimitation of its frontiers or boundaries.

The term "country" as used in the text of this report also refers, as appropriate, to territories or areas.

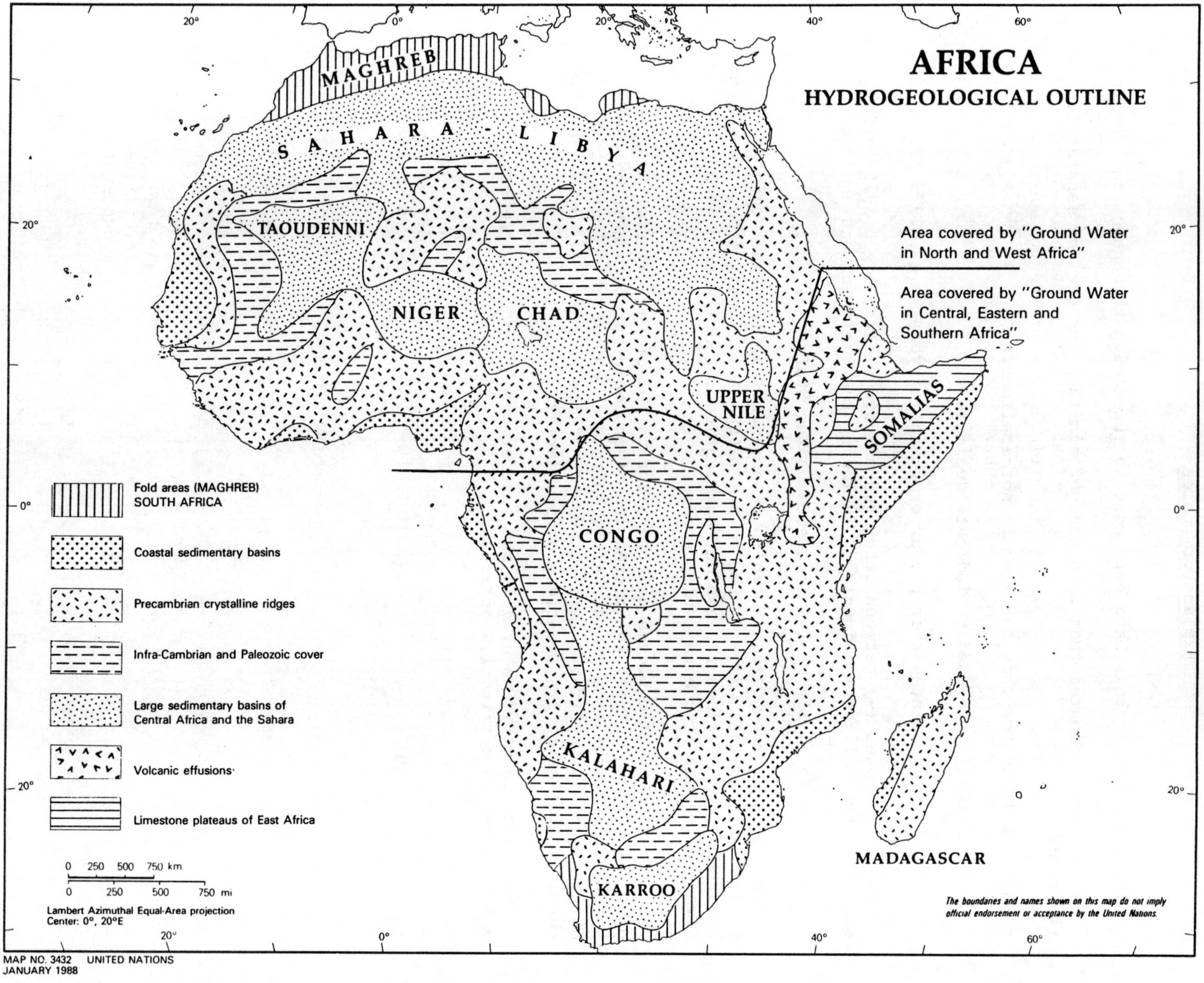
AFRICA
HYDROGEOLOGICAL OUTLINE
Area covered by "Ground Water in North and West Africa"
Area covered by "Ground Water in Central, Eastern and Southern Africa"
MAGHREB
SAHARA - LIBYA
TAOUDENNI
NIGER
CHAD
UPPER NILE
SOMALIAS
CONGO
KALAHARI
KARROO
MADAGASCAR
Fold areas (MAGHREB) SOUTH AFRICA
Coastal sedimentary basins
Precambrian crystalline ridges
Infra-Cambrian and Paleozoic cover
Large sedimentary basins of Central Africa and the Sahara
Volcanic effusions
Limestone plateaus of East Africa
0 250 500 750 km
0 250 500 750 mi
Lambert Azimuthal Equal-Area projection
Center: 0°, 20°E
The boundaries and names shown on this map do not imply official endorsement or acceptance by the United Nations.
20°
0°
40°
60°
MAP NO. 3432 UNITED NATIONS
JANUARY 1988

PART ONE

OVERVIEW

This volume deals with ground water from the standpoint of the physical conditions of the accumulation of this natural resource, the state of knowledge about its potential, its exploitation and the uses to which it is put. It deals with all the African countries north of the Equator, except for Ethiopia and Somalia which belong geographically to East Africa, to be covered in a second volume on all the countries of central-equatorial and southern Africa, including Madagascar and the neighbouring island countries and territories.

I. LARGE AQUIFER SYSTEMS

This vast territory of 17.2 million km^2 with 300 million inhabitants can be subdivided on the basis of geological, morphological and climatic considerations into a number of large aquifer systems in which the ground-water resources can be reasonably well distinguished from the standpoint of their accumulation, their fossil or renewable state and their accessibility.

i) To the north-west, the mountains and plateaus of the Atlas and the Rif and the Mediterranean and Atlantic coasts in the north and west. This is the "Maghreb" of Morocco, Algeria and Tunisia, north of the Sahara. From the geological standpoint these are mainly sedimentary formations strongly affected by Alpine orogeny. The region has contrasting landscapes: it has different climates ranging from the Mediterranean or subhumid type to the semi-arid or even arid type: Moroccan plains north of the Atlas, Algerian high plateaus; here the ground water is intensely exploited to an average extent of 80 or 90 per cent of the renewable resources or even higher in some places, especially in the semi-arid and arid areas.

ii) In the north-east, the Mediterranean fringe constitutes a kind of extension of the Atlas but much more modest in its relief, extent and altitude. The mountains receive quantities of rain which can recharge the neighbouring aquifers, but the renewable resources are small and generally overexploited.

iii) To the south of these areas lies the Saharan region and the deserts which form its eastern extension - the Libyan and Nubian deserts; this is an enormous, generally flat, monotonous territory where the rainfall is infrequent, irregular and very meagre, except over some mountainous areas. It is made up of sedimentary basins mainly of continental origin but with some lagoonal and marine basins in which the beds generally lie in regular horizontal or subhorizontal strata. Two sandstone formations constitute large aquifers of the fossil and Mesozoic types: the "intercalated continental" in Algeria and Libya and the "Nubian sandstones" in Libya-Egypt-Sudan. To the west (western Algeria and Mauritania), the formations are of hard Paleozoic rocks with low permeability in which the ground-water resources are much smaller, except locally.

iv) The crystaline Precambrian basement rock underlying these sedimentary basins emerges in great masses at the west-east axis of the Sahara: Tiris-Adrar, Yetti-Karet (Mauritania-Algeria). The Hoggar mountains which rise to almost 3,000 m in volcanic peaks (Algeria) flanked in the south by the ranges of Adrar des Iforas (Mali) and Aïr (Niger), Tibesti (Chad), Ouaddai (Chad), Darfour (Sudan) and the majority of the territory to the east of the Nile as far as the Red Sea. As far as ground water is concerned, this is a mainly barren region, with the exception of a number of alluvial deposits at the foot of the mountains. When present - a fairly rare occurrence - the ground water is far from abundant and in many cases heavily mineralized.

v) To the south of this ancient backbone the general situation of the aquifers is fairly similar to the one found in the north; here too there are enormous sedimentary basins subject to a desert climate, with hard and unproductive Paleozoic strata in the west (Tagant, Mauritania), except for a number of limestone layers, and with sandstone strata in the east: intercalated continental (Azaouad in Mali), the Aïr sandstone in Niger and the Nubian sandstone further east, which contain fossil aquifers.

vi) The Chad basin, occupied in its centre by the eponymous lake which is shallow and has declined in size over the last decade, is formed by a complex of sediments of various ages, mainly recent, Quaternary and Cenozoic, in which the ground-water resources are considerable: in places artesian, but with relatively low unit yields per well, for the clay strata are frequent and extensive.

vii) The basement-rock areas of West Africa which cover the majority of the territory of Guinea, Sierra Leone, Liberia, Côte d'Ivoire, Ghana, Togo, Benin, Burkina Faso and Cameroon, as well as large areas of Mali and Nigeria. The rocks are exposed to Sudano-Sahelian climatic conditions and are water-bearing in their altered and fractured parts. At its northern edge the crystalline shield of western Africa is flanked by a sandstone rim of Precambrian or Paleozoic age which constitutes a major aquifer in Mali and Burkina Faso. The unit yields obtainable from the wells or boreholes are not large except in a few cases (Bobo Dioulasso sandstone) but they are usually sufficient for village and livestock needs. The sedimentary basins in the central part of Niger, along the axis of which run the River Niger and its main tributary the Benoue, which has its source in Cameroon, are made up mainly of gray argillaceous Cretaceous formations containing artesian aquifers.

viii) The coastal sedimentary basins are very different in extent, the largest being the Senegalese-Mauritanian basin which runs southwards into Guinea-Bissau. Then come the bevel-shaped coastal basin of Nigeria which narrows towards the east (Cameroon) and towards the west (Benin, Togo, Ghana), and the very narrow but economically important coastal basins along the shoreline of Guinea, Sierra Leone, Liberia, Côte d'Ivoire and Ghana. These basins contain recent, Quaternary and Cenozoic sediments with very productive sandstone and limestone layers. They are intensely exploited - sometimes overexploited.

II. CLIMATIC CONDITIONS: EFFECTS ON THE RECHARGE OF THE AQUIFERS

The territory is subject to very varied climatic conditions in which latitude plays an essential role. From the anticyclone of the Azores, a high-pressure centre, the trade winds blow towards the Equator and are deflected westwards by the rotation of the earth. In January a cool dry wind - the Harmattan - blows from the Sahara towards the west coast of Africa from Mauritania to the Niger delta. At this period the whole of Africa south of the equator is subject to a low-pressure system (below 700 mm). In July, in contrast, a high-pressure system prevails over southern Africa and a cyclonic depression is centred over the plateaus of Iran. As a result, the winds tend to blow towards the east and a monsoon from the south-west brings heavy rains to the western coast.

As a general rule the winds blow from the sea to the land, bringing rain; but there are notable exceptions: the Harmattan and in mid-year some local winds from the Maghreb which blow towards the Mediterranean, and some regular winds which blow towards the north-east of Africa in the direction of the Arabian Peninsula. The mountains halt the wet winds.

In January, the regions to the south of the 20th parallel N (from Nouakchott to Port Sudan) have average temperatures below 20°C. In July, the whole of the continent north of the equator (except for the coastal zones) has temperatures above 30°C, sometimes 32°C.

The temperature ranges are very small in the equatorial regions (1°C) but increase in step with distance from the equator; they are from 20°C to 30°C in the Sahara.

The rainfall is irregular with wide variations from season to season and year to year.

In the extreme north of the continent the Maghreb and certain coastal parts of Libya and Egypt and, in the extreme south, the Cape region have rainfall of the Mediterranean type (winter rains).

The very wet equatorial regions to the south of 10° latitude N have two rainy seasons when the sun is high above the horizon, generally from March to June and from September to November. From the 10th to the 15th parallel N the tropical regions have only one rainy season, from May to October. Lastly, the subtropical desert region, i.e. the whole of the north of the continent with the exception of the Mediterranean zone, receives only occasional and irregular showers.

The annual rainfall is two to six metres along the coast of West Africa from Conakry to Abidjan and from the Niger delta to Libreville in Gabon; one to two metres in some mountainous regions of the Maghreb and south of the line from Dakar to Mogadishu; 500 to 1,000 mm in the High Atlas, in the coastal regions of Algeria and Tunisia and in a strip 300 to 500 km wide to the north of the line mentioned above; less than one metre to the north of the line from Nouakchott to Port Sudan, with the exception of the Maghreb, the majority of this region receiving less than 20 mm.

Climatic zones

The climatic zones, characterized by very different vegetation types are as follows:

- Mediterranean zone with dry summers (hot season) northern Maghreb.

- Steppe zone with the following subdivisions:

Pre-Saharan regions south of the Maghreb with drier summers. This climate is sometimes described as "semi-arid Mediterranean". The rainfall is less abundant and the temperature range broader than in the Mediterranean zone.

Regions to the south of the Sahara with semi-arid tropical climate of the Senegalese or Sahelian type. They receive more abundant rainfall in the hot season from June to September.

- Wet savannah zone or zone of tropical Sudanese climate. The wet season grows longer the closer to the equator, but in some places the uninterrupted dry season can last from four to five months. The belt of wet savannah is 500 km wide on average.

- Desert zone with Saharan climate (Sahara).

- Equatorial forest zone with very wet climate and two rainy seasons or continual rain. It includes, over a width of 300 km, the region of the Gulf of Guinea from Freetown to Accra and from Lagos to Douala, southern Cameroon and the Congo basin as far as the rift valleys.

- Coastal fringe zone, a narrow coastal strip in which the climate is heavily influenced by the sometimes very powerful coastal currents. The Canaries current, flowing north to south from Tangiers to approximately the 20th parallel N, is cold; the Guinea current, flowing west to east from Dakar to the equator, is warm.

Aridity and evaporation

The climatic zones can also be classified according to the index of humidity or aridity (Thornthwaite), which takes into account both the temperature and rainfall and its distribution and which expresses a characteristic ratio between potential evapotranspiration and the amount of rainfall.

The surface aquifers (lakes) undergo large variations in level owing to the imbalance in some years between the headings "evaporation" and "recharge". This is particularly true of Lake Chad. It is also true of the unconfined ground-water aquifers when the piezometric surface is shallow in comparison with the soil. Evaporation produces - and can be measured by - concentrations of salts in the aquifers. The question of the depth to which evaporation takes place is disputed. However, all authors agree that this effect operates for several metres (five metres on average and as deep as eight to ten metres). Some authors speak of much greater depths.

Conclusion

The amount of rainfall available to recharge the ground-water aquifers depends on three main climatic factors: the annual rainfall, its distribution in time or the "heaviness of the precipitation", and the value of the potential evapotranspiration, which is essentially a function of latitude, altitude and temperature.

In some cases each of these three factors singly can have a decisive influence.

In all the regions in which the rainfall exceeds roughly 1 to 1.2 metres a year, neither the heaviness of the rain nor the evapotranspiration value should be taken into account, for a large part of the rainfall is almost always available for infiltration, in some places after runoff. In this case the decisive factor is the amount of the rainfall.

In the case of rainfall below 250 mm, it is the heaviness of the precipitation which is important. It is interesting to note that in conditions of increasing aridity - decline in rainfall accompanied by an increase in the evaporation potential - the heaviness of the showers increases to the point where most of the annual precipitation sometimes falls in a few hours. Accordingly, in the Sahara some daily figures can produce a surplus - which can persist over several days - of rainfall over potential evapotranspiration; this gives the water time to infiltrate and thus recharge local aquifers in particular cases.

The following table compares some potential evapotranspiration values with the rainfall at a number of climatological stations in Africa.

	Annual rainfall (cm)	Potential evapotranspiration (cm)	Quotient (percentages)
Arid and hyper-arid zone (rainfall below 250 mm):			
In Salah (Sahara)	0.5	140	0.3
Biskra (southern Algeria)	18	133	
Moudjeria (Mauritania)	17	187	3
Coastal regions			
Nouadhibou (Mauritania)	4	116	4
Tarfaya (Morocco)	11	85	13
Zone with rainfall between 250 and 1,000 mm:			
Kayes (Mali)	74	187	30
Algiers: wet Mediterranean climate	76	92	83

Thus in some regions a large or even overwhelming part of the rainfall is almost immediately lost through evaporation. The heading "evapotranspiration" in the water balances is often the largest. Some authors offer the following figures for the various regions of Africa: evapotranspiration, 40 to 98 per cent; infiltration, 2 to 40 per cent; runoff, 2 to 12 per cent.

In regions with rainfall between 250 mm and one metre (steppe and dry savannah) the potential evapotranspiration is the decisive factor, for the rainfall is spread out better in time. During the rainy season, which can vary widely from one year to the next, the potential evapotranspiration can still have a large value. However, a very variable remainder is almost always available for runoff and infiltration. In contrast, during the dry season which can last from three to six months, some regions of Africa have climatic conditions of the semi-arid or arid type while receiving more annual rainfall than some countries in the wet temperate zone of Europe. During the dry season, the evaporation effect can be considerable in surface and shallow aquifers.

Some recent studies on the treatment of rainfall data in the Sahelian countries tend to show that the rainfall falls into two distinct categories:

- A "monsoon" system with moderate rainfall fairly well distributed in time and little variation from year to year. To some extent this rainfall can help to maintain the vegetation but most of the water evaporates after having soaked the supper layers of the soil.

- A system of very heavy, brief and frequent shows which produces large amounts of surface runoff and deep infiltration. This type of rainfall is essential for the renewal of surface- and ground-water resources. It is the decline in the heaviness or frequency of these showers which causes "drought" when the main consequence is a drop in the level of water in the wells, which can even dry up completely. Lastly, a drop of 50 per cent in the amount of total annual rainfall as a result of less frequent showers can mean so surface runoff or recharging of the aquifers.

III. PRODUCTIVITY OF THE AQUIFERS

The values given below are by way of example. Additional data will be found in the country papers (see Part Two).

Extensive sand formations

In Africa sand dunes cover large areas north of the 14th parallel. Little is known about their role as aquifers in the Sahara. But it is known that the sands themselves, despite their great permeability, cannot provide a large reservoir in many cases since they quickly lose, through runoff or evaporation, the rainwater which they absorb.

Country	Location	Geology	Flow rate per installation
Mauritania	Plain of Kaffa	Sand dunes	5 to 10
	Plain of Assaba	Sand dunes	5 to 10
Senegal	Malika	Sand with clay	26
Cape Verde	Tiaroye	Sand dunes	50

Alluvial fill, deltas, chott deposits, Quaternary formations of the Chad basin and coastal sedimentary basins

Country	Location	Geology	Flow rate per installation (m^3/h) a/	Drawdown (m)
	Fluvial alluviums. These aquifers are among the most important and serve large populations.			
Algeria	Wadi Biskra	Sands-gravels	-	-
Morocco	Doukkala	Sands-gravels	10 to 1,000 m^3/day	-
	Tafilalet	Gravelly alluviums	-	-
	Sous	Gravelly alluviums	Up to 360	1
Mauritania	Wadi Seguelil	Gravelly alluviums	10	
Egypt	Nile	Coarse Pliocene-Pleistocene gravels	1000	3
	Coastal or continental alluviums			
Côte d'Ivoire	Treichville lagoon	Coarse sands	210	-
Guinea	River Nunez	Alluviums	20 to 50 (subartesian boreholes: 7)	-
Togo	Coastal zones	Argillaceous sands	3 to 5	-
Cameroon	Flats	Fill formations	10 to 80	-
	Coastal sedimentary basins			
Cote d'Ivoire	Abidjan	Paleocretaceous sands and limestones	18	80
Benin-Togo	Coastal region	Cretaceous sands	1 to 35 m^3/h/m; (average: 8 to 15)	-
Togo	Afagnagan	Cretaceous sands	18	10
Libya	Syrte	Miocene limestones and sands	25	42
Morocco	Agadir	Pliocene limestones and sandstones	5 to 20 m^3/h/m	-
	Plains of Doukkala and Berrechid	-	10 to 100	-
Senegal	Basin (total)	Maestrichtian sandstone	15 to 120 (artesian)	-
Tunisia	Zarzis-Djerba	Upper Miocene	50 (artesian)	-

a/ In the column "Flow rate per installation", the underlined values indicate a specific yield.

Terminal continental sandstones and conglomerates (Late or Post-Cretacean

Terminal continental

Burkina Faso	Bobo Dioulasso	-	90
Mali	Gondo	-	50 to 100 (up to 300/day)
Mauritania	Trarza	Sandy intercalations	1 to 4
	Bennichab	Sandy intercalations	30
	Nouakchott	Sandy intercalations	15
Senegal	Casamance	Argillaceous sand-sandstone	6 $m^3/h/m$
Togo	Lomé-Agouévé	Variegated sandstone	5 to 40 $m^3/h/m$

Intercalated continental, Nubian sandstone and other continental Precretaceous or Cretaceous continental sandstone

Algeria	Chardaia	Intercalated continental argillaceous sandstone	variable
Cameroon	Bénoué-Garoua	Cretaceous sandstone	10 to 20, up to 50
Nigeria	Sokoto	Consolidated Eocene sands	heavy flows (variable)
Egypt	Casis de Kharga	-	3,000 to 4,000 per day (artesian)

Limestone tableland of the hammadas of northern Africa (Pliocene-Pleistocene)

The hammadas cover vast areas south of the Atlas; their surface is generally made up of a subhorizontal plate of hard Pliocene-Villafranchian lacustrine limestones with varying degrees of sandiness, often overlying softer sand-clay formations. The scant rainfall which infiltrates in the hammadas quickly circulates through a karstic system, flows towards peripheral or central depressions and is rapidly lost through evaporation. The few wells found in the hammadas are fed from dune or alluvial formations. Water-drilling operations have generally not produced positive results.

Karstified limestone aquifers of the Jurassic, Cretaceous (North African Cenomanian-Turonian plate) and Eocene periods

Country	Location	Geology	Flow rate per installation (m^3/h)	Drawdown (m)
Algeria-Morocco	High plateaus	Jurassic limestones	150 (artesian)	
Morocco	-	Liassic limestones	Up to 500 (artesian)	
	Doukkala	Upper-Jurassic marly limestones	10 to 100	
	Bahira	Dolomitic limestones	150 to 200	
	Sous	Cenomanian-Turonian limestones	Up to 1,200	
		Cretaceous sandy-marly limestones	1 to 10	
Mauritania	Trarza	Eocene limestones	0.1 to 1	
Senegal	Pout-Ndiass	Paleocene limestones	Up to 4	
Tunisia	Djebel Zaghouan	Liassic limestones	2,000 (in 6 springs)	

These few examples show that the karstified limestones of North-West Africa can yield rates of flow often in excess of 50 m^3 per hour, sometimes as high as 100 and even several hundred in certain cases.

Tectonized zones of northern Africa with complex structures of marl-sandstone, marl-limestone, flysch, etc., of the Jurassic and Cretaceous periods.

The ground-water resources are very local; they are found mainly in fractured zones with thin limestone or sandstone seams alternating with schists, marl-limestones, clays, etc. The available yields are very variable.

Dolomitic-limestone massifs and plateaus of the Upper Precambrian and Cambrian periods

The dolomitic-limestone sedimentary formation (of Upper Precambrian and Cambrian age) is often very thick and constitutes a major ground-water reservoir:

Dolomitic limestones of Tin Hrassan (Burkina Faso) in arid zones. Transmissivity: 5×10^{-4} m^2/s. Flow: 4 m^3/h, with a downdraw of 10 m. Storage coefficient: 1.8×10^{-3};

Fissured dolomitic limestones of Atar (Mauritania). Flow: 70 m^3/h, with a downdraw of 4 m. Such a flow is exceptional for an arid area; it is produced by a river bed infiltrated by flood waters;

Precambrian and Cambrian limestones of the Anti Atlas (Morocco). A number of overflow springs have flows of 20 to 40 m^3/h and up to 250 m^3/h;

Mention must also be made of the dolomites of Tiara (Burkina Faso) and Gondo with its karstic sink-holes (Mali), for which no figures are available.

Precambrian and Paleozoic hard sandstones, schist-sandstones and quartzites

Country	Location	Geology	Flow rate per installation (m^3/h)
Mauritania	Hodh	Cambrian pelitic sandstones	Up to 0.2 to 0.5
		Brazer sandstone	2 (maximum)
	Ayoun el Atrous	Precambrian sand-stones	0.2 to 0.3
Togo	Bombouaka	Sandstone	0.3 to 7
	Dapango	Sandstone	3 to 7 (maximum)
Togo-Benin	-	Atakora quartzites	2 to 3, up to 7

Schists (mainly Precambrian and Paleozoic) and clays

When they are not totally impermeable these formations do contain some meagre water resources, mainly in fracture zones. Some examples of available yields per installation are given below:

Country	Location	Geology	Flow rate per installation (m^3/h)
Ghana	-	Volta schists	Very low
Guinea	-	Black Gothlandian slates	Very low in fractures and seams
Burkina Faso	Banfora	Schist-sandstone	12 (exceptional)
Mali	Nara	Cambrian schists	Very low
	Azaoud-Timbuktu	Metamorphized Pre-Cambrian schists	0.5
Mauritania	Atar	Schists under alluvium	20
Togo	Sansanne-Mango	Schists	0.3 to 1 per day
	Buen	Marly-sandy schists	0.5 to 10 per day

Examples of available yields per well and borehole in crystalline zones

Country	Location	Geology	Flow rate per installation (m^3/h)	Drawdown (m)
Cote d'Ivoire	Yamoussoukro	Fractured granites	6	-
	Daloa	Granitogneiss	2 to 4 (up to 12)	12
Benin	Parakou	Fractured granites in tectonic depressions	7 to 8	-
Benin-Togo	-	Birrimian schists, quartz seams	3 to 7 (exceptional)	-
Ghana	-	Granites and granodiorites with quartz seams	5 to 20 m^3/day	-
Burkina Faso	Various	Mica schists	less than 1 m^3/day	-
	Various	Granitogneiss	1 to 4	10 to 20
Mauritania	Fort Detrick	Mica schists and gneiss with pegmatitic seams	20 m^3/day	-
	South-east	Diorites	0.5	-
Chad	Ouaddai	Granitic sands	2	-
Togo	Elavagnon	Mica schists and graniteogneiss	2 to 5	6 to 20
	Kande	Chlorite schists quartz seams	7 to 12	9 to 15
	Dapango	Alkaline granite-gneiss	1 to 5 m^3/day	-
	Palime	Granites and granodiorites with amphibolites and quartz	5 to 20 m^3/day	-

In summary, a flow rate of 5 m^3/h is a good one for granites and granitogneiss; a rate of 1 m^3/h is considerable for mica schists and metamorphic schists. Better yields are obtained in the quartzy zones.

Crystalline and metamorphic rocks (basement formations, granites and gneiss

Since they have virtually no porosity, the crystalline rocks are impermeable except in faulty, fractured or altered zones. The best yields are usually obtained when a relatively thick altered stratum overlays a fault zone.

The nature and structure of the altered stratum vary according to the parent rock. This stratum can be almost entirely argillaceous and therefore barren.

Volcanic rocks

Lavas, especially basalts, dolerites and certain basal rocks which sometimes give high yields can be put in a seperate category; a few examples are given below:

Fissured dolerites in arid zones - Ayoun el Atrous (Mauritania): less than 0.1 m^3/h; non-fractured: 0.2 to 0.3 m^3/h;

Basal rocks of Akjoujt (Mauritania): 30 to 45 m^3/h, with a drawdown of 13 m;

Basal rocks of Conakry (Guinea): 13 to 72 m^3/h (very rainy tropical climate), with a drawdown of 20 to 50 m;

Green rocks of Kongolikan (Burkina Faso), fractured: 3 m^3/h.

Conclusion

There is almost nowhere in Africa where ground water is not found at one depth or another. The highest flows are provided by clay-free alluviums, continental or marine Cretaceous sandstones and karstic limestones.

Most of the ground water is acceptable for human consumption and therefore for livestock as well.

In arid zones the ground water is usually of calcium/magnesium bicarbonate facies at the higher level, i.e. near the regions where the surface runoff infiltrates. It then acquires a higher sulphate content and finally increased amounts of chlorine and sodium at the end of the course in the regions where the evaporation effect is high and operates directly on shallow aquifers. This is particularly the case in pre-Saharan North Africa for the sabkhas (continental depressions).

Some geological formations, especially of Permian-Triassic or Cretaceous age and lagoonal origin, contain mineral salts which pass in solution into the ground water. This is particularly the case in North Africa.

In the coastal sedimentary basins, often made up of permeable formations, pumping causes sea-water intrusion which tends to contaminate the fresh water aquifers.

In the Precambrian basement rock in tropical rain country the water is usually not very mineralized or aggressive.

Mineral-water and thermo-mineral springs abound in the African continent in the fracture zones. They constitute a major potential resource which has been explored and exploited in only a few places.

IV. EXPLOITATION OF THE GROUND WATER

In Africa as in the rest of the ancient world ancient, densely populated civilizations with advanced social organization and a sophisticated way of life were associated with the big rivers. These rivers furnished abundant water, rich soil, and fish and game in their valleys and deltas, as well as means of transport and places of refuge.

Away from the big rivers the surface-water resources are scanty especially in the dry season. In tropical Africa they are limited to waterholes. In the northern Africa the ancient inhabitants, the Berbers, usually established themselves in the mountainous regions near the sources of permanent rivers. It was the Arabs, the occupiers of the plains and Saharan oases, who developed the use of ground water through the construction of wells and infiltration galleries, employing the original techniques of Central Asia and the Middle East. Various methods of dewatering were also imported from those regions.

However, until recent times and with the exception of Arabized Africa, ground water was drawn off only from shallow holes dug in alluvial beds devoid of surface water in the dry season. These crude wells are in general use in the pre-Saharan regions. They are rarely more than a metre deep and provide temporary water points still frequented by nomads; they usually last only a short time, for flood water in any amount destroys them.

Traditional wells and drains

The digging of wells and construction of traditional drains - underground galleries linking aligned wells - was practised mainly in arid countries under Arab or Turkish influence in northern Africa, including the oases of the Sahara, Libya and Nubia and some of the Southern fringes of these deserts.

The wells were excavated with simple digging tools in soft earth of good consistency. Sometimes the walls or vaults were reinforced in places with timbering or brickwork, either dry-stone or with lime mortar. Some of these wells, especially in arid piedmont areas, attain considerable depths, sometimes 100 metres and more.

The well systems described in the Bible and very numerous in Iran, where they are called kanats, are widespread in northern Africa where the total length of the galleries amounts to several thousand kilometres. In Egypt and the Sahara they are called foggaras; in Morocco, rhettaras. This system makes it possible to obtain the ground water from the soil without using dewatering methods. The galleries are first built as trenches which climb underground until they intersect over a certain length the saturated formations to be drained. The length of the galleries is limited by the maximum depth of the "head well", which depends on the techniques used and the nature of the terrain.

These drains can only be built in formations of suitable consistency where the digging is easy: lacustrine formations, soft sandstone, tuff, consolidated alluvium, etc. The aquifer must also be relatively shallow and lie under land which slopes sufficiently for the galleries to discharge in the open air; but the slope must not be too steep, for the head wells must be of a reasonable depth. These foggaras are found in the beds of certain wadis and their environs: middle or adjacent beds on the flanks of gently sloping valleys and at the foot of dejection cones spreading from piedmonts. Some drains penetrate rock formations and reach aquifers whose flow is blocked downstream by natural obstacles.

The construction, cleaning and maintenance of these drains - arduous and dangerous work - is now very difficult. Many of the installations are deteriorating and collapsing for lack of maintenance. In small aquifers with irregular recharge the drains can cause a permanent discharge - often unused - which quickly leads to total depletion: this is particularly the case in the plain of Haouz (Marrakesh) and the plain of Sous (Morocco).

The traditional means of raising the water from the wells vary according to region, raw materials, depths and uses. For shallow irrigation wells (norias) bucket wheels operated by animal traction are widely used. For greater depths a simpler procedure is often employed; it involves a treadmill worked by an animal (cow or camel) which hauls up a leather water-bucket by means of a system of ropes and pulleys. This method raises hardly more than a few cubic metres a day.

The deepest wells are drawn by hand, for they are used only to supply the population and livestock. Beam wells are a traditional feature of the landscape in the Nile Valley. They are also found in Sudan and in all the sub-Saharan countries from Chad to Mauritania.

Wells drilled and dug by modern methods

In the deserts the discovery of ground water by deep drilling is essential for oil exploration works, especially for the mixing of drilling mud and the raising of oil by injecting water under pressure. The general geological studies and the geophysical studies carried out for this purpose have led to the identification of deep confined aquifers which have then been exploited by means of artesian boreholes. Thus, even before the proclamation of their independence the African territories under British administration benefitted from the experience acquired in oil exploration in the Middle East during the second quarter of the twentieth century and from the progress made in the same period by British and Swedish manufacturers of drilling equipment for the exploitation of ground water. In French Africa drilling for water also underwent a great expansion, especially from the time when oil exploration activities were started in the Sahara, i.e. from the 1950s.

The ground water was first exploited by borehole in the arid zones of northern Africa: Algeria, Morocco, Tunisia, Egypt and northern Nigeria where there are vast stocks of ground water. This ground water sometimes has natural outlets in topographically low-lying areas such as the chotts, where it is subject to direct evaporation; in other cases these depressions offer favourable conditions for the drilling of artesian wells. Artesian wells have been dug in chotts in Tunisia from the end of the 19th century using big augers operated by groups of workers. Mention must also be made of the many artesian bore-holes drilled in the 1940s and 1950s in the New Valley, i.e. in the depressions of El Kharga and El Dakhla in Egypt's Western Desert.

Many small boreholes have also been drilled in all the countries of the semi-arid or arid zone in order to supply from shallow aquifers the administrative or economic urban and rural centres and modern agricultural enterprises. These works were first carried out in northern Africa; they were then extended to the wetter areas and as far as the equator, for the wet tropical countries also need ground water to supply their towns and villages.

The number of water-drilling rigs in Africa has increased rapidly over the past decade, especially in the arid countries. These rigs are used by a number of African and foreign companies and by State services such as departments of water development, or equipment, etc.

In most cases the boreholes are not equipped with motorized pumps. In rural areas many types of hand-operated or animal-traction pumps have been tried out. Some of these pumps are particularly simple and tough, for example the India Mark II developed with the help of UNICEF, which is now manufactured in Africa, in particular in Mali.

In addition to drilled wells, there are many wells dug by hand on the initiative of the administration in areas where they could not be constructed by the methods traditionally used by the local people (shovels and picks). In areas of hard rock, particularly Paleozoic schists and sandstones, compressed air tools and explosives are used to excavate the wells. These operations are usually costly.

In many African countries in the wet tropical zone the formations usually contain very loose clay seams which make it impossible to dig wells by hand, for the walls collapse even before the digger reaches the water-bearing strata underlying the clays. In such cases an appropriate lining must be used; this is always tricky and sometimes expensive or difficult, which means that the wells must be drilled.

The construction of wells is also very difficult in areas of sand-clay sediments where the installation of a prefabricated reinforced-concrete lining is always essential.

The installation of motorized pumps is justified only when the water requirement is large, and account must be taken of economic and social factors, the chemical quality of the water and the height of the lift. The communities or bcdies concerned must also have the technical and financial means to maintain and repair the installations.

During the last 15 years the digging and drilling of wells has undergone spectacular development in the region, partly as a result of the International Drinking Water Supply and Sanitation Decade and partly because of the periods of exceptional drought (1973-1975 and 1983-1985) which affected the arid and Sahelian zones north and south of the Sahara.

Thus, for a Sahelian country in which about 20 wells a year were dug in 1965/66 the number of wells drilled had increased to four or five hundred in 1985.

Ground water is intensively used to supply urban and industrial areas, especially in arid regions and coastal zones. This is particularly true of Tangiers, Fez, Meknes, Marrakesh, Agadir, Constantine, Tripoli, Benghazi, Port Sudan, Ibadan, Cotonou, Lomé, Bobo Dioulasso, Abidjan, Bissau, Banjul, Dakar and Nouakchott.

The exploitation of ground water in Africa is intended mainly to meet the water needs of the towns, villages and pastoral areas and those of industrial and mining enterprises. In contrast, irrigation with ground water is limited either by its cost and the expenditure of convertible currency involved in the purchase of pumps, motors and fuel, or by the exhaustion of the aquifers in arid regions. Apart from the countries of North Africa - from Morocco to Egypt - the areas irrigated by ground water are still very small. However, small market-garden centres have been spontaneously created around the hand pumps installed in villages and this kind of small-scale operation is tending to increase (Mali).

CONCLUSION

The sharp increase in the use of ground water in Africa goes hand in hand with the continent's rapid entry into the modern world. This use is important for all sectors of the economy but was first concentrated in the towns, the mining centres and some priority farming regions. It is now tending to be extended to the small centres in the most isolated tropical and desert regions. A considerable effort to this end is being made both by Governments and by international and bilateral technical co-operation bodies. This sharp increase in the use of ground water is almost always one of the fundamental conditions for economic and social development, for it is an essential factor in the life or survival of many existing centres of population and a fundamental condition for the establishment of new centres.

However the development of ground water is beset with many difficulties. Firstly, the areas with the best aquifers from the standpoint of the capacity of the rocks to absorb, hold and discharge large quantities of water are the desert zones where there is little or no recharge from rainfall and the coastal zones subject to deep intrusion of sea water in the direction of the wells. In contrast, the rainy tropical areas have rocks which are poorly suited to the absorption and storage of water supplied by rainfall and surface runoff.

Furthermore, ground-water prospecting and the drilling and digging of wells are usually difficult and expensive operations owing to the weakness of the infrastructures, the unfavourable natural conditions, the remoteness of the zones to be reached and the wide dispersal of the villages, as well as the lack of equipment, qualified personnel, project-uptake facilities, and investment and maintenance funds.

Lastly, and this is not the least problem, African villagers do not always have the motivation, the basic technical capacity and the material resources required for the satisfactory operation, maintenance and repair of the manual pumps supplied to them. Substantial progress has nevertheless been made in recent years in several fields: training of technical personnel at various levels, including management and decision-making; rational planning of drilling operations; introduction of appropriate technologies for the construction and restoration of wells and for the movement of the water; introduction of relatively cheap and effective methods of prospecting (particularly remote-sensing and geophysical techniques); computerization of data and inventories; manufacture of equipment - especially hand pumps - in Africa itself; grassroots animation and education of villagers and creation of African water-drilling enterprises.

However, much remains to be done to ensure that the ground-water resources of North Africa are managed to best effect, i.e. without wastage or long-term threat to the existence of these resources in terms of both quantity and quality. This comment applies equally to the intensely exploited coastal zones, especially at Nouakchott, Dakar and Lomé.

Nor are the objectives of the International Drinking Water Supply and Sanitation Decade about to be achieved for the villages of the countries of Western and Central Africa south of the Sahara. However, it can be hoped that towards the end of the century the necessary infrastructures - wells and boreholes - and the corresponding elementary superstructures will be in place in all the villages and that the maintenance of the pumps, if not their replacement when they are worn out, will be undertaken mainly by the villagers themselves. The organizations of the United Nations system - as can be seen from the list of projects in the annex - will have contributed to this vast undertaking in a very considerable and in many cases decisive manner.

PART TWO - COUNTRY PAPERS

ALGERIA

Area: 2,381,700 km^2

Population: 20.5 million (United Nations estimate, 1983)

I. GEOGRAPHICAL SUMMARY

Algeria is both a Mediterranean country with the Tell and the steppe in the north covering only 280,000 km^2. and a desert country with the Sahara in the south covering nine-tenths of the territory. Northern Algeria, flanked by the Mediterranean coast, is a region of high land with an average altitude of 900 m made up of four separate zones: the high plateaus, the interior plains, the mountain zones and the coastal zone; from south to north first are found large lagoonal formations in a very dry region (chotts) and zahres), followed by alternating zones of mountains and interior plains, and finally the valleys of the watercourses and a narrow coastal strip running east to west for 1,000 km. The Sahara Desert, of which Algeria occupies about a half, has five types of relief: the crystalline or volcanic mountains, the hammada (a flat plateau at the foot of mountains), the reg (a stony and arid plain), the erg (sand dunes) and the closed basins, sabkhas or dayas.

Climate

There are six main climatic zones; from north to south:

- The coastal zone; the coastal zone itself and a subcoastal zone;
- The Tell Atlas: an east-west band between the coastal zone in the north and the high steppes in the south;
- The high plains of the Tell, mainly in the east of the country, and the high plains of Constantine and Aures;
- The high steppes, a zone of chotts and high plateaus between the Tell Atlas and the Saharan Atlas;
- The Saharan Atlas, the northern edge of the Sahara;
- The Saharan zones.

Temperatures

Apart from the coastal zone which is under the direct moderating influence of the sea, the daily and annual temperature ranges are high, especially in the plains of the Tell Atlas (El Asnam), only 50 km from the sea, where the climage is of the continental type with a harsh winter and a very hot dry summer; this is also the case in the high steppes, but here the altitude keeps the annual average temperature down.

The annual averages at several stations in each of the six climatic zones are given in the following table:

Climatic zone	Cold season average (centigrade)	Hot season average (centigrade	Annual average (centigrade)
Tell Atlas (plains)	14	23	18.5
High plains of the Tell	14	26	20
High steppes	12.5	25.5	19
Saharan Atlas	15.5	28	21.7
Northern Sahara	15	28	21.7
	15	29.5	22.3

The annual ranges, i.e. the differences between the averages of the coldest and the hottest months, also increase steadily from north to south:

	Degrees centigrade
- Coast	14
- Tell Atlas (20°C in the plain of Cheliff)	19
- High plains of the Tell	19
- High steppes	21
- Saharan Atlas	22
- Northern Sahara	24

Rainfall

The average annual rainfall also follows in general terms the north-south zonal system, but with big discrepancies between east and west in each zone, the Oran region (west) being generally less wet than the Constantine region (east). The annual rainfall in Algeria ranges from 2,000 mm in the high mountains bordering the Mediterranean to less than 100 mm in the northern Sahara.

The lowest rainfall is found in the Saharan regions where the notion of an annual average has little meaning, given the very large variations from year to year and the total lack of rain in some years. Thus, the average rainfall over thirty years at Bechar is 91 mm. This represents the arithmetical average of a series of values between 30 and 180 mm.

This phenomenon is even more marked in the Sahara: at Timimoun the annual rainfall ranges between zero and 140 mm.

Average rainfall
(mm/year)

Climatic zone	West	Central	East
Coast	400	700	900
Tell Atlas (plains)	500	450	700
Tell Atlas (mountains)	600	700 - 1,000	800 - 1,600
High plains of the Tell	-	-	400
High steppes	250	250	
Saharan Atlas	200	200	400 - 700
Northern Sahara	50	50	150

The wettest regions are in the northern part of the Tell Atlas, a zone of mountains rising directly from the sea (Atlas of Blida, Djurjura, Petite Kabylie).

Almost everywhere the dry season is from May to September inclusive. It becomes increasingly less distinguishable as the annual rainfall declines. In an average year Algeria receives about 65 billion m^3 of rainfall.

Northern Algeria has a Mediterranean climate, but with continental features in the areas far from the sea and protected from its influence by the coastal mountains. There is almost no rainfall in summer and it is unevenly distributed in time and space during the rest of the year. The eastern Tell, with its fairly high elevations, is very wet. The Tell of Oran, in contrast, is hot and dry. The high plains of Constantine are well watered and surrounded by mountains from which the waters flow towards the plains. The high plains of Algier and Oran have a relatively low and irregular rainfall. The Sahara is extremely arid. It is situated entirely to the south of the 200 mm isohyet. Of the whole of the territory of Algeria, only the central and eastern coastal zones and a few very mountainous areas in the interior receive more than 600 mm of rainfall per year.

Surface water

Algeria's flow systems are determined by the abrupt and fragmented nature of the relief, the narrowness of the Tell band, and the climate; the watercourses are irregular, the commonest type being the wadi, a succession of streams rather than rivers, which are rarely perennial. In winter the flood waters are violent and carry down large quantities of solid material; in summer the majority of the wadis are difficult to regulate.

Dams are expensive and they accumulate a lot of sedimentation. Surface-water flows occur mainly in the northern part of the country over an area of about 300,000 km^2 north of the line Bechar-Laghouat-Biskra. Typical features of these flows are:

- The extreme irregularity of the inflows from season to season and year to year;

- The violence and swiftness of the flood waters;

- The large amount of solid material carried down.

These characteristics make it difficult to exploit the hydrometric network and to interpret the measures taken, so that there are big gaps in the series of observations.

Hydrographic network - made up of 17 drainage basins which can be placed in three categories:

i) The basins flowing into the Mediterranean

These are the basins of the wadis which drain the Tell Atlas and the Sahelian coastal basins. They cover 130,000 km^2 and receive an annual rainfall of between 400 and 1,500 mm.

They deliver to the Mediterranean about 12 billion m^3/year.

From west to east they are: the Tafna, the Macta, the coastal basins of Constantine ,the Kebir Rhumel, the Seybouse and the Medjerda.

ii) The endorheic basins of the high plains

They cover some 100,000 km^2 and receive an average annual rainfall of 300 mm. Much of the surface water of these basins is lost through evaporation in the chotts.

From west to east they are: the basins of Chott Chergui, Chott Zahrez and Chott Hodna and of the high plateaus of Constantine.

Their potential is estimated at 750 hm^3/year.

iii) The Saharan basins

The surface water found in the Sahara comes from the wadis which drain the Saharan slopes and the Great Atlas in Morocco (Chott Melrhir, southern slopes of Mts. Ksour and Guir). The basins cover 100,000 km^2, receive between 100 and 300 mm of rainfall and deliver to the Sahara an annual average of about 700 hm^3 in the form of very irregular floodwaters.

Hydrometric network - The first stations in Algeria's hydrometric network were established in 1924.

This network has been steadily developed:

- In 1950 it had 40 stations operated by four hydrological teams;

- In 1968 immediately before the three-year plan it had 77 stations operated by seven teams.

In 1982 the hydrometric network had 220 stations operated by 25 teams which unfortunately provided somewhat imprecise data for a long series of annual observations.

Surface-water potential - The figures given in the table below are based on analysis of data taken from the hydrometric data bank established by DEMRH (Department of Environmental Studies and Hydraulic Research) and they constitute a stock of information equal to 3,000 station-years.

For the tributary basins of the Mediterranean which are controlled before they reach the sea DEMRH has tried as far as possible to reconstitute the natural discharges by adding to the observed yields the amounts held back by the reservoir dams.

However, low-water river intakes have not been included.

ALGERIA
SURFACE-WATER POTENTIAL

Drainage basin	Sub-basins	Millions of m^3/year	Drainage basin total in millions of m^3/year
Tafna			315
Macta			250
Oran coastal			50
Cheliff	Above Boughzoul	240	
	Below Boughzoul	1,400	1640
Algiers coastal	Mazafran	350	
	Harrach	280	
	Sebaou	980	1610
Isser			495
Soummam			684
Constantine coastal	East Kebir	800	
	West Kebir	330	
	Saf Saf	240	
	Guébli	500	
	Jijel coastal	1,400	3270
Kebir Rhumel			550
Seybouse			496
Medjerda	Medjerda	190	
	Mellegue	80	270
Chott Cheroui			240
Zarhez			60
Chott Hodna			250
High plains of Constantine			200
Chott Melrhir	W. Djedi	100	
	South Aures	180	280
Saharan Atlas			150
Guir			200
TOTAL:			11,010

II. GEOLOGY

Summary geological background

Algeria has two main geological regions. The African tableland of the Sahara and the Mediterranean fold region: the Atlas Mountains. The geology of the Atlas has evolved through the following phases:

In the Paleozoic era, during the period from the Ordovician to the Middle Carboniferous, the land was covered by the sea. Then, especially in the Permian-Triassic era, it passed through alternate continental and lagoonal phases. A process of marine sedimentation began at the beginning of the Jurassic and lasted until the end of the Cretaceous period. The banks of the sea were displaced north-south and south-north, and the depths of the sea declined from north to south. The sedimentation processes in the Cenozoic era underwent many changes.

The main tectonic phases were as follows:

The beginning of the Mesozoic era saw the formation of the marginal trench of the southern Atlas which separates the northern and southern parts; geosynclinal development began in the north, and platform conditions persisted in the south.

During the Mesozoic and Cenozoic eras the Algerian Atlas was subject to the influence of Hercynian (two phases) and Alpine (five phases) tectogenesis.

The last phase of Alpine folding, of Astian age, played an important role in the rejuvenation of the relief and the development of fissuration and karstic phenomena, as well as in the creation of saliferous domes and lodes. These processes have left their imprint on the present geological structure.

Stratigraphy

The Precambrian granitogneiss basement rock outcrops in the western Sahara (Eglab) and central Sahara (Hoggar). It is overlain with Paleozoic sedimentary formations, mainly of marine origin and Cambrian to Carboniferous age, which are characterized by Hercynian orogeny. These formations outcrop: in the west, at the periphery of the Tindouf basin; in the centre, north of Hoggar, in the "Tassili enclosure"; in the south-west, at the eastern edges of the Taoudenit basin; in the north-west, in the basins of the Saoura and the Bechar.

In the most depressed parts of these basins the Paleozoic sediments are overlain by three sedimentary systems: "intercalcated continental", Cretaceous and Tertiary. The intercalcated continental is a complete thick series (up to 1,000 m) of Post-Carboniferous to Precenomanian age. These are detrital formations of sand, sandstone and conglomerate with lenticular intercalations of marl. These facies are similar to those of the nearby Albian continental formations in the Saharan Atlas and on the high plateaus, with a thickness of 50 to 1,000 m. Cretaceous (marine) and Tertiary formations occupy the centre of the basins; they constitute enormous tablelands - the hammadas - and they are often totally covered by the sandy formations of the ergs.

A major accident separates the African tableland of the Sahara from the fold area of the Maghreb (or Mediterranean), which is also called "Berbérie" by geologists. The Berbérie is a geosyncline affected by two orogenic phases in the Eocene and Miocene periods. The geological formations found there are mainly of Jurassic age in the western part of the Atlas and of Cretaceous age in the east. Bordering the Mediterranean small crystalline and Paleozoic massifs are found in the Kabylies on the edge of the geosyncline.

Here the Jurassic formations can be 3,000 to 4,000 m thick. They include deep-sea clay sediments and dolomitic limestones which form thick tablelands and folds in the region of Tlemcen and the Saida mountains.

Three-quarters of the mountains are composed of Cretaceous formations - limestones, schistoid marls, red sandstones, clays; they reach their maximum thickness (3,000 to 6,000 m) in Hodna. The high plateaus are generally covered by continental formations of Cretaceous age.

The Tertiary period (thickness 1,000 to 3,000 m) is represented by phosphated limestones and paleogenic sandstones and marl-schists and by various neogenic formations of lacustrine and continental origin which border the depressions, especially the interior plains (Chéliff, Mitidja).

The tectonics of the Berberie is complex: the fols, drifts and warps are very numerous and closely packed, especially in the Tell Atlas.

III. GROUND WATER

All types of aquifer are found in Algeria by virtue of the diversity of the geological formations and structures.

A trial classification of these aquifers based primarily on geological, lithological and morphological criteria is given below. First it is necessary to describe one of the dominant features of the country's groundwater resources: the great potential which allows intensive exploitation by groups of boreholes is associated with two main types of aquifer which are, in order of importance:

- A number of extensive limestone formations, unconfined or confined (Chott Chergui, Saida platea, Tolga aquifer);
- The great subsided plains covered with a thick layer of alluvial fill which are wellfed both by precipitation and infiltration of water from the wadis which cross them (plains of Algiers, Bel Abbes, Mascara, Maghnia, Annaba and Chéliff).

Apart from these two main types and the big aquifers of the continental formations of the Sahara, the exploitation of ground water is severely limited by several factors:

- The very abrupt relief of northern Algeria;

- The fragmentation and compartmentalization of the reservoirs due to erosion and tectonics;

- The aquifers' lack of depth;

- The low unit yield of the boreholes, which means that many must be drilled in one place, with the associated high costs (this is true of the Pliocene Quaternary fill of the high steppes);

- The risk of salination of the aquifers bordering the sea or in the vicinity of closed depressions (chotts and sabkhas).

Aquifers - trial classification

There are six standard categories, in order of increasing actual potential:

Type 1 - small isolated aquifers drained by weak contact and adjacent springs with their source in superficial alluvial or colluvial deposits, usually perched and eroded, dating from the Pliocene or Quaternary periods.

The discharges of the (isolated farm) wells are low, as is their total potential.

The reserve capacity is nil and the springs usually run dry in the dry season.

Type 2 - Aquifers in relatively thick alluvial plateau formations or in the fill of endorheic depressions in the high plains.

They are usually large in extent and thickness, especially in the high plains. However, with rare exceptions, the unit yields of the wells are low and they do not furnish large volumes at any one point. The water must therefore be exploited by a large number of scattered wells. The total potential is high owing to the large amount of rainfall. All these aquifers are drained either close to the soil or through the hydrographic network (the case in the coastal aquifers), or at deeper levels towards the evaporation surfaces of the closed depressions of the high plains of central Algeria known as chotts.

Type 3 - Narrow alluvial aquifers lying along major watercourses and in direct hydraulic contract with them.

The transmissivity, and therefore the yields of the installations, is high and the reserve stocks low. These aquifers can be intensively exploited, but the exact scale of their connections with the wadi must be known and the reserve capacity of the alluvial mass must be used to the maximum in the case of seasonal agricultural exploitation.

Particularly good examples of this type of aquifer are found in the valleys of Kabylie: Wadis Isser, Sebaou and Soumman.

Type 4 - Extensive limestone or dolomitic massifs with karstic circulation.

These aquifers usually receive heavy infiltration, in the order of 20 per cent of the rainfall, but their exploitation by boreholes is difficult owing to the heterogeneity of the fracturation which produces extremely variable unit yields. Thus they can be directly exploited only in favourable sites, in particular tectonized or subsided zones. Moreover, their unconfined part is often in a zone of abrupt relief and difficult access. The confined part, at the edges of outcrops, is usually worth exploiting. However, this exploitation is limited in practice, for tectonic accidents rapidly produce major displacements and, in view of the depths to be reached, the cost of the boreholes quickly becomes prohibitive.

In most cases this type of aquifer is well drained by powerful springs which can only very rarely be regulated by pumping. However, two aquifers are important exceptions to this general situation:

- The buried karst of the Ziban region west of Biskra which is exploited only in its confined part by artesian boreholes;

- The dolomitic aquifer of Chott Chergui, where all the resources can be exploited in one place in a favourable zone where the aquifer is overlain by a lens of extremely permeable limestone which serves as a drain for the whole of the dolomitic aquifer.

Type 5 - Aquifers based in the alluvial fill of the great subsided plains.

These are by far the most important aquifers in northern Algeria. The plain of Algiers where the sand and gravel fill reaches thicknesses of over 100 m is the best example. In general terms, these aquifers are unconfined for most of their extent. They are usually found in the Tell zone, where the rainfall is heavy, and are in constant hydraulic contact with major wadis.

The unit yields of the boreholes are variable owing to the heterogeneity of the fill. However, some zones can be tapped over a large area. For example, the two exploitation zones of Bas Mazagran should be supplying, through 35 boreholes, close to 2 m^3/s to the city of Algiers by 1990.

The reserve capacity of these aquifers is very high: their large stocks make it possible not only to control the dewatering rates over the year but also to compensate for the irregularity of recharge from year to year.

Type 6 - Very extensive and thick aquifers with extremely high geological reserves in comparison with the exploitation stocks.

This type of aquifer is represented by the "terminal" and "intercalated" continental complexes of the Sahara. In this type of aquifer the only real problem of exploitation is the management of a stock which is very large in comparison with the annual rate of recharge and which must be exploited in a planned manner so as to ensure economically acceptable use at all points. A knowledge of the geometry of the reservoirs and the economic conditions of exploitation is essential for the determination of the various items in the water balance.

Ground-water resources

For the purposes of evaluation of ground-water resources, northern Algeria has been subdivided into 107 hydrogeologically homogeneous regions. These regions are usually drainage basins or sub-basins. Some isolated mountainous regions, such as small coastal basins where agricultural and urban development is insignificant, have been excluded.

It is rare for there to be no correlation between hydrographic and hydrogeological basins. Some notable exceptions are the upper Tafna basin, the aquifer of the dolomitic plateau of Saida, the upper Rhumel and the chott region.

Almost all the big alluvial plains and the extensive limestone massifs have been the subject of full hydrogeological studies. In terms of exploitable volume, this means almost 70 per cent of the total resources of northern Algeria.

For other zones with clearly lower potentials, the fragmentary information collected from isolated drilling and geophysical prospecting is still not sufficient for regional summaries to be made. For these zones and for the sparsely populated and little developed mountainous regions, global estimates have been made for a number of hydrogeological basins on the basis of rainfall-infiltration ratios. This method, whenever it has been possible to verify it by means of complete studies, has given good results in the determination of the total infiltration in a basin. It is based on the following principle: for a clearly delimited hydrological basin, without underground exchanges with neighbouring basins, the total flows measured at the downstream hydrometric station correspond closely to the total floodwater runoff and the base flows, which represent the volumes of water passing through the aquifers at any given moment. In the case of a basin subject to hydraulic exploitation, the figure includes the station's upstream withdrawals whether from surface water or from the aquifers.

In addition to a number of hydrological basins which have been properly studied, in their varied climatic, geological and geomorphological conditions, it is possible to draw experimental series of curves linking, for example, total infiltration (base flows) to rainfall for different permeability values of the basins.

Conversely, given knowledge of the average rainfall in a basin and taking into account the dominant lithological and morphological features, it is possible to estimate the basin's infiltration.

The ground-water potential is equal to the infiltration under the following conditions:

- Perfect hydrogeological shelf at the station (or at least knowledge of the leakage rates);

- Identity of underground and surface hydrological basins;

- Aquifers in balance (constant stocks);

- Negligible losses through evaporation.

This method gives the best results for the mountainous regions where there are no evaporation zones of any importance or where the aquifers are small in extent. Moreover it is in these regions where it is needed, for they rarely have hydrometric stations.

The exploitable potentials often represent only a small part of the total potentials, which are shown in the following tables:

GROUND-WATER RESOURCES

Northern Algeria

(hm^3/year)

Drainage basins	Resources	Uses	Remainder
Tafna	44	23	21
Oran coastal	60	47	13
Oran high plateaus	60	60	0
Cheliff (below Boughzoul)	121	73	48
Cheliff (above Boughzoul)	82	82	
Zarhez	20	20	
Algiers coastal	343	388	45
Constantine high plateaus	56	35	21
Isser	31	31	0
Sebaou	50	50	0
Soumam	100	100	
Hodna	133	133	
Kebir Rhumel	80	58	22
Seybouse	20	10	10
Constantine coastal	88	56	32
Chott Melrhir	233	215	18
Macta	182	100	82
Medjerda	25	15	10
TOTAL	1 728	1 496	322

Region	Drainage basins	Potential (hm^3/year)
NORTHERN ALGERIA	Tafna	44
	Macta	182
	Oran coastal	60
	Oran high plateaus	60
	Cheliff	203
	Zahrez	20
	Algiers coastal	343
	Isser	31
	Sebaou	50
	Soummam	100
	Hodna	133
	Constantine high plains	56
	Constantine coastal	88
	Kebir Rhumel	80
	Seybouse	20
	Chott Melrhir	215
	Medjerda	25
SAHARA	(Albian)	1032
	(Tertiary)	610
TOTAL		3352

The tables below show that Algeria's renewable ground-water resources amount to 3.3 billion m^3/year, distributed as follows:

Region	Resources (hm^3/year)	Use (hm^3/year)	Remainder (hm^3/year)	Date (estimage)
Northern Algeria	1 700	1 500	350	1974
Sahara	1 600	600	1 000	1970

Thus the rate of resource utilization is about 85 per cent for northern Algeria and 37 per cent for the Sahara. It has in fact increased since these last surveys were made.

Water resources of the Algerian Sahara

The ground-water resources of the Sahara are found in two big geological complexes.

- The intercalated continental clay-sandstone complex found throughout the northern Sahara; its very great extent and average thickness of several hundred metres make it a considerable reservoir. The annual recharge of this aquifer, although low in terms of the exploitable geological reserves, is evaluated at 8.5 m^3/s. It is effected through distribution from the wadis flowing down from the Saharan Atlas and through direct infiltration of rainfall on the Great Western Erg and the plateau of Tademait.

The natural outlets are, in the west and south-west, the springs of Touat-Gourara-Tidikelt, which explains the foggaras and the evaporation zones which mark these regions. In the north-east, towards the coastal zone of Tunisia, the outlets are in the fault areas of El Hamma and Medenine.

The artificial outlets are the foggaras of Touat-Gourara-Tidikelt in the west and south-west, and some one hundred boreholes, of which seven are very deep (about 1,500 m).

This aquifer's water balance has been studied on a mathematical model. The present exploitation is 6.5 m^3/s.

- The terminal continental complex which has several levels: Senonian Carbonic (especially Maestrichtian), Lower Carbonic Eocene, and sandy Miocene-Pliocene.

The corresponding aquifers are in more or less close hydraulic contact depending on the location and can be considered as a single recharging aquifer under the Miocene-Pliocene clays. The boreholes are artesian in the centre of the basin. The aquifer is unconfined at the edges. It is generally shallow or of moderate depth (100 to 400 m).

As in the "intercalated continental", the recharge is large in absolute terms but small in terms of the exploitable stocks. It is effected through infiltration of the runoff at the periphery of the basin (Saharan Atlas, Mzab, Dahar) and through direct infiltration of the exceptional rainfall on the Great Eastern Erg, which rests directly on the permeable horizons of the terminal complex.

This aquifer's natural outlets are the springs at the topographical low points, which have now mostly dried up as a result of intensive exploitation, and the evaporation zones represented by the chotts of Ouargla, Wadi Rhir, and Chott Melrhir.

The main artificial outlets are the more than 2,000 boreholes in Algerian territory.

The water balance of the terminal complex is as follows:

- Recharge: 18.5 m^3/s

- Total dewatering: 21.5 m^3/s

- Drawoffs from stocks: 3 m^3/s (between 1950 and 1970)

The water drawn off in the Algerian Sahara and actually used amounts to 8.5 m^3/s.

It thus appears that in 1970 a large part of the drawoff already came from aquifer stocks. These stocks are considerable. They are estimated at $6 \times 10^{13} m^3$, which in theory means continuous yield of 100 m^3/s for 2,000 years. However, overexploitation of this kind would lead to a steady lowering of the piezometric surface in amounts varying according to location, with the maximum amounts close to the big pumping centres and the minimum (possibly close to zero) in certain regions remote from the drawoff points.

From the economic standpoint this steady lowering of pumping levels will mean a growth in investments over time (increase in the number of boreholes and higher pumping costs) and a rise in the unit cost of the water.

The decision to exploit the water resources of these regions depends on economic criteria: the drawoff corresponding to an acceptable investment value and growth rate and acceptable operating costs for a given geographical distribution of present and future drawoff points. It was therefore essential to determine the evolution of the drawdowns at all points of the aquifers. This was the reason for the establishment of the two hydrodynamic models of the ERESS project (Study of the Water Resources of the Northern Sahara).

The exploitation assumptions have been determined on the basis of two different evaluations of the rates of agricultural development through irrigation:

	1971-1980	1981-1990	1991-2000
High assumption	600 ha/year	1,500 ha/year	2,400 ha/year
Low assumption	600 ha/year	900 ha/year	1,200 ha/year

The drawoff increases will be the following for these two assumptions:

Aquifer	Present yield (m^3/s)	Yield forecast for 2000 (m^3/s)	
		Low assumption	High assumption
Intercalated continental	6.5	23	33
Terminal continental	8.5	18	25
TOTAL	15	41	58

The drawoff increase between 1970 and 2000, a very large total amount has been simulated differently according to zone, so as not to cause drawdowns above 40 m, at least in the terminal continental aquifers. The amounts at present exploited in the region of Wadi Rhir have therefore been increased only a little as this zone is particularly sensitive. In contrast, the drawoffs in the El Hadjira region, which receives little attention at present, move from 0.1 m^3/s to 1.6 m^3/s (low assumption) and even to 2.5 m^3/s (high assumption) for the terminal continental aquifer alone.

For the regions in which this aquifer has been extensively exploited, the amounts of water have been simulated on the basis of the intercalated continental.

GROUND-WATER RESOURCES

ALBANIAN SAHARA

Region	Resources m^3/s	Use $m/^3/s$	Remainder m^3/s
El Oued	1,275	0	
Bou Aroua	0,510	0	
El Meghaier	0,700	0,100	
Djamaa	1,659	0,442	
Touggourt	1,170	0,310	
El Hadjira	1,000	0,000	
Ouargla	1,200	0,450	
Hassi Messaoud	1,200	0,230	
Chardala	1,206	0,347	
Zelfana	1,200	0,290	
El Golea	2,900	0,500	
In Salah	3,935	1,123	
Gourar (foggara)	2,148	0,887	
Gourar (non-foggara)	3,840	0,000	
Touat (foggara)	4,363	2,085	
TOTALS m^3/s	28,300	6,760	26,500
$hm^3/year$	892	214	835

GROUND-WATER RESOURCES

TERTIARY SAHARA

Region	Resources	Use m^3/s	Remainder m^3/s
El Oued	1,313		
Bou Aroua	1,200		
El Meghaier	2,461		
Djamaa	3,755		
Touggourt	3,567		
El Hadjira	1,630		
Ouargla	1,095		
Hassi Messaoud	4,350		
TOTALS m^3/s	19,371	12,600	6,670
hm^3/year	610	390	220

IV. CONCLUSION - GROUND-WATER DEVELOPMENT

Ground water has been used in Algeria since time immemorial: chanelled spring waters (remnants of aqueducts from the Roman and later eras); hand-dug wells, sometimes very deep; underground galleries similar to the kanats of Central Asia, here called "foggaras" (periphery of the Tademait); and even earlier shallow wells in Wadi Rhir.

The exploitation of ground water by means of boreholes underwent great development after the Second World War. There has been rapid hydraulic development since Independence, especially since the Secretariat of State for Water was elevated to the rank of ministry in 1977. Regional study services and drilling enterprises have been established. A very large drilling programme is being carried out both in the north and in the Sahara. The exploitation of ground water is expected to reach 3.5 billion m^3 per year by 2000.

Hydrogeological studies in Algiers are carried out under the responsibility of a "Hydrogeological Service" which is a special unit of the Department of Environmental Studies and Hydraulic Research (DEMRH) attached to the Ministry of Water and Land and Environmental Development.

Hydrogeology was formerly the responsibility of the Service for Scientific Hydraulic Studies created in 1942.

There is a large amount of hydrogeological documentation on Algeria. It includes several hundred monographs and more than 2,000 reports and unpublished technical notes. Hydrogeological maps have also been published.

V. REFERENCES

Canteleev L.Y., Gouloubev, S. M.	Les eaux souterraines de l'Algérie, Ministry of Water Development of the USSR, Moscow "NEDRA", 1978.
Emsellem Y.	L'eau en Méditerranée - Rapport final. Plan Bleu, Expertise No. 2, June 1982.
DEMRH (Achi K.)	Water Resources in Algeria. The first Arab Symposium on Water Resources - Vol. I, Regional Hydrogeology in the Arab World - Part III ACSAD - Damascus 1980.
Cote M.	L'eau en Algérie. Mise en oeuvre et utilisation, in Colloque sur l'eau - University of Constantine, 10-13 April 1983.

- - - - -

The following are also cited:

UNESCO	Etude des resources en eau du Sahara septentrional (Algérie-Tunisie) - final report of projet REG-100, Paris 1972.
Achi K.	Les ressources en eau souterraines de l'Algérie - 1979.
Margat J.	Les resources en eau du bassin méditerranéen, Aix-en-Provence, 1982, No. 2, pp. 15-19.

BENIN

Area: 112,622 km^2

Population: 3.72 million (United Nations estimate, 1983)

I. INTRODUCTION

Benin is a coastal State of West Africa extending for almost 700 km between the 6th and 12th parallels between the Atlantic Ocean and the River Niger.

The territory is 110 km wide as far as the 9th parallel N and then widens to 325 km; this makes the two northern provinces the country's largest regions, occupying 73 per cent of the total area: Borgou, with its provincial capital at Parakou (57,000 km^2 or 45 per cent of the country's total area); and Atakora with its provincial capital at Natitingou (31,200 km^2 or 28 per cent of the total area). The other four provinces are Zou, in the centre of the country with its capital at Abomey (18,700 km^2 or 16.5 per cent of the area); and the three southern provinces: Ouémé with its capital at Porto-Novo, the former capital of the country (4,700 km^2 or 4.2 per cent of the area); Mono, with its capital at Lokossa (3,800 km^2 or 3.4 per cent of the area); and Atlantique, the smallest province but with its capital at Cotonou - the most important town of the People's Republic of Benin by reason of its population and economic (port town) and political roles, the majority of the ministries having their headquarters here. The province is only 3,222 km^2 in size, i.e. 2.9 per cent of the country's total area.

The six provinces have roughly equal populations (500,000 to 700,000 inhabitants each).

Taken as a whole, Benin has a low relief with three characteristic types: a table morphology which is very visible in most of the sedimentary basins; an undulating morphology overlying the whole of the Precambrian basement rock; and a fairly pronounced chaotic morphology at the edges of the quartzite chains of the Atakora mountains.

These geomorphological formations correspond to five natural regions: the low sandy coastal strip, two to five kilometres wide and bounded in the north by the towns of Cotonou and Porto-Novo; the intermediate zone of the barre country which forms a gently undulating plateau covered with ferruginous clays; the silica-clay plateau extending from north of Abomey to the Atakora foothills; the plains of the Niger with their silica-clay cover; and the Atakora mountains with their chains running north-east to south-west, from 500 to 800 m high and 5 to 50 km long. Benin has two big geological formations: the Precambrian African shield and the sedimentary basins. The shield covers 80 per cent of the country; the sedimentary basins include formations ranging from Mesozoic to Quaternary in age and they occupy the north-east and southern parts of the country where they correspond in the north to the southern part of the Niger's sedimentary basin and in the south to the western part of the sedimentary and coastal basin of Nigeria.

Climate

The climate is tropical Sudanese. The following climatic conditions are found from north to south:

- A dry season from December to February throughout the country but growing progressively longer from south to north;

- A rainy season from March to November throughout the country, but especially in the southern regions. From July the climate undergoes two opposite developments: the rainfall decreases south of a zone between the 8th and 9th parallels N, and in August the rainfall is minimal on the coast: in the north, in contrast, the maximum rainfall is in August. In the south there is then a second rainy season which from September and especially in October is the equal of the rainy season in the north:

- In November the rainfall decreases throughout the country.

Lastly, the first rainy season in the coastal areas resembles the single rainy season in the northern areas both in the nature of the rainfall and in its abundance and frequency: high precipitation levels occur on only a few days. The second rainy season is different in nature, with frequent but less heavy rains. There are also differences of distribution from west to east. In the north-west the rain comes earlier at the beginning of the season (March) and stays later at the end (October). This is due to the monsoon which blows in the north in March and brings rainfall in the order of 500 mm; it is halted by the Atakora mountains, which thus receive more rainfall on their western slopes. Later (April) the monsoon ceiling reaches several thousand metres and the Atakora with its average height of 700 m no longer presents an obstacle to the monsoon's movement eastwards and north-eastwards.

The same thing happens in the reverse direction at the end of the rainy season; this time the cause is the Harmattan, a sand wind from the north which pushes back the intertropical front.

On the coast Grand-Popo in the west has noticeably less rainfall than Porto-Novo in the east (900 mm against 1,300 mm). This difference is due to a cool current which flows along the coast and lessens the effect of the monsoon.

Some figures on temperature and relative humidity provided by the Agency for Air Traffic Safety (ASECNA) are given below:

	Average temperature (centigrade)	Average relative humidity (percentages)
Cotonou	m 23 August	65 (December)
	M 32 March	98 (October-November)
Savé	m 20 August	31 (February)
	M 36 February	98 (May-October)
Matitingou	m 18 December	18 (January)
	M 37 March	99 (July-September)
Kandi	m 16 December	99 (September)
	M 39 April	17 (February)

m: minimum; M: maximum.

Surface water

The Precambrian massif of the Atakora is the hydrological centre of Benin; towards the north the peneplain is covered by the sandstone formations of the Cretaceous and intercalated continental periods, but only close to the Niger. The southern slope, in contrast, disappears under the sedimentary formations of the Upper Cretaceous and terminal continental periods.

The tributaries of the Niger flow northwards, the Oueme and its tributaries southwards. When they reach the sedimentary plain the Mono and the Kouffo spread out into vast deltas terminating in the system of lagoons which has formed behond the offshore bar. These lagoons originally contained fresh water but they have become saline as a result of the works at the port of Cotonou.

Only the Pendjari first flows northwards then turns back on itself and emerges from the Atakora mountains to join the Volta by way of the Oti.

The River Niger flows for 150 km between Benin and Niger, forming their common frontier. At Malanville in the north-east of Benin it has an average flood flow of 2,100 m^3/s for three months and a low-water flow of 140 m^3/s, according to data from more than 20 years of observations. Along the frontier it is fed by the Mekrou, the Alibori, the Sota and a few small backwaters. These three tributaries drain a third of Benin and produce a flood crest in the Niger in September, while an earlier crest produced by the Upper Niger takes five months to arrive at Malanville in March.

SOME EVAPORATION DATA (ASECNA)

		Jan.	Feb.	Mar.	Apr.	May	Jun.	Jul.	Aug.	Sept.	Oct.	Nov.	Dec.	Total	mm/m	mm/day
COTONOU (30 years)	PAN "A"	132	158	179	185	132	124	127	160	154	160	145	140	1,786	149	4.9
	PICHE	57	72	87	77	66	57	55	69	58	63	81	67	809	675	2.2
SAVE (10 years)	PAN "A"	154	170	219	198	175	127	85	80	104	124	153	139	1,728	144	4.8
	PICHE	140	133	136	110	86	58	42	42	48	58	111	105	1,069	89	2.9
NATITINGOU (30 years)	PAN "A"	202	206	210	172	120	122	80	70	97	114	152	170	1,715	143	4.7
	PICHE	225	216	194	135	75	66	34	29	43	58	152	206	1,433	119	3.9
KANDI (not specified)	PAN "A"	247	276	318	290	232	262	145	126	135	172	220	232	2,655	221	7.3
	PICHE	260	305	319	244	126	88	44	42	45	98	209	215	1,995	166	5.5

PAN: evaporation in the pan.

PICHE: evaporation in the "Piche evaporomoter".

Geology

The main feature is the Atakora mountains which cross the country in the north-west; this ancient mountain formation is geologically of Precambrian age and forms a succession of terraced indurated surfaces tilted towards the north.

The basement rock of schist, granite and gneiss is the main feature of the geology as far as Dassa-Zoume. There this rock sinks southwards and is overlain by a gently undulating plateau of Tertiary age covered with red sand-clay soils: this is the fertile barre country. In the centre of the sedimentary stratum there is a marshy depression of black clay which extends from Togo to Nigeria.

On the ocean coast a line of offshore bars isolated a number of fresh-water lagoons until the construction works at the port of Cotonou made the water fairly brackish. Step faults give the sedimentary layers an average slope of about 7.5 per cent.

The geological series of Benin's sedimentary coastal basin is as follows:

Coastal sands	
Quaternary:	Sandy alluviums or argillaceous sands. (River flats, normally flooded);
Terminal continental:	Sandy continental sedimentary series with seams containing various amounts of clay or sandstone; one hundred metres thick along the coast.

II. GROUND WATER

Background

In 1948 the Compagnie Générale de Géophysique (CGG) carried out a long series of local geophysical prospecting tests for ground water, with a view to supplying water to the centres of Oti, Savalou, Savé, Tanguieta, Parakou, Dassa-Zoumé, Djougou, Boukoumbé, Natitingou, Nikki and others.

In 1950 BURGEAP made studoes of the sedimentary basins, followed up by the Piccard and Slansky missions (1952-1954). In 1963 the United Nations sent the Lessard mission which studied the north-east of the country in particular, and then the Benham mission in 1964 which studied the centre and the north-east. In 1968 the Wakuti company made a series of mechanical (394) and geophysical soundings in 277 villages (210 positive and 30 per cent negative); it carried out the first large-scale study with 288 km of drilling and 743 electrical soundings (AB/2 = 100 - 300 m). In 1969. on the basis of the Wakuti study, 393 wells were dug by hand with financing from the European Communities.

The works were completed on schedule and to a high standard; it was thus surprising to find that 57 per cent of the installations ran dry in periods of drought. The Office of Geological Studies was therefore requested to make a survey; this showed that the success rate varied according to rainfall and could be as high as 80 to 90 per cent in certain granites but fell to 50 per cent in the schists of Oti and in certain metamorphic rocks. The geophysical studies were resumed in 1969 as part of a UNDP/FAO project by the Compagnia Mediterranea de Prospezioni (CMP) which in conjunction with a consultant, Mr. J.L. Astier, pushed the electric and seismic soundings to greater depths and trained Benin's first national geophysicist. The UNDP/FAO project produced a "summary of information" in 1971.

The United Nations Water Conference (Mar del Plata, 1977), followed in 1980 by the launch of the International Drinking Water Supply and Sanitation Decade (IDWSSD) underlined the great backwardness in the sector. The United Nations instigated and co-ordinated multilateral and bilateral actions in an operational project which began in 1979. Major programmes were launched with the participation of UNICEF, IBRD, OPEC, the United Nations Capital Development Fund, the Conseil de l'Entente, Cooperation Suisse and the European Development Fund (EDF).

The Hydraulics Department has been considerably strengthened since 1981. In 1982 it had seven young engineers who had studied in Canada, China and Romania. The United Nations then initiated a three-year programme for two groups of ten technical assistants specializing in hydraulics.

The available equipment was then overhauled and brought into use, in particular some Bucyrus-Erie light percussion drills and an amount of geophysical prospecting equipment.

Two drilling workshops equipped with an Atlas Copco Mobil Drill with down-the-hole hammer and an Ingersoll Rand TH 55 (hammer and rotary) were then supplied by UNICEF. Two drilling enterprises (FORAKY, a subcontractor of EDF, and INTRAFOR-COFOR of the Conseil de l'Entente) opened offices in Benin.

A set of aerial photographs on the scale 1:50,000 (or 1:65,000) was taken between 1950 (Kandi, Gaya) and 1975 (Natitingou, Djougou); the photographs are of good quality but with the passage of time some of the sites of the villages have changed. The topographical cover is old, and the names of the villages indicated are often different from their present ones. The total-cover scale is 1:200,000 with the 40 m contour lines shown; one-third of the country has been covered on scale 1:50,000.

Geophysical and hydrological studies have so far been limited to the environs of the largest towns and to the coastal zone. However, since 1983 systematic use has been made of geophysical techniques, especially for granites.

EDF V had planned for 1985 the production of a map for ground-water research and exploitation in Benin and the extension of the map of possible locations for hand-dug wells (published in the report of EDF II) to the sedimentary system.

There are 200 copies of these five-colour maps on scale 1:500,000 for the basement rock and 1:200,000 for the sedimentary rock, showing fuller details, which give the main data on the aquifers (depth and thickness, limits, hydrochemistry) collected at the sites of the many holes which have been drilled.

Description of the main aquifers

Benin can be divided into two aquifer systems: the crystalline basement rock in the north and the sedimentary formations in the south.

The crystalline basement rock is made up of intrusive and metamorphic rocks such as quartzites, sandstones and schists, which are impermeable in their mass. However, two types of aquifer are found here: the first type in the clay-sand alteration stratum, which is usually 10-30 m thick and has been exploited since time immemorial by large-diameter hand-dug wells. But the permeability is very low, at best $K = 8 \times 10^{-6}$ m/s (BURGEAP/CIEH, 1972). (CIEH: Inter-State Committee for Hydraulic Studies).

The underlying altered rock and the fissured zones constitute a discontinuous and much larger reservoir. The combination of alterites with good storage capacity resting on very permeable fissured rock offers excellent drilling conditions. Fissured and fractured rocks and the intersections of regional faults can be identified by interpretation of the aerial photographs. The drillings carried out on this basis produce positive holes in a maximum of 60 per cent of the cases. In the granites geophysical prospecting by drilling and electrical sounding increase the success rate by 10 per cent.

The depths of the hole - 30 to 70 m - is no problem for the rigs equipped with down-the-hole hammers, which is the most suitable equipment for very hard formations. The permeability is at least 10 times higher and low-recharge aquifers have flows higher than 2 m^3/h in granites in 50 to 60 per cent of the cases and higher than 5 m^3/h in the schists surrounding fragile rocks in 65 to 80 per cent of the cases. The following aquifers are found in the sedimentary layer:

Coastal sands: Small fresh-water aquifers often floating on salt water and therefore to be exploited with care;

"Terminal continental": A common aquifer in Africa, sandy, with variable clay or sandstone strata (0 - 100 m);

Eocene: Zoogenic, nummulitic fissured limestone with phosphated strata and strata of clay-marl or sandstone-limestone. These strata can contain up to 60 m of black clay in Mono;

Cretaceous: Maestrichtian sands with some clay or marl strata (60 to 200 m). There have been cases of artesian boreholes in Mono with flows up to 50 m^3/h.

The thickness of the whole series increases rapidly on outcrop contact with the basement rock and in a southerly direction. The latitude 6°30'N forms the northern limit of drawoff by hand pump. At Bopa (north of Lake Ahémé) the limestone and sand aquifers can be 80 to 150 m deep.

A few kilometres lower at Possotome, the terminal continental is found at 40 to 100 m and its piezometric level exceeds 50 to 60 m, the limit of manual drawoff. Further south, fresh water is found in the form of lenses floating on salt water.

According to the hydraulic surveys made by UNDP/DCTD in 1982-1983, Benin then had 600 water points acceptable as to quantity and quality and in an operational state. An emergency programme was launched to reach the objective of 7,000 new water points to produce 10 litres per day per inhabitant. It must be noted that at least 30 per cent of the rural population is unfit for work because of water-borne diseases.

The available data are still too limited to provide statistics or a table of typical yields.

NUMBER OF WATER POINTS

Province	EDF IV	EDF V	UNICEF I	UNICEF II	ENTENTE	SUISSE	BOAD	HYDRAULICS	TOTAL	REQUIREMENTS (percentages)
ATACORA		190			80				270	15
ATLANTIQUE							120	30	150	16
BORGOU				86	190				278	24
MONO	40	50							90	12
OUEME		65			60				125	20
BASEMENT			73	44		35			152	16
SEDIMENTARY			65				80		145	
								TOTAL	1,210	= 17

German, Dutch and French volunteers are also working in small programmes for hand-dug wells, but the results obtained so far have not been very encouraging.

III. GROUND-WATER EXPLOITATION

The following are the State services responsible for water and hydraulics :

- Public Works, Construction and Housing - Hydraulics Department;

- Industry, Mines and Energy - Electricity and Water Corporation of Benin (SBEE): urban distribution and sale of water;

- State Farms and Rural Development - Agricultural Engineering, CARDER (Centre for Regional Action and Rural Development): rural surface hydraulics;

- Health: management of the sanitation sector of water-drilling projects, in particular the integrated "health wells" project at Ouesse;

- Transport and Civil Aviation - ASECNA Service: meteorology;

- Interior - Hunting, Waters and Forests Service;

- Social Affairs: village education in water-point hygiene.

The Hydraulics Department was responsible for 85 per cent of the metrage drilled in 1982 (almost 5,500 m drilled); 10 per cent (650 m) was drilled by SBEE and the remaining five per cent by the other ministries (these were mainly wells dug by hand with the assistance of German, Dutch and French volunteers; these wells often run dry in the low-water season).

UNICEF supplied an Atlas Copco Mobil Drill rig(down-the-hole hammer) and an Ingersoll Rand TH 55 Cyclone (down-the-hole hammer convertible to rotary drill), capable of drilling a 10-inch diameter hole 300 m deep. This rig is to be handed over to the Hydraulics Department, which has two Bucyrus-Erie 22W percussion rigs in use in the coastal zone.

The SBEE has a Gotcker rotary rig which in principle works only in the coastal sedimentary zones and a new Rotamec down-the-hole hammer).

In the EDF IV and V projects the drilling operations were assigned to the Sobea-Foraky group which uses two Ingersoll Rand TH 55 rigs. In the Entente project Intrafor-Cofor is using two Bonne Esperance rigs. The plan is to drill 15,000 m/year in 1985-1990.

With a view to standardizing the dewatering,the Hydraulics Department has decided to accept only two types of hand pump: the Abi-Vergnet (EDF/Entente) and the India Mark II (UNICEF).

The national drillers capable of carrying out all phases of a 60-70 m rotary or down-the-hole hammer drilling operation were still few in number in 1985 (four or five), but their skills and numbers are being increased and there should be about 30 of them in five years time.

Cotonou is served by six boreholes located 15 km from the town on the Godomey plateau. The aquifer's capacity is about 20,000 m^3/day and the present boreholes deliver about 17,500 m^3/day.

Porto-Novo is served by three boreholes located at Ouando, about 4 km from the town. The aquifer's potential is about 4,000 m^3/day.

A network links Abomey and Bohicon from Cana, 7 km from Bohicon. The capacity of the boreholes is 2,700 m^3/day.

Lokossa has two boreholes with a capacity of 1,920 m^3/day.

The price of a cubic metre of water is CFAF 80 at Cotonou and Porto-Novo and CFAF 90 at Parakou; but the water in the villages, whether privately supplied or not, costs more than a franc a litre, i.e. more than CFAF 1,000 per cubic metre.

If the requirements are estimated at 20 litres per inhabitant, there will have to be 7,000 boreholes in 2000 and 15,000 in 2010.

The construction works to increase the capacity of the port of Cotonou are one of the problems affecting the distribution and protection of drinking water: the interior lagoons which used to contain fresh water are becoming increasingly saline, creating problems for the villages built there on piles. These villages can be supplied locally by careful exploitation of the lens of fresh water floating on the salt water; but overexploitation could cause an additional intrusion of salt water.

The integrated use of surface and ground water is envisaged in the studies on the Oueme.

IV. CONCLUSION

Benin has some difficult problems where water is concerned. The precipitation/evaporation balance shows a slight surplus in the north where the crystalline formations capture water accessible only by down-the-hole hammer drilling; the drilling of these boreholes has to be preceded by complex studies involving the interpretation of aerial photographs and geophysical prospecting. The use of these techniques can only hold the risk of failure down to an acceptable level. In the south the potential resources are larger but the population is dense and in many areas the ground water lies too deep to be drawn off by hand. The coastal zone is threatened by salt-water intrusion, especially as a result of the recent construction works at the port of Cotonou. However, social welfare and economic activity demand that the whole of the population should have access to water of an acceptable quality, especially when it is remembered that a large part of the village population (30 per cent according to some estimates) can engage in only limited activity because they suffer from water-borne diseases, the commonest being caused by the "Guinea worm". The new drilling techniques are capable of solving the problem. However, the cost of the borehole and the price of the ground water remain high.

An evaluation made by the Department of Geohydraulics in September 1982 for the Travaux à l'Entreprise project of the West African Development Bank (BOAD) envisages 200 positive boreholes with an average failure rate of five per cent for the sedimentary layer, 30 per cent for the basement and 10 per cent overall. Two pumps (a total of 360) have been installed at the majority of the drilling sites and they will deliver 0.6 m^3/h for 10 hours a day for 365 days a year, i.e. 2,190 m^3/per year per pump.

The cost elements are as follows (the amortization periods are indicated):

	Amortization period	CFA francs
- Drilling works per year per pump	(20 years)	146,200
- Pumps	(10 years)	49,400
- Basement-rock pumps	(10 years)	10,715
- Pump installation	(10 years)	14,310
- Technical assistance of pump supplier	(10 years)	10,000
- Studies	(20 years)	37,650
- Activization of provincial services		
Hydraulics Department and laboratory analyses		8,900
- Well-head construction		3,100
- Training programme	(20 years)	7,700
- Hydraulics Department operations (per borehole)		
Two vehicles at 3 million (3 years)		
Salaries: Two team chiefs at 70,000 per month		
Two technical assistants at 50,000 per month		
- Overheads 300,000 per month		20,000
- Maintenance and spare parts		40,000
Repair mechanic		20,000
		375,975

For a drawoff rate of 2,190 m^3 per year per pump this means CFAF 172 (US$ 0.38) per cubic metre.

The Presidential Decree stipulating the powers, organization and operation of the Hydraulics Department dates from only August 1981; water research and exploitation is therefore only just beginning in Benin but it is progressing rapidly. Moreover, with the support of the United Nations project for assistance to the services of the Hydraulics Department, drilling activities have made truly remarkable advances.

The Hydraulics Department has decided to make an immediate start on the evaluation of the country's water resources and it has begun to mobilize the necessary means with respect to geophysical and hydrological operations, laboratory analyses, etc.

V. REFERENCES

Aicard-Slansky, M. 1:500,000 Geological Maps, sheets for Porto-Novo west and east, 1959.

Archambault, A. *Les Eaux Souterraines de l'Afrique occidentale*, BURGEAP-BCEOM, Paris, 1953.

Besairie, H. *Contribution à l'étude stratigraphique et structurale des formations sédimentaires du Bas-Dahomey*, 2 vols., 1943.

Boudin, G. Final report, EDF II Construction Project, Cotonou, 1969.

Bouton Report on the situation of the village water supply programme, UNDP project BEN-78-004.

Bouzid, M. *Donnes hydrogéologiques sur le bassin sédimentaire côtier du Dahomey*, FAO, 1967 (Project DAH3).

BRGM Water resources planning map of Benin, CIEH, 1979.

BURGEAP *Etudes préliminaires du sédimentaire côtier, 1950*.

BURGEAP *Programme d'hydraulique villageoise du Conseil de l'Entente*, 1982.

BURGEAP *La recherche d'eau dans les terrains cristallins africains*, CIEH, Ouagadougou, 1972.

Ferrand *Prospection hydrogéologique de la bordure nord du bassin sédimentaire du Bas-Dahomey.* Hydraulics Service, Cotonou, 1979.

GEOHYDRAULIQUE *Méthodes d'études et de recherche de l'eau souterraine des roches cristallines*, CIEH, Ougadougou, 1978.

GEOHYDRAULIQUE *Idem-travaux complémentaires*, CIEH, 1979.

GEOHYDRAULIQUE *Enquête hydrogéologique: puits du 2ème FED*.

GEOHYDRAULIQUE *Etudes de factibilité*, 5eme FED, Cotonou, 1981.

GEOHYDRAULIQUE *Etudes de factibilité*, BOAD, September, 1982.

UNITED NATIONS Ground Water in Africa, New York, 1971.

UNITED NATIONS Report of the Lessard technical assistance mission, 1963; Benham mission 1964; Dijon mission 1977.

Slansky   Contribution à l'étude géologique du bassin sédimentaire côtier du Dahomey et du Togo, 241 pages, 1961.

UNICEF   Projet hydraulique villageoise (Zou), C Hubert, Cotonou, 1981.

Wakuti Co.   Final report, 1969 drilling operations.

Note: The majority of these documents are kept at Cotonou in the Hydraulics Department.

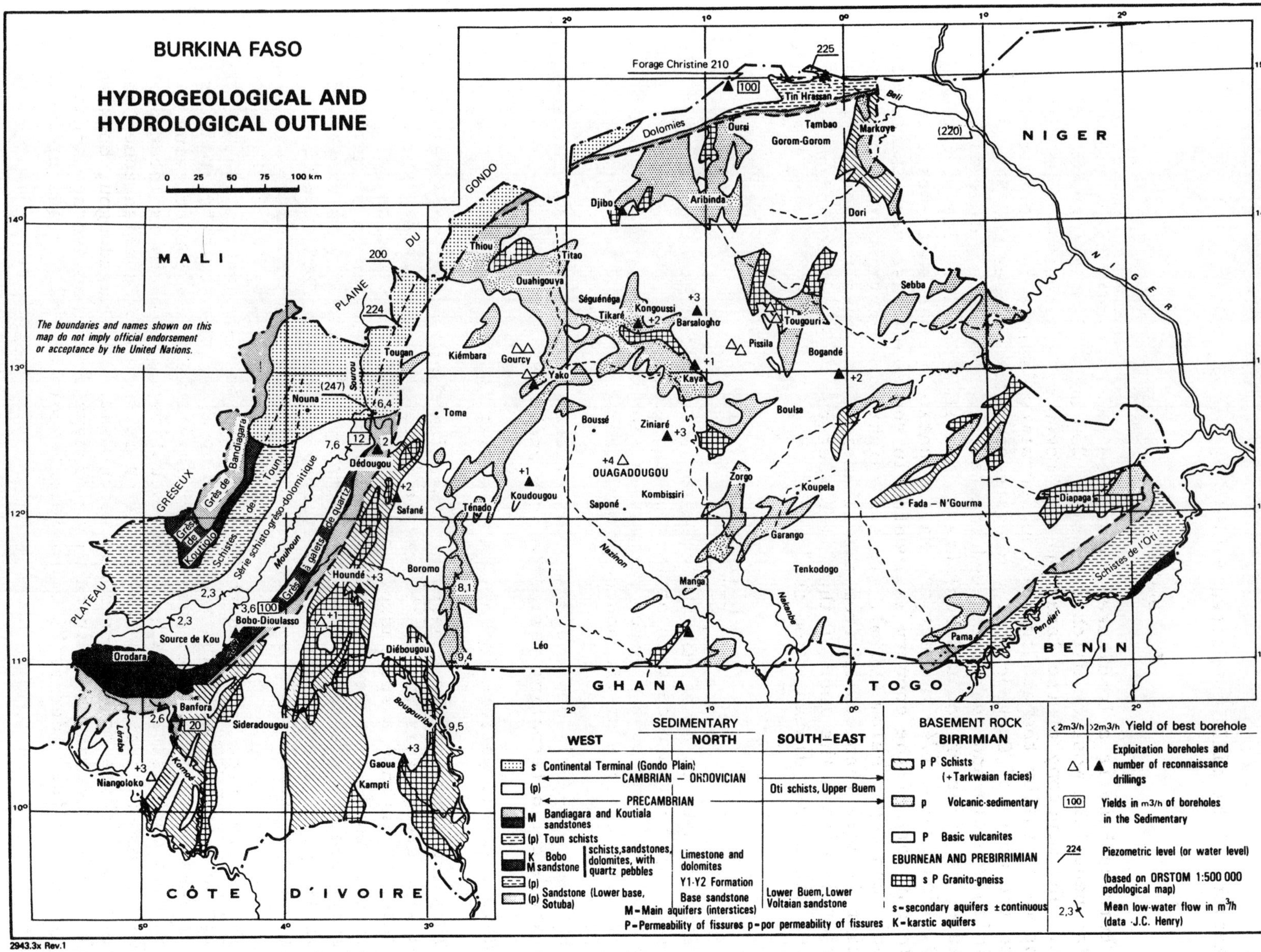
BURKINA FASO
HYDROGEOLOGICAL AND HYDROLOGICAL OUTLINE
0 25 50 75 100 km
The boundaries and names shown on this map do not imply official endorsement or acceptance by the United Nations.
MALI
NIGER
BENIN
TOGO
GHANA
CÔTE D'IVOIRE
GONDO DU PLAINE
PLATEAU GRÉSEUX
Forage Christine 210
225
100
Tin Hrassan
Dolomies
Oursi
Tambao
Gorom-Gorom
Markoye
Beli
(220)
Djibo
Aribinda
Dori
Thiou
Titao
Ouahigouya
200
224
Tougan
Kiémbara
Gourcy
Yako
Séguénéga
Tikaré
Kongoussi
+2
+3
Barsalogho
Pissila
Tougouri
Sebba
Bogandé
+2
Kaya
+1
Nouna
(247)
Sourou
6,4
Toma
7,6
12
2
Dédougou
+2
Safané
Boussé
Ziniaré
+3
+4
OUAGADOUGOU
+1
Koudougou
Ténado
Saponé
Kombissiri
Zorgo
Koupela
Boulsa
Fada – N'Gourma
Diapaga
Garango
Tenkodogo
Schistes de l'Oti
Pendjari
Pama
Nazinon
Nakanbe
Manga
Léo
Boromo
8,1
9,4
Diébougou
Bougouriba
9,5
Gaoua
+3
Kampti
Grès de Bandiagara
Schistes de Toun
Grès de Koutiala
Série schisto-grèso-dolomitique
Mouhoun
Grès à galets de quartz
Houndé
+3
+1
2,3
2,3
3,6
100
Bobo-Dioulasso
Source de Kou
Orodara
Banfora
2,6
20
Sideradougou
Léraba
Komoé
+3
Niangoloko
SEDIMENTARY
WEST
NORTH
SOUTH–EAST
s Continental Terminal (Gondo Plain)
CAMBRIAN – ORDOVICIAN
(p)
Oti schists, Upper Buem
PRECAMBRIAN
M Bandiagara and Koutiala sandstones
(p) Toun schists
K Bobo
M sandstone
schists, sandstones, dolomites, with quartz pebbles
Limestone and dolomites
Y1-Y2 Formation
(p)
(p) Sandstone (Lower base, Sotuba)
Base sandstone
Lower Buem, Lower Voltaian sandstone
M – Main aquifers (interstices)
P – Permeability of fissures p – por permeability of fissures
BASEMENT ROCK
BIRRIMIAN
p P Schists (+ Tarkwaian facies)
p Volcanic-sedimentary
P Basic vulcanites
EBURNEAN AND PREBIRRIMIAN
s P Granito-gneiss
s – secondary aquifers ± continuous
K – karstic aquifers
< 2m3/h
> 2m3/h
Yield of best borehole
Exploitation boreholes and number of reconnaissance drillings
100 Yields in m3/h of boreholes in the Sedimentary
224 Piezometric level (or water level)
(based on ORSTOM 1:500 000 pedological map)
2,3 Mean low-water flow in m3/h (data J.C. Henry)
2943.3x Rev.1

BURKINA FASO

Area: 274,200 km^2

Population: 6.61 million (United Nations estimate, 1983)

I. BACKGROUND

Burkina Faso is a landlocked country in West Africa: a vast, flat and regular peneplain most of which is between 250 and 350 metres above sea level. The monotony of the landscape is broken by a number of mountain chains which follow the geological configuration, mainly in the south-west and to the north of Ouagadougou: elongated parallel ridges made up of Birrimian schists and quartzites and sometimes capped with a thick lateritic cuirasse, or piles of rounded boulders of granitic origin. The high point is Mt. Téna Kourou (749 m) near the Mali frontier in the south-west, and the lowest part of the country is located in the south-east near the borders of Benin and Togo (below 200 m). In a straight line, Burkina Faso's southern frontier is more than 600 km from the Atlantic coast, and the northern frontier is within the loop of the Niger, between a few dozen and a few hundred kilometres distant from the river.

Climate

From October to April/May the climate is dominated by a dry flow of air from the high pressure centres of the Sahara, giving hot days and cool nights, while from May/June to September there is a flow of moist air from the oceanic high pressure centres of the southern hemisphere.

Between mid-April and mid-June and between mid-September and mid-November there are brief alternating periods of dry and moist air flows.

This over-all pattern is influenced by latitude.

Thus the north of the country or "Sahelian zone" has a short rainy period with great variation in the amount of rainfall, and a dry period aggravated by intense evapotranspiration, with wide daily and annual temperature ranges. At Dori on the 650 mm isohyet it rains on 54 days a year on average and the potential evapotranspiration is generally in excess of two metres for an average annual temperature of between 28.5°C and 29°C (maximum temperature 45.4°C and minimum 6.8°C).

The central zone comprises the larger part of the country, leaving the Sahelian zone in the north and a narrow strip along the southern frontier which widens out only in the south-west. This zone has a longer rainy season which can last six months in an exceptional year. It is located between the 650 mm isohyet (which separates it from the Sahelian zone) and the 1,000 mm isohyet to the south and south-west. At Ouagadougou, between the 800 and 900 mm isohyets, it rains on 75 days a year on average annual temperature in the order of 28°C (maximum temperature 42.6°C and minimum 9.5°C).

The southern zone, to the south and south-west of the 1,000 mm isohyet, has rainfall for about half the year and a fairly low annual temperature range. At Bobo Dioulasso, between the 1,100 and 1,200 mm isohyets, it rains on 93 days a year on average, the potential evapotranspiration is between 1,700 and 1,800 mm, and the average annual temperature is below 27°C (maximum temperature 41.6°C and minimum 10°C).

Surface water

Burkina Faso is crossed from north-west to south-east by the watershed between the Niger basin and those of the Voltas. The centre, the south and the west are drained by the White, Red and Black Voltas and by the Comoé and its tributary the Léraba. The extreme south-east is drained by the Pendjari, itself a tributary of the Voltas.

The north and the east are drained by the right-bank tributaries of the Niger, which are temporary rivers linking a "string of waterholes" (some of which remain wet throughout the year) and they are of great importance for livestock raising. The only permanent rivers are the Black Volta, the Comoé and the Pendjari, of which the Black Volta is by far the best fed, with an average flow at Banza of 13 m^3/s and a minimum flow in March/April of about 2.8 m^3/s. The minimum flow of the Comoé falls to below 2 m^3/s from March to June and that of the Pendjari is in the order of 500 l/s in March/April.

Many of the valleys are dammed in order to accumulate the water in the rainy season. The country is dotted with more than 700 dams of varying size, depending on the topography, the amount of recharge and the techniques used in their construction (sometimes very rudimentary). The Loumbila dam which supplies Ouagadougou with drinking water has an area of 1,000 ha, with a maximum depth of 6.5 m, while some of the small dams collect no more than a few thousand cubic metres of water, with depths of about one metre, and they dry up completely in the course of the year. The maintenance of these installations is a major concern in view of the risk which they offer to downstream installations and low-lying areas.

The country has a number of natural lakes. These are Lake Bam, which occupies the fossil bed of an ancient tributary of the White Volta; it is about 20 km long and about a kilometre wide and it dries up about every seven years (FAO report cited); Lake Dem, which at its low-water level covers a few hundred hectares; Lake Kou, with an area of about 100 ha at low water; and the "hippo pool" fed by the Black Volta, and the Tengréla and Karfiguéla waterholes, located near Banfora.

Geology

The crystalline formations of the African Precambrian basement rock cover more than 80 per cent of the area of Burkina Faso.

In the Gondo plain in the extreme north-west, the tertiary continental deposits of the "terminal continental" formation rest in unconformity on the Precambrian formations "A".

- The Precambrian crystalline formations "C" and "D", granites and gneiss, schists, quartzites and green rocks, are generally impermeable. However, the altered or fractured zones constitute important local aquifers which can be used for village water supplies.

- The Precambrian sediments "A" have at their base sandstone-quartzites and conglomerates, then mainly sandstone formations interbedded with schists and rare dolomitic limestones in the west of the country which give rise to perennial springs in places (for example, Kou), then heterogeneous formations of argillaceous schists, quartzitic sandstones, interbedded limestones and dolomites, brecchias, and marbled intra-formational conglomerates, which represent the peripheral facies of the subsidence basin of the Gourma running down from the north (Mali and Niger). These formations are aquifers in places. Many sandstone facies (Gobinangou sandstone) are found in the south-east of the country, with typical tillitic strata overlain by an important formation: the series of Pn edjaride pelitic schists and lydites with calcareous and phosphated strata. These formations have hydrogeological potential.

- The terminal continental formation of Gondo, consisting of sands and clays, is a continuous aquifer connected with the underlying dolomites.

II. GROUND WATER

Bodies concerned

The Department of Hydraulics and Rural Development (DHER), which was set up in July 1984, is responsible, inter alia, for:

- The inventory of the country's water resources;

- The hydrological hydrometric and hydrogeological studies required for the evaluation of the country's water resources;

- The studies required for the determination of a policy for water resources development and the corresponding installations;

- The preparation of national programmes for the development of water resources for domestic, agricultural and livestock, and energy needs, and for the development of rural areas;

- The planning, preparation, monitoring and evaluation of water infrastructure programmes;

- The co-ordination of all water resources activities;

- The drafting of laws and regulations governing the use of water resources.

DHER has three quasi-State bodies under its technical control:

a) The National Water Office (ONE) which is responsible for "the management of all undertakings and all operations directly or indirectly affecting the storage, treatment and purification, transport and distribution of untreated water and drinking water. It may also be required to attend to the treatment of waste water before its disposal or to recycle this water in the supply system".

b) The National Office of Dams and Irrigation (ONBI), which is responsible for "building rural dams and supplying water to the plains. It is also responsible for the establishment, management and maintenance of agricultural production infrastructures: dams, rice fields, pumping stations, fruit and vegetable packaging and preserving centres, etc., in liaison with training bodies and producers' organizations...".

c) The Rural Development Fund (FDR), which is responsible for "the mobilization, co-ordination and management of funds for the promotion of social and economic development for the rural population with its participation". The FDR is therefore active in the surface- and ground-water sectors.

In addition, the Volta Valleys Development Authority (AVV), responsible to the Ministry of Rural Development, plans and implements water programmes for the repopulation of the valleys of the Voltas and for the utilization of the fertile land. In addition to the Ministry of Rural Development, there are several other ministries with important responsibilities in the area of water:

The Ministry of Public Health is responsible for the quality of drinking water and the protection of drinking water supplies, as well as for health education;

The Ministry of Equipment and Communications is responsible for urban development and rules on matters of individual sanitation installations;

The Ministry of Trade, Industrial Development and Mines regulates the industrial use of water and is responsible for the campaign against the pollution of the environment by industrial wastes.

The Ministry of Finances contributes to water policy in that it controls the use and distribution of public funds for all State actions with respect to drinking water and sanitation.

The Ministry of Planning and Co-operation is responsible in the sector for the overall planning and mobilization of the financing needed for the country's economic and social activities.

The Technical Water Committee was set up in 1978 as an inter-ministerial co-ordination body; it also acts as a national action commission for all matters of drinking water supply and sanitation.

Lastly, the Water Code came into force on 1 April 1983.

Background of Hydrogeological Research

It was not until 1954 that the first hydrogeological studies were made; they consisted mainly of inventory work and the supervision of prospecting operations. Some 700 wells were dug between 1952 and 1961. Regional studies were then undertaken, some with the assistance of a United Nations expert.

In December 1979 the Inter-State Committee for Hydraulic Studies (CIEH) published a summary of hydrogeological knowledge about the territory, which listed 53 studies in chronological order, and included the summary table which appears below.

Village water supply programmes are now being carried out throughout the country. These programmes have been preceded by local studies on the establishment of water points. In bedrock areas, i.e. in the great majority of cases, it is a question of identifying deep fissures. The techniques used generally include field surveys, an interpretative study of aerial photographs and geophysical prospecting exercises, In these exercises use is generally made of geo-electrical and aerial geophysical methods and soundings. However, research is being carried out to refine these methods (especially with respect to multi-directional aerial prospecting) and to introduce seismic, microgravimetric, magnetic and electromagnetic techniques. Specialized national personnel are usually assigned to the projects (hydrogeologists, drillers and drilling assistants, well-diggers), except in the case of the geophysical unit which is attached to the Ground Water Service at Ouagadougou and has three electrical prospecting and one seismic prospecting sets of equipment. However, this unit lacks the resources and qualified personnel to cope with the very large volume of work involved in the establishment of several hundred water points per year throughout the country.

In practice the water points are first established on the basis of geomorphological surveys using aerial photographs. In the event of failure or when a priori the area is very unfavourable, the geophysical team is called in. According to the CIEH report on the state of hydrogeological knowledge of the country, the main ground-water resources are:

- Aquifers in lateritic cuirasses: when the cuirasses are submerged, these water-bearing strata constitute a major resource. Their good consistency means that they can be tapped by unlined wells (especially in the area of Koudougou);

- Sand aquifers: traditional wells are limited by the poor consistency of the formations: they cannot tap the water of the granular sands lying at the base of the alteration profile and they can exploit only the unproductive argillaceous sands. The granular sands, when they are present, can be reached only by modern wells;

- Alluvial aquifers: they are very local and are exploited by sump holes;

- Aquifers in fractures of the crystalline basement rock. Drilling with "down-the-hole hammer rigs" has made it possible to exploit these strata, which were not suitable for modern wells or wells dug by excavators such as the "Calweld". The use of this new technique has made it possible to obtain better yields from the installations constructed under the Volta Valleys (AVV) and village water supply (DHER) programmes;

- Sandstone aquifers in the west and south-east of the country (sedimentary basin);

- Sand aquifers in the terminal continental formation of the Goudo plain. Deep wells must be drilled to exploit the deep permeable strata.

Summary of Studies

	Author	Year	Purpose	Content	Proposals
Hydrogeology of Nord-Dori	BURGEAP	1954	R P	C P	
Material on the Hydrogeology of Upper Volta	BURGEAP	1954	R		
Material on Installations in Nord-Dori	BURGEAP	1957			F M
Resources and Needs of the cercles of Boulsa, Bogandé, Dori Sud	GERHAV	1962	R	C P B	P B
Water resources of Ouahigouya	SCET	1963	R B P	D	B
Ougadougou water supply	SCET	1964	R P	Ge C	
Data for the equipment of five cercles in southern Upper Volta	BURGEAP	1964	R B P	C P	P Ge
Critical data on the SATEC wells campaign	BURGEAP/SATEC	1965	P	D	
Hydrogeology of Nord-Voltaique: Markoye, Darkoye, Kouyera	DGM	1965	R P	F	
Hydrogeology of Ténado cercle	BRGM	1965	R	C P	F P Ge
Hydrogeology of Boussé and Ouagadougou cercles	BRGM	1965	R	P	F P
Preliminary hydrogeology of Bangao depression	DGM	1966	R	F	M
Water resources of Yako cercle	BRGM	1966	R	C P	F P Ge
Hydrogeological survey - supply for 20 urban centres	BRGM	1966	R P	F Ge C	F P
Hydrogeological survey of Koudougou cercle	BRGM	1967	R B P	C P	F P Ge
Ground water resources in Toma cercle	BRGM	1968	R B P	C P D	F P Ge
Mining and ground water resources in Upper Volta	COMTEC	1968	R B P	F GesC P D	F P
Water resources in Tougan cercle	BRGM	1969	R B P	C	F P
Hydrogeological study - cattle trail	BRGM	1969	R B Pa		P
Ground water resources in Nouna cercle	BRGM	1969	R B P	C P D	F P S
Ground water resources in Zorgo cercle	BRGM	1970	R B P	C P	F P Ge
Hydrogeological studies in Oudalan	UNDP	1971	R P	see Hydrogéo	1971
Hydrogeological study in the north of Upper Volta	HYDROGEO	1971	P	F Ge Ci	
Ground-water resources in Kaya and Pissila cercles	BRGM	1972	R B P	C P	F Ge
Hydrogeological study - Tambao water supply	HYDROGEO	1972	P	Ge	B
Summary of carbonate formations in Oudalan	BRGM	1973	P	C	Ge
Ground-water resources in Boulsa cercle	BRGM	1973	R B P	C P	F P
Ground-water resources in Diébougou cercle	BURGEAP	1973	R B P	C	P B
Ground-water resources in Dédougou cercle	BRGM	1973	R B P	C P	F P Ge
Hydraulic programme in the Sahel	SCET	1973	R	P	F P B Ge
Hydrogeological studies: First phase AVV	LCHF	1974	R P I	F Ge	
Resources and needs in Diapaga and Fada N'Gourma cercles	BRGM	1975	R B P	C P	F P B
Hydrogeological studies - Ranch de Léo installation	LCHF	1975	R Pa	F Ge C D	F B
Ground-water resources of Yatenga ORD	BURGEAP	1975	R B P	C P	F P Ge
Hydrogeological studies: Second phase AVV	LCHF	1975	R P I	F Ge	

Summary of Studies (continued)

	Author	Year	Purpose	Content	Proposals
Sahel Emergency Mission - 26 urban centres	BURGEAP	1975	P	F Ge D B	
Hydraulic resources - Centre-East of Upper Volta	UNIV.ST.MONTPE.	1975	R	F C D	
Hydrogeology of Haute-Sissili	UNIV. GRENOBLE	1975	R Pa	F Ge Ci D	
Hydrogeological mission in Bani region	DHER	1976	R P		F
Report of the 1975-1976 drilling campaign - North Upper Volta	DIWI WALTER	1976	P	F D	
Hydrogeological studies: third phase AVV	LCHF	1976	R P I	F Ge	
Water resources in Kongoussi and Barsalogho sub-prefectures	BURGEAP	1976	R B P	C P D B	F P Ge
Progress of work on the Dori-Djibo road	SEGIRACO	1976			M
Hydrogeology of Boromo, Fada, Koupéla, Nouna centres	DHER	1976	R P	P	Ge
Hydrogeological studies: fourth phase AVV	LCHF	1977	R P I	F Ge	
Water resources - Est-Banfora livestock reception zone	DHER	1977	R Pa		
Hydrogeological studies: fifth phase AVV	LCHF	1978	R P I	F Ge	
Water resources inventory of Comoé ORD	IWACO	1978	R B P	GesC P	F P Ges S
Ground-water resources in the Bougouriba valley	UNDP	1978	R P	F Ge	
Hydrogeological studies for the cattle trail - Ouest-Volta	DPEOV	1978	R Pa		P
Water supply for nine secondary centres	DHER	1978	R P	F Ge C D	Ge
Ground-water resources in the Bougouriba valley	UNDP	1979	R P	F	
Fracture systems in Upper Volta - Landsat photographs	FAO	1979	R Landsat satellite imagery study		

Key

R: Estimate of resources
B: Estimate of needs
P: Human water supply
Pa: Livestock water supply
I: Irrigation

Works proposed following the study

F: Drilling
P: Wells
B. Dam

Contents

F: Drilling works
Ge(s): Electrical geophysics (and seismic)
C: Chemical analysis
Ci: Chemical and isotopic analysis
P: Water-point card index
D: Discharge test
B: Dams

Ge: Electrical geophysics
M: Development of water holes
S: Tapping of springs

Not enough data are available for an evaluation of the yield of the aquifers. The discharge of village water points rarely exceeds a few M^3/h and no pumping trials have been undertaken. However, the widespread decline in piezometric levels, which seems to have been going on for some years and has caused the exhaustion of some wells, remains a source of concern.

At present, the aquifers do not seem threatened by over-exploitation owing to the low dewatering capacity both by traditional wells and by hand or foot pumps. The damming of surface water usually has to be undertaken for supplies to urban centres.

With respect to the physical and chemical characteristics of the water, the CIEH report states that the water usually has a low mineral content and is aggressive and of good chemical quality. The submerged cuirasse strata have high a iron content. Bacteriological pollution is common when the aquifers are tapped by sump holes or in the vicinity of big urban centres (poor bacteriological quality of the aquifer in the alterites close to the Ouagadougou dams).

In view of the lack of information on ground-water resources, the Government has sought financing for a detailed inventory of the country's water potential, including:

- Village surveys on present supplies to the people - prior to the launching of village water-supply projects;

- An estimate of the reserves and buffer stocks in the drawoff zones;

- An estimate, by means of piezometric checks, of the resources of the crystalline bedrock;

- An analytical study of fracturation based on aerial photography and field surveys;

- The development of simple geophysical techniques suited to the country's geology, and the implementation of accurate drilling campaigns.

Mention must also be made of the current study of the country's water balance financed by the Netherlands and carried out by IWACO, and of the establishment by DHER's Ground Water Service of a computer file of ground-water points, with assistance from the UNDP/DTCD project for strengthening DHER.

III. EXPLOITATION OF GROUND WATER

Two of DHER's services are responsible for the development and exploitation of ground water: the Ground Water Service for the village water supply programmes, and the Urban and Industrial Water Service for the water-supply programmes for towns and industrial establishments. As organized at present, the Ground Water Service includes the study and programming office with its hydrogeology and water-points index sections, the office of works, with its geophysical and drilling sections, the projects office, and the water analysis laboratory. The Urban and Industrial Water Service includes the water supply office, with a section dealing with urban and semi-urban water supplies and a section for industrial water supplies, and the sanitation office, with its waste-water and rainwater sections. In 1984 these services had the following staff:

- Ground Water Service: five engineers, eight senior technicians, one works supervisor;

- Urban and Industrial Water Service: four engineers, four senior technicians;

- Staff assigned to projects: five engineers, 11 senior technicians, 65 permanent and 500 temporary employees providing administrative support and manpower for the works units.

The engineers and senior technicians usually have suitable qualifications, notably from the Inter-State Rural Development School at Ouagadougou and the School for Senior Technicians in Hydraulics and Rural Development at Komboinsé, or even from abroad. However, there are still not enough of them.

Training and further-training courses are organized for field staff either centrally or as part of the projects. However, the training needs remain very large.

The logistical means available to DHER for the implementation of the works consist of a fleet of 105 light and 66 heavy vehicles; there were also six drilling units in 1984.

There are seven repair workshops/garages; at Ouagadougou, Bobo Dioulasso, Dédougou, Banfora, Ouahigouya, Koudougou, and Kaya.

The diversity of the equipment and the shortage of spare parts mean inadequate maintenance and repair. The result is that the equipment is out of use for long periods and generally underused.

The sinking of the boreholes and wells is divided among State and private bodies which have the resources summarized in the following table taken from the June 1983 CIEH report cited in the bibliography.

Annual Installation Capacity for Wells and Boreholes in Burkina Faso

Bodies	Boreholes: Machine	Boreholes: Boreholes Type	Boreholes: Theoretical annual capacity	Wells: Machines	Wells: No. of teams	Wells: Annual capacity
	2 SH 70	R MFT	160			
	1 Dando	B	30			
DHER	1 Dando	R B	60		7	250
	1 IR.TH 60	R MFT	100			
BUVOGHMI	2 Failing	R MFT	200			
AVV	2 Stenuick HS 66 D	MFT	200	1 Calweld	-	80
National army	1 Schafer	MFT	50			
Aforcom	2 Stenuick HS 66 D	MFT	250	-	-	
SADE						150
NGO */				-	-	100
Total	13		1,150	1	7	480

R: Rotary; B: Percussion; MFT: Down-the-hole hammer.

The operations carried out by the Administration (DHER) in the two campaigns of 1977-1978 and 1978-1979 were as follows:

Boreholes

Campaign	Total: Number	Total: Metrage	Positive: Number	Positive: Metrage	Percentage failures
1977-1978	78		51		35
1978-1979	69	2, 81	52	2,200	25
1979-1980	47	1,770	33	1,190	30
Average	65	2,375	45	1,650	30

*/ NGO activities are not negligable but they are not integrated in the national programme despite the efforts of the Permanent Secretary for Non-Governmental Organizations.

Wells

	New wells			Deepened wells		
	Total	Positive	Percentage failures	Total	Positive	Percentage failures
1976-1977	172	148	14	16	14	12
1977-1978	112	91	19	76	69	9
1978-1979	148	132	10	42	36	14

It will be noted that the actual drilling is below the theoretical capacity, i.e. 40 per cent for boreholes and 60 per cent for wells. This is due in part to the unreliability of the equipment and the difficult hydrogeological conditions.

The zones of rolling alteration require rigs equipped with advanced casing systems for drilling by down-the-hole hammer with eccentric bits. As the existing machinery does not have this equipment, use is made of temporary casing installed by cutting, but the drilling rate suffers.

It should be note that the units equipped with mud pumps very rarely make use of this advantage.

The sinking of wells encounters the same difficulties with respect to hydrogeological conditions and inadequate equipment. The capacity of the derricks (900 kg) is too low to permit digging by excavator. A capacity of 1,500 kg and a faster lifting speed would be useful in certain loose formations. Again for reasons of capacity, the filter tubes cannot be manufactured on the surface and then lowered, which would mean they could be of better quality with regular distribution of the draining channels; this would speed up operations by making the tubes immediately available to the drilling crews so that they would not have to wait for the concrete to set, as is the case when the filter column is poured down the well.

Use of ground water

The available yields remain low for both wells and boreholes. The wells supply only 5-10 m^3/day and they are often dry during certain periods. The rural boreholes are equipped with manual pumps and the yield rarely exceeds one m^3/h.

It is estimated that there were 3,500 modern water points in Burkina Faso in July 1980 and 5,500 in July 1984. A thousand water points were established during the 1983/84 campaign.

To meet the objectives of the International Drinking Water Supply and Sanitation Decade, a total of almost 10,000 would be required by 1985 to deliver 10 litres per day per inhabitant, and almost 20,000 by 1990 to deliver 25 litres per day per inhabitant.

Eighteen urban centres are at present equipped with a drinking-water network supplying a total population of 320,000. Half of the supply is provided through private connections and the other half through public standpipes. The average rate of cover of the population of these centres is 50 per cent. The total investment during the five years 1978-1983 for the 18 centres amounted to almost CFAF 7.5 billion.

The industrial water consumption included in the urban statistics accounts at present for five to six per cent of the total consumption at Ouagadougou and Bobo Dioulasso. In exceptional cases it is much higher. For example, the VOLTEX textiles complex at Koudougou consumes 370,000 m^3/year.

The present situation with regard to agricultural water supplies can be summarized as follows:

- Downstream of the dams: a little over 5,000 ha, including 3,900 for the Banfora sugar complex alone.

- Pumping or diversion: about 3,000 ha developed, mainly planted with rice, vegetables and fruit;

- Improved flatlands: a few hundred hectares planted with rice during the rainy season;

- Unimproved flatlands: between 3,000 and 4,000 ha fully developed, used for rice growing in the rainy season.

The agricultural water-supply projects which have been identified or are under study involve a total area of about 90,000 ha. The largest of these projects will also be used for electricity production.

The main limitation on the intensive exploitation of ground water stems from the nature of the deposits. The unit yields are generally low for the 80 per cent of the country's area made up of crystalline formations. Moreover, for some years there has been a decline in the piezometric levels because the rainfall has been insufficient fully to recharge the aquifers.

In many cases the water in the surface reservoirs is used to supply urban areas, either directly after treatment or by means of wells or boreholes tapping the aquifer underlying the dammed lakes. However, favourable conditions for the artificial recharge of ground-water aquifers are rarely found in Burkina Faso.

IV. CONCLUSION

At present in Burkina Faso, as in the past, ground water is used mainly to supply rural dwellers; it is being increasingly exploited by means of boreholes equipped with hand or foot pumps.

The cost of this kind of borehole, including the development of the wellhead area and the village education work, is in the order of CFAF 4 million. The cost of maintaining and repairing the pumps is estimated at CFAF 50,000 per year on average. On the basis of a water point for every 500 inhabitants, this means an annual expenditure of CFAF 100 per inhabitant, with the basic investment usually financed by external aid; from the economic standpoint, the provision of drinking water to the rural population is essentially a social charge. However, further development of the water points might be envisaged in future if the yields exceed the requirements of the human population; the surplus will be used for small-scale activities (potteries, brickworks), the irrigation of small gardens and the watering of sedentary livestock. Nevertheless, there will be no spectacular development of the resources owing to the nature of the aquifers (mainly discontinuous) which limits the available yield per water point.

V. REFERENCES

United Nations: Ground Water in Africa, United Nations, 1971.

Explanatory notice to the 1:1,000,000 geological map of the Republic of Upper Volta, 1975.

United Nations: Water and Hydraulics. Technical report for a water policy.

Final report, project UNDP-72-039, United Nations, 1977

C. Diluca: Etat des connaissances hydrogéologiques en Haute-Volta, CIEH, 1979.

CILSS/OECD/Club of the Sahel/CIEH: L'hydraulique villageoise dans les pays membres du CILSS. Enquête et propositions en vue d'une gestion rationnelle de l'eau. Situation en Haute-Volta.

Ten-Year Plan, International Drinking-Water Supply and Sanitation Decade. Volumes I, II and III, 1983.

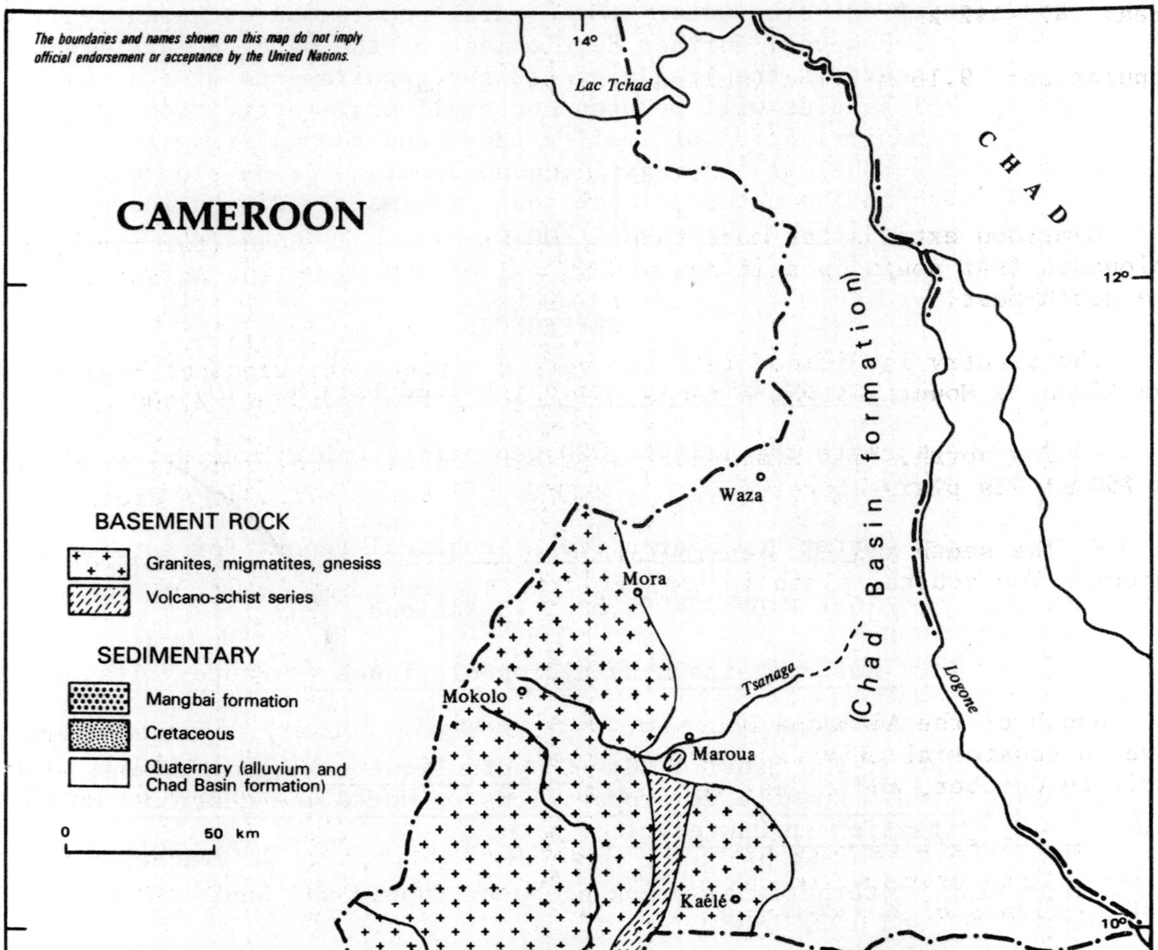

Source: GROUND-WATER EXPLORATION AND EXPLOITATION
Agency Terminal Report DP/UN/CMR-74-013/1 New York, 1979

CAMEROON

Area: 475,442 km^2

Population: 9.16 million (united Nations estimate, 1983)

I. BACKGROUND

Cameroon extends for more than 1,500 km from Lake Chad (far north) to Moloundou (far south) and it has a seaboard of 350 km on the Atlantic in the south-west.

The country is divided into two very different geographical regions by the Adamaoua Mountains which range in altitude from 1,500 to 2,500 m:

- The north, from Ngaoundéré to Lake Chad (altitude ranging from 300 to 750 m) has plains covered with savannah and some mountainous areas;

- The south and the centre have plateaus covered with equatorial forest. The coastal plain is dominated by Mt. Cameroon (4,070 m).

Climate

South of the Adamaoua Mountains the low-lying densely forested areas have an equatorial climate with two separate seasons: heavy rainfall from April to October, and a less wet season from November to the end of March.

The rainfall is very heavy from June to September. It reaches 3,963 mm at Douala and even as high as 10,500 mm at Bibundi on the southern and western flanks of Mt. Cameroon.

The average annual temperature is 26°C and there is little variation between a monthly minimum of 23°C and a maximum of 30°C.

The climate evolves towards the tropical type in the direction of the high plateaus of the west and of Bamenda, i.e. on high land. The rainfall is less heavy than in the forest zone, and the temperatures are more moderate with wider seasonal ranges (average annual temperature 20°C, monthly minimum 15°C, and monthly maximum 25°C).

North of the Adamaoua plateau the climate is of the normal tropical type with a very dry season from November to May and a rainy season from June to September.

The rainfall declines towards the north. It is 1,420 mm at Poli, only 977 mm at Garoua and 500-600 mm at Kousséri. At Garoua the climate is of the Sudanese type; it becomes Sahelian-Sudanese from latitute 9°N.

The temperatures are high with large seasonal variations. Maroua, for example, has recorded 11.2°C in December and 44.9°C in May.

Surface water

As a direct effect of the relief and the climate, the perennial rivers and streams are numerous south of the Adamaoua, whereas most of the streams in the north, called "mayos", are seasonal.

The following are the main watercourses in the south:

Watercourse	Flow	Length
Wouri	-	250 km
Moungo	-	Navigable for 110 km
Dibamba		
Sanaga	The average flow is 2,300 m^3/S; 6,500 m^3/s in October; it drops to 400-500 m^3/s at its lowest level	
Ntem	-	400 km
Mbam	-	400 km

The north has only two important perennial watercourses:

- The Logone, the main tributary of the Chari which flows into Lake Chad;

- The Bénoué, with 350 km of its course in Cameroon, flows into the Niger after having received Mayo Kebi as its main tributary, with an average flow of 400 m^3 s, which can decline in April to 2 m^3/s and increase at flood crests to 6,000 m^3/s at Garoua.

From September to October these two rivers flood vast areas; these flood waters act as reserve stocks during the low-water season.

Geology

Crystalline and metamorphic formations outcrop over 90 per cent of the territory of Cameroon. The sedimentary areas occupy the remaining 10 per cent, i.e. about 45,000 km^2.

The following are the main geological units:

The Precambrian basement rock (subdivided into three systems): the Lower Precambrian, a complex of granites, micaschists and migmatites; the Middle Precambrian (south-east, north of the Adamaoua around Poli), made up of a series of schists-amphibolites with intercalations of quartzitic sandstones and phyllites; the generally non-metamorphized Upper Precambrian, made up of tillites of sandstone and schist intersected by granitic intrusions (Lower Dja series).

The sedimentary formations are of Mesozoic and Cenozoic age.

Several clearly delimited sedimentary basins are found in the north and west of the country.

The coastal basin of Douala is composed mainly of marine sandstones and limestones of Cretaceous age which in places achieve thicknesses of 1,000 to 2,000 m. The base sandstone is overlain by Paleocene, Miocene and even Quaternary sands.

The Lake Chad basin is composed of a succession of mainly sandy-argillaceous formations from very old to very recent in age:

- The sandstone-clay interbedded continental;
- The terminal continental - 200 to 400 m of sands with interbedded sandstones;
- The Lower Pliocene represented by sands;
- The Upper Pliocene including multicoloured clays;
- The Early Quaternary with clays and sands, sometimes in lenses;
- The clay-sand alluviums of the Lake Quaternary and recent alluviums.

The Garoua basin includes Cretaceous formations with the following main subdivisions:

- Lower series of clays and sandstones in the south-east;
- Upper series of sandstones in the west of the basin.

Mention must also be made of the smaller sedimentary basins of Koum, Babouri-Figuil, Mayo Oulo, Mayo Sorawel, Mamfé, Bamusso and Djerem.

A series of parallel fractures running north-east/south-west, with faults whose displacements sometimes reach 1,000 m in the axis of the trench, throw up large volcanic massifs of various ages and compositions from the Adamaoua to the Atlantic. Many thermal springs emerge from these volcanoes.

II. GROUND-WATER RESEARCH

The Hydrogeological Service, a national service in the Department of Mines and Geology, is responsible for the inventory of ground-water resources throughout the country, the monitoring of aquifer fluctuations, the collection of the results of all water-drilling operations and, where necessary, the issue of drilling or pumping permits.

It also acts as adviser to the users of ground water, the final goal being the rational and planned exploitation of ground water.

In 1972 the ground-water project financed largely by UNDP strengthened the Hydrogeological Service by training local drillers and furnishing drilling equipment (drills, compressors, etc.).

Background

The Agricultural Engineering Service was set up in 1952 to take charge of livestock and village water supply in the northern part of Cameroon.

In 1953 the Department of Mines and Geology put Mr. P. Schwoerer, a geological engineer, in charge of the "Hydro-North" mission, which was designed to study hydrogeological problems and offer advice to executing agencies such as the Agricultural Engineering Service.

This mission was replaced in 1961 by the hydrogeological research mission and 1962 saw the establishment at Garoua of the "Water Office", which subsequently became the Hydrogeological Service.

From 1962 to 1964 the Fonds d'aide et de coopération (FAC) financed the study of three hydrogeological systems:

- The big flats, the spreading zone of seasonal rivers or mayos;
- The isolated and discontinuous surface aquifers;
- The dune system (from Limani to Yagoua).

A number of ad hoc studies were carried out from 1965 to 1969, including in particular some regional studies such as those on:

- The Mandara Mountains (C. Dassibat, BRGM), which led to the sinking of wells and construction of a number of dams in the Mandara Mountains;

- The Cretaceous basins (Wakuti Co.). The Middle and Upper Cretaceous basin of the Bénoué and the Lower Cretaceous basins of Hama Koussou, Babouri-Figuil and Mayo Oulo;

- The artesian aquifer of the Chad plain (geophysical prospecting) carried out by Lerici Co.;

- The Logone aquifer (M. Biscaldi, BRGM).

From 1972 to 1975, as part of the ground-water research and exploitation project financed by UNDP and executed by the United Nations, the Office for Hydrogeological Studies (Pisa, Italy) studied a number of priority areas (about 30 sites) for the installation of wells and boreholes with a view to the exploitation of the ground water, with a training sector for national drillers.

This project was extended until 1977 with a second phase consisting mainly of a borehole programme.

From 1978 to 1983 the FSAR project (Special Fund for Rural Actions) took over the UNDP project and implemented a programme of 430 boreholes.

Activities

The Hydrogeological Service sinks wells and boreholes after interpretation of the photo-geological documents and field surveys. It also monitors the drilling works and prepares water-chemistry maps on the basis of several chemical analyses.

It must be made clear that all the hydrogeological work in connection with ground-water research undertaken by the National Service has been concentrated almost exclusively in the northern part of the country, which has a sub-Sahelian climate. Studies have just begun (1985) in other regions, especially in Mbam (Arlab Studies Office) and were to be extended to the rest of the territory (1985).

Equipment and Specialized Personnel

From 1954 to 1955 the "Hydro-North" mission had three light drilling rigs which could reach a depth of 15 m in soft terrain and one Rotary Caly which could reach 80 m, also in soft terrain.

From 1960 to 1964 the hydrogeological research mission and subsequently the Water Office gradually acquired:

- Three Boilot winches mounted on trailers and driven by Bernard diesel motors; this equipment could operate down to a depth of 50 m in soft terrain;

- Four portable Platy and Banka hand drills for shallow alluviums;

- One APEC OK Bonne Esperance drilling rig, convertible from rotary to percussion, down to 200 m;

- Two small Toy and Explorer continuous coring rigs.

With the exception of one Boilot winch, this equipment is not in use.

The Ground-Water Project

Three drilling units supplied under UNDP and World Bank projects are at present being used in a national project: the Ground-Water Project (GWP); the equipment includes:

One Failing 250 rig which can reach 80 to 100 m in soft terrain; this is a rotary rig but used with a compressor it can also operate as a down-the-hole hammer;

- One Stenuick rig which operates only as a down-the-hole hammer with five-inch diameter; it can reach 80 m in crystalline formations;

- One Ingersoll Rand TH 60 convertible rig (rotary and down-the-hole hammer); this is the most recent acquisition and it can exceed 400 m with three-inch diameter in any formation.

Well-digging equipment has been furnished under the UNDP project. It includes compressors, pneumatic hammers and motor winches.

The Hydrogeological Service has a well equipped laboratory with research and testing equipment such as limnimetric probes to measure the water levels in the holes, screens for granulometric studies, submersible pumps for pumping tests, and a number of pH-meters, conductivity meters, etc. It also uses Johnson electro-geophysical and diagraphy equipment.

The personnel includes two engineers, two geologists, four technicians and a number of drillers.

The ground-water project is directed by a geologist assisted by an administrator.

Prospecting methods

At the beginning of the implementation of the FSAR project a geophysical prospecting campaign was carried out in the most difficult areas (bedrock, thin alterite stratum under zones of basement uplift, dune zone) by a BRGM geophysicist working in collaboration with two Cameroon senior geophysical technicians. This campaign added to the electro-geophysical data produced by the Lerici, Hydrogéo and Hydro-plan campaigns.

A BRGM hydrogeologist then worked with the team of geologists on the aerial photographs, the results of the geophysical campaign and the data collected in the course of field surveys for installations in areas of discontinuous aquifers (dune zone, Mandara Mountains) and in crystalline formations (syenite of Mindif, Kaélé, etc.).

In Cameroon geomorphological observations and structural geology form the basis of hydrogeological prospecting. The available aerial photographs do not cover all the villages concerned. The most frequently used research tools are the inventory of water points and the structural map showing faults, discontinuities, fractures and vegetation. They generally give fairly satisfactory results.

III. GROUND-WATER RESOURCES

Cameroon has three hydrogeological systems:

- The crystalline basement rock;
- The small sedimentary basins;
- The areas of widespread aquifers, mainly Logone-Chari-Chad.

Precambrian and metamorphic crystalline basement rock

Crystalline rocks cover nine-tenths of the total area of Cameroon. There are also many more areas with outcrops of bedrock (Mindif, Guider) and areas in which the rocks are covered with a more or less thick layer of alluviums and alterities.

In these regions the aquifers are usually small and limited to favourable sites where the rock is fractured and faulted. The best yields are obtained in areas where the fissured rocks are overlain with a thick stratum of alteration or alluvium (big flats of the mayos).

The yields are usually 1-3 m^3/h in the region of the Mandara Mountains and can attain 30 m^3/h in places in the Kribi region (Hévéacam drilling on a covered fault).

Small Cretaceous sedimentary basins

The sedimentary basins of Hama Kossou, Babouri-Figuil, Sorawel and Mayo Oulo contain aquifers from which wells and boreholes can draw a maximum of 5 m^3éh per unit.

The basin of Mayo Oulo has an area of 120 km^2; it contains two aquifers: one phreatic and one artesian, tapped by a GWP borehole in 1982 at Sorawel between 120 and 148 m and giving a yield of 500 l/h.

The Koum basin has an area of 1,200 km^2; water is found in the sandstones at Gouga, but its characteristics are not known.

Widespread aquifers

In the Bénoué basin, which has Lower and Upper Cretaceous formations, the permeability coefficient (K) is between 10^{-5} and 10^{-4}m/s (Garoua sandstone). The transmissivity coefficient (T) is between 3×10^{-3} and 2×10^{-2} m^2/s.

The storage coefficient (S) is 2.5×10^{-2}.

The sub-outcrop aquifer of the Quaternary basin of Chad is fed deep underground by the watercourses (Chari, Logone, Serbewel). The transmissivity (T) ranges from 4×10^{-5} m^2/s to 7×10^{-4} m^2/s.

In the Douala basin there are four aquifers located in:

- The base sandstones exploited by shallow boreholes in outcrop zones;

- The Paleocene sands which cover 80 km^2;

- The Post-Eocene and Oligocene formations;

- The Quaternary formations exploited by industries at Bassa to the extent of 1,400 m^3/day.

In the area of Mayos Tsanaga and Kalia the piedmont aquifers have variable yields of 40 to 80 m^3/h; they are exploited to supply Maroua. $T = 8 \times 10^{-3}$ to 4×10^{-2} $S = 2.5 \times 10^{-2}$.

The depth of the aquifers and the installations (in order of magnitude) is as follows:

Geological period; aquifer depth	Geographical location	Depth of wells and boreholes (m)
Cretaceous Bénoué sandstones - 30 m in the west, rising almost to the surface near the Bénoué	Northern Cameroon	40 to 200
Paleocene sand aquifer, 20-37 m	Douala Basin	60 to 200
Quaternary terminal continental and intercalated, 7-36 m	Chad Plain	30 to 600
Small Cretaceous basins	Hama Kossou, Babouri-Figuil, Mayo Oulo	120 to 180
Piedmont aquifer - Quaternary alluviums on crystalline bedrock	Northern Cameroon	12 to 39

IV. CHEMICAL COMPOSITION OF THE GROUND WATER

The chemistry of the water varies throughout the country. In the Bénoué basin the water has a very low mineral content and a low pH content (5.1) and it is therefore very acid.

The basic composition is as follows (mg/1):

Dry residue	= 50	HCO_3	= 24.4
Ca	= 2.4	Cl	= 1.2
Mg	= 0.9	SO_4	= 9
Na	= 1.3	NO_3	= 2.1

In the Chad basin (Logone, Chari, Chad) the salt content varies greatly from south to north near Lake Chad. The water generally has a low mineral content.

Hôpital de Yagoua pH = 6.2 Composition in mg/l

Calculated dry residue	= 102		
Total salinity	= 129		
Ca	= 16	HCO_3	= 44.6
Mg	= 4.4	NO_3	= 0.2
Fe	= 0.004	PO_4	= 0.2

Tildé

pH	= 6.5		
CaO	= 53	Cl	= 3
MgO	= 17	SO_3	= 14
K_2O	= 57	SiO_2	= 54
Na_2O	= 27	NO_3	= 20

Douala Basin (Paleocene)

Water with very low mineral content (dry residue below 170 mg/l)

pH = 4 to 6

CO_2 = 40 mg/l

Fe = 0.7 to 3.5 mg/l

In the crystalline rocks the water is usually more mineralized than in the other areas, as can be seen in from the following three examples:

	Boula-Ibi (granites)	Mindif (syenites)	Badjengo (embrechites)
pH	7.6	7.4	7.1
Cl	40	68	8
P_2O_5	-	-	3
SO_3	14	24	8
SiO_2	50	50	74
CaO	184	202	85
MgO	66	35	29
K_2O	20	-	8
Na_2O	77	78	25

The water in the sedimentary formations, although of good chemical quality, usually has a high iron content, for example 1.4 mg/l in the Akassou Yagoua borehole and 3.7 mg/l at Douala.

Although many local studies have been carried out, it is still not possible to evaluate the ground-water resources of Cameroon or to draw up a water balance. The periods of drought in recent years, which have caused wells to dry up, have underlined the importance of making observations of the fluctuations in aquifers and studying the measures to be taken for their recharge.

The dune zone, i.e. the vast and wide belt of dunes from Limani to Yagoua, poses serious problems for ground-water research. The drilling operations in this zone have so far all produced very disappointing results, despite the joint use of aerial photography and geophysical techniques. The intention is therefore to extend the prospecting into the underlying crystalline formations between 30 and 45 m deep.

V. EXPLOITATION OF GROUND WATER

Bodies concerned

Ground-water exploitation is the responsibility on the one hand of the Department of Agricultural Engineering and Village Hydraulics, which digs the wells and waterholes to supply the villagers and their livestock, and on the other hand of the National Water Company of Cameroon (SNEC) which has a monopoly of water distribution in the urban centres. SNEC is a semi-public company represented in each province by a regional office. It commissions and finances the exploitation boreholes to supply towns such as Douala, Garoua, Maroua and Kousséri. The Department of Agricultural Engineering is also represented in each province.

Cameroon has two private drilling companies:

- FORACO (Forage rationnel construction)

- SOTRAHY (Société de travaux hydrauliques)

and also a drilling service which operates mainly in rural areas: the Ground-Water Project (GWP).

This project is an offshoot of UNDP project CMR 74/013. It has the three drilling rigs mentioned above, namely a Failing 250, a Stenuick and an Ingersoll-Rand.

In several months of field operations per year GWP has drilled 84 to 120 boreholes 40 m in depth, i.e. a total of 3,500-5,000 m.

The personnel includes four drillers and three assistant drillers trained on the site. The technical qualifications of the drilling personnel needs to be somewhat improved, especially with respect to the use of muds, the installation of filter blocks and the repair of the machinery.

SOTRAHY is a small drilling company which has been operating in Cameroon since 1979. Depending on the year and the number of orders, it has drilled between 800 and 2,000 m a year (exceptionally, 5,000 m).

FORACO is a French drilling company operating in several African countries. The metrage drilled in Cameroon is very variable and depends on orders: from 600 to more than 3,600 m/year.

Users

The main users of ground water are SNEC to supply certain urban centres, the Ministry of Livestock for the watering of animals and the Department of Agricultural Engineering and the National Rural Development Fund (FONADER) for village water supplies.

The main towns supplied with ground water are:

Garoua: with 10 boreholes the town of Garoua consumes 15,000 m^3/day, to which must be added the production of a number of private boreholes for industrial use such as those of Cameroon Breweries (1,000 m^3/day), SOCAPROD, the Veterinary Laboratory and the Office of Scientific Research. A total amount of 1,700 m^3/day is drawn off from the Garoua aquifer.

Maroua: 3,500 m^3/day are exploited by SNEC with four boreholes, to which must be added a number of private boreholes, making a total of 5,000 to 6,000 m^3/day;

Guider: a drain in Mayo Guider gives a variable yield of 500-750 m^3/day in high-water periods, decreasing at low water (April-May) to 250 m^3/day, which is quite inadequate for the town's needs;

Kousséri: two artesian boreholes producing 500 m^3/day feed part of the town; six shallower borehoes with yields of 6-10 m^3/h installed by GWP and equipped with hand pumps with a capacity of 700 l/h supply the peripheral districts; the total extraction is about 640-700 m^3/day;

Waza: to meet the hotel requirements in the National Park, two boreholes operating alternately each delivers 10 m^3/h, a total of 240 m^3/day;

Douala: at present 12 SNEC boreholes produce 60,000 m^3/day. In addition to these major exploitations, mineralized water is drawn off from the Adamaoua mainly to water livestock, and ground water is drawn off for bottling; for example, at the northern tip of the Douala sedimentary basin Cameroon Breweries market water drawn from a borehole. Water from the Muyuka basalts will also be exploited by borehole in the near future and sold as table water.

In the rural areas and especially in north Cameroon all the villages have at least one well dug by the Department of Agricultural Engineering. The yields are very variable: from a few m^3/h to nil in the dry season.

At present the region has about 340 boreholes equipped with hand pumps discharging 700-800 l/h distributed among the Departments of Bénoué, Mayo Danai, Margui Wandala, Logone et Chari, and Diamaré.

Ground water is not yet used for irrigation purposes. This is planned at Afade (Kousséri region) for rice growing. The medium-term requirements (1990-1995) are estimated as follows: 30,000 m^3/day for Garoua, 100,000-200,000 m^3/day for Douala (100 litres per day per inhabitant), 5,000-10,000 m^3/day for Maroua, and 1,500-2,000 m^3/day for Yagoua.

The exploitation of ground water is not envisaged for other towns.

Problems

At Douala the problem of salt-water invasion exists above a certain level of exploitation and over a fairly well defined area.

At Garoua where the stocks are limited (45,000-50,000 m^3/day) and at Maroua uncontrolled exploitation may lead to the rapid exhaustion of the aquifer. Already at Garoua there has been a significant decline in the aquifer (10 m) in the area exploited by the oldest boreholes.

Water-balance studies must therefore be carried out immediately (recharge, exploitation) to determine the protected areas reserved exclusively for urban supply. It might even be decided that certain aquifers could be exploited only for this purpose.

VI. CONCLUSION

Importance of ground water

In Cameroon the importance of ground water varies from region to region. In the provinces of the north and the far north (Guider, Poli, Tcholliré) ground water is the only resource available during the long dry season, i.e. from October to June, for the mayos (seasonal rivers) and the waterholes are then dry. On the other hand, south of the Adamaoua the water is more abundant: most of the villages are supplied from surface water (springs, rivers, etc.). Only the risk of pollution during the low-water period threaten the supply to Douala, which already has integrated use of ground and surface water.

Ground water is vital for the northern part of Cameroon. The establishment of permanent water points close to concentrations of population enables the people to give more time to agricultural or domestic tasks by reducing the distances they have to go to obtain water. It must be realized that when the water point is 10 km from the village, the walk there and back to fetch water takes between five and seven hours for buckets or containers of 15 to 30 litres. It takes less than an hour when the wells or boreholes are close to home. Furthermore, the installation of modern water points reduces the incidence of water-borne diseases such as bilharziosis. The improvement of sanitation standards promotes increased productivity and can lead eventually to better living conditions.

Cost of ground water

The cost of the water is very variable. It depends on the cost price of the water point (borehole or well), its capacity and the means of drawoff (foot or hand pump or bucket and rope). The unit cost per cubic metre is calculated on the basis of the cost price of the installation, its probable life and the various costs of maintenance and drawoff. For the existing 400 or so water points (usually boreholes equipped with foot or hand pumps) a distinction must be made between satisfactory water points and boreholes which are used for about 10 hours a day and those with low yields which are used almost 24 hours a day.

On the basis of an amortization period of five years, the average cost of the water is about CFAF 150 per cubic metre (about US $0.40 in April 1986).

The cost is roughly the same for the 20-metre wells dug by Agricultural Engineering.

Given the increasingly severe drought, especially in the north of the country, it is becoming ever more necessary to extend the search for and development of ground water to meet human and livestock requirements.

Operation FSAR II (Special Fund for Rural Actions) initiated in 1984 envisages the installation of 1,000 village boreholes equipped with hand pumps, the creation of maintenance facilities at village and department level and the installation of piezometers in the far-north province.

The Hydrogeological Service is to be strengthened in staff and equipment and it will be responsible at the national level for the inventory, protection, exploration and exploitation of ground water, for which operations it has its own drilling equipment furnished under the Ground-Water Project.

VII. REFERENCES

Tillement (1962) — Hydrogéologie du Nord-Cameroun, doctoral engineering thesis, Univ. Cl. Bernard-Lyon.

C. Dassibat (1967) — Etude hydrogéologique des monts Mandara, BRGM report YAO 67 A6.

Lerici C.M. (1967) — Etudes hydrogéologiques au Nord-Cameroun, Fondation Ing. C.M. Lerici de politecnico, final report 427.

Rister J.J. and Dassibat C. (1965) — *Mission de recherche hydrogéologique-plaine de piémont FAC/BRGM*, BRGM report YAO 65 A 13.

R. Biscaldi (1970) — *Hydrogéologie de la nappe phréatique du Logone-Chari-Tchad, rapport de fin de mission 1969-1970*, BRGM report YAO 003.

P. Peaudecerf (1970) — *Etudes hydrogéologiques de la nappe phréatique du Logone-Chari-Tchad, compléments hydrodynamiques*, BRGM report 70 YAO 005.

United Nations, Final Report (1975) — *Recherche et exploitation pilote des eaux souterraines dans le Nord-Cameroun*.

G. Martin (1978) — *Alimentation en eau des villes de Garoua-Maroua et Guider, études et travaux, mai 1977*, BRGM report 78 AGE 002.

G. Martin (1979) — *Alimentation en eau de la ville de Douala à partir des eaux souterraines*, 79 AGE 021.

Théophile Mba Mpondo (1980) — *Sur les possibilités d'alimentation de la zone résidentielle de Mindif par les forages*.

CIEHéBRGM (1976) — *Carte de planification des ressources en eau souterraine des Etats membres du CIEH de l'Afrique soudano-sahélienne*.

ARLAB (1985) — Subcontractor of the National Fund for Rural Development. *Programme d'urgence d'hydraulique villageoise dans les provinces du nord et de l'extrême-nord, phase I, final report*.

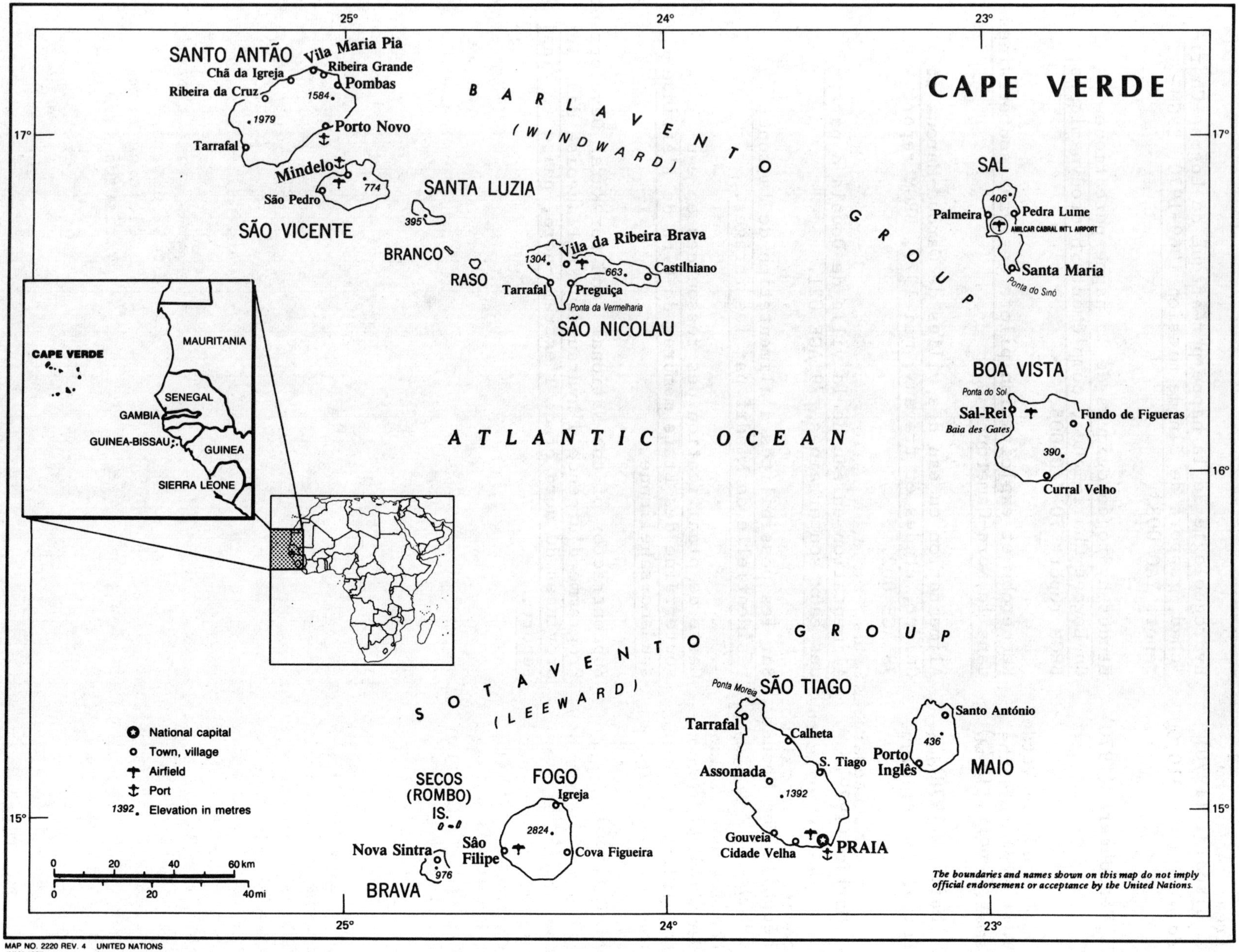

MAP NO. 2220 REV. 4 UNITED NATIONS
JULY 1984

CAPE VERDE

Area: 4,033 km^2

Population: 310,000 (United Nations estimate, 1983)

I. BACKGROUND

Physical geography

The Cape Verde Archipelago is located about 700 km off the coast of West Africa on the latitude of Dakar. It has 10 main islands and five smaller islands which form two groups:

- the Windward Islands, including Santo Antao, Sao Vicente, Santa Luzia, Sao Nicolau, Sal and Boa Vista;

- the Leeward Islands, including Maio, Sao Tiago, Fogo and Brava.

Sao Tiago is the biggest island, with an area of 991 km^2. It has the country's capital which is also the largest town: Praia.

From the geomorphological standpoint there are two types of island: the low flat islands of Maio, Boa Vista and Sal, and the high islands of Santo Antao, Sao Nicolau, Sao Tiago, Fogo and Brava, with elevated topography and pronounced relief with slopes of more than 45 degrees in many places. The maximum altitude (2,289 m) is reached at the peak of Fogo volcano only six kilometres from the sea. Here are found almost all the forms of landscape typical of the volcanic activity which produced the islands: basaltic mesas deeply gashed by river erosion, and sharp peaks and crests, all evidence of ancient volcanic formations, but also well preserved volcanic cones produced by more recent eruptions. Fogo is remarkable for its enormous "caldron" from which emerges the present 1,000-metre volcano. The steep coastlines of the high islands, with cliffs of 100 m and higher, contrast sharply with the broad sandy plains of the low islands.

Climate

Cape Verde's climate is of the dry subtropical type; it corresponds to the mainland climate in the Sahelian zone on the southern edge of the Sahara. Almost all the islands are subject to the trade winds blowing from the north-east and east which are usually dry and cool. The rainy season is short and lasts from July to October. The precipitation is caused by the northwards displacement of the zone of intertropical convergence, but the islands are usually affected only by the northern sub-zones with their low moisture content. This means that the rainfall is almost always torrential and irregularly distributed in space and time, delcining from south to north, as can be seen from the annual rainfall statistics (1950-1973) for the low islands:

Maio - 219 mm

Boa Vista - 128 mm

Sal - 84 mm

In the high islands these values are affected by the relief: they can double or even triple. The following average amounts of rainfall were measured on Sao Tiago (1949-1970):

Praia (altitude: 64 m) - 260 mm

Trindade (205 m) - 355 mm

Curralinho (950 m) - 906 mm

There can be some light rain in the dry season, more especially at high levels and on the windward slopes. This occurs when cold air from the north meets the high land.

The temperatures are moderated by the small size of the islands and the influence of the ocean. The monthly averages rarely reach 30°C. Praia has recorded a maximum daytime temperature of 36.2°C (1941-1960) and a minimum of 16.2°C. The average monthly temperatures range from 26.8°C in September to 22.2°C in February.

Owing to the hot sunshine and the strong winds, the average evaporation (Piche) is high, especially in the coastal areas:

Praia, Sao Tiago (1941-1960) - 2,730 mm/year

Mindelo, Sao Vicente (1931-1960) - 1,684 mm/year.

On the upper slopes the evaporation does not exceed 110 mm per month. The evaporation calculated for 1981 by the Penman method was high:

Praia (64 m above sea level), Sao Tiago - 2,226 mm

Trindade (205 m), Sao Tiago - 1,773 mm

Sao Jorge (3.9 m), Sao Tiago - 1,304 mm

Population

Nine of the 10 main islands are inhabited. In 1980 the total population was 296,093, distributed as follows:

Island	Total	Population Urban (percentage)	Rural (percentage)
Boa Vista	3,397	39	61
Brava	6,984	26	74
Fogo	31,115	13	87
Maio	4,103	35	65
Sal	6,006	70	30
Sao Tiago	145,923	33	67
Santo Antao	43,198	20	80
Sao Nicolau	13,575	15	85
Sao Vicente	41,792	88	12
Total	296,093	36	64

Thus the population is mainly rural, except on Sao Vicente and Sal where almost all the people live in the towns. The main towns are Praia (Sao Tiago) with 37,480 inhabitants and Mindelo (Sao Vicente) with 36,265 inhabitants. The average age is 25. Half the population is under 20.

One peculiar demographic feature is the lack of population in the 30-40 age group. If the annual growth rate (three per cent) and the emigration rate remain at their present levels the population will be 485,000 in 2000. However, according to the National Development Plan, it would be undesirable for it to exceed 420,000.

Surface water

Cape Verde has no lakes or permanent watercourses; the small size of the hydrographic basins, usually narrow and long, and the steepness of the slopes produce flash floods in the valleys and large quantities of solid matter in the water. A small amount of runoff can still occur some days after the rains.

There is hardly any means of measuring the rate of flow of the country's watercourses. But some estimates of surface runoff have been made (United Nations, 1980):

Island	mm/year	Rainfall (percentage)
Sao Tiago	108	34
Fogo	182	40
Brava	122	35
Maio	15	15
Boa Vista	10	13
Sal	10	13
Santo Antao	125	36
Sao Nicolau	37	25
Sao Vicente	10	13

Experiments are being carried out at present (1984) in two basins on Sao Tiago to measure, *inter alia*, the rates of flow and runoff. A first partial and approximate measurement has indicated a value of 14 per cent.

Geology

Although a considerable amount of geological work has been carried out in the Archipelago during the past 20 years, it has not advanced beyond the stage of a general survey.

Like most islands in the Atlantic, the Cape Verde Archipelago is of volcanic origin. Basaltic and pyroclastic rocks cover about 83 per cent of the surface, and phonolites and associated rocks about nine per cent. The petrographic sequence is as follows: 1) basaltic lavas; 2) phonolitic heavily alkaline lavas; 3) recent and contemporary basaltic lavas.

Sedimentary rocks are found on all the islands but over small areas and especially in the peripheral zones. Outcrops are found in notable quantity only on Boa Vista and Maio. They are usually thin, except on Maio where the outcrop of calcareous rocks attains a thickness of 435 m. Calcareous rocks are widespread (90 per cent). Oligocene-Miocene conglomerates are also found. The sedimentary rocks, although not extensive, are of major importance for study of the stratigraphy by virtue of the fossils which they contain.

Geological units

The oldest formations are the marine sediments, mainly limestones and neritic marls of the Lower Cretaceous (Neocomian) period and sometimes of the Upper Jurassic (Portlandian) period, which outcrop on Maio. Between the Cretaceous and Eocene periods complex and prolonged eruptive activity produced what has been called the "ancient internal irruptive complex". This complex forms the central core of the islands. With the exception of Fogo, Brava and Sal where the outcrops are small or non-existent, this complex is readily discernible. It is the formation with the most varied lithology with intrusive granular rocks such as gabbros, syenites, essexites and carbonatites, veins, breccias, basaltic tuffs and small intrusions of phonolites. Breccias and basaltic veins predominate, and sometimes the veins make up 50 per cent of the formation. These are compact rocks, heavily altered on the surface.

Following a long period of erosion, volcanic activity occurred at the beginning of the Miocene era with a very explosive phase which formed in particular the submarine breccias of the Flamengos formation on Sao Tiago with a small and variable proportion of pillow lavas, the breccial conglomerate (continental) of the dos Orgaos formation, also on Sao Tiago, and a few continental breccias on Maio.

After a period of erosion in the Miocene and Pliocene eras, a system of volcanic, mostly basaltic rocks came to occupy a vast area of the islands and form the main reliefs. Sometimes called the "intermediate series", these rocks are known on Sao Tiago as the "eruptive complex of Pic d'Antonia". It is possible to distinguish in this system:

i) A lower explosive part where brecceas and pyroclastic material predominate; it achieves its maximum development in the lower vein complex of Santo Antao and Sao Vicente; there is also a submarine phase of abundant pillow lavas associated with tuffs; it appears mainly on Sao Tiago but also on Sao Nicolau;

ii) The relatively thick upper part can attain several hundred metres; it is composed mainly of basaltic beds alternating with ancient alluviums, pyroclastic deposits and base breccias.

The last great upheavals of the Pliocene era produced the Assomada formation (Sao Tiago) which includes thick basaltic strata in unconformity with the previous series.

With the exception of Fogo, the Archipelago's only active volcano, the volcanic activity ended with the eruptions which produced the recent series of Pliocene-Quaternary age in which uncompacted pyroclastic elements predominate, forming in the majority of cases well preserved volcanic cones with some associated flows. The latter are commoner on Santo Antao.

Owing to the morphological factors (narrow, deep and steeply-sided valleys), alluviums are rare, except on Santo Antao, Sao Tiago and Sao Vicente where they are fairly widespread. In most cases they are poorly sorted and coarse, especially in the lower part, and the sand and limestone content increases in an upward direction.

Tectonic structures

Little is known about these structures; the majority of those which have been studied are pseudo-tectonic or volcano-tectonic. Only three main faults have been identified (Mitchell-Thomé, 1972): a) an incurved north-south fault on the western edge of Campo da Preguiça, Sao Nicolau; b) an inverse fault parallel to the north coast on Santo Antao, responsible for the lifting and northwards tilt of the block between Ribeira Grande and the coast; c) lastly, a large fault running north-north-west/south-south-east in the south-east of Maio, with a displacement of 450 m; this would be part of a deep-lying diapiric structure. The results of gravimetric measurements seem to confirm the existence of a rift caused by the intrusion of essexites under Maio. This is the only island where aerial photography has revealed a dense fault system running in four main directions: north-south and east-west and diagonally. The present work (1984) on Sao Tiago indicates the same main directions and points to the existence of a minor fault running north-east/south-west in the south-east of the island.

The basalts are intensely jointed but there are no detailed studies on this topic. Very few noteworthy unconformities have been identified. The main one is the 75° angular unconformity between the horizontal Tertiary limestones and the Mesozoic limestones. Erosional and sometimes angular unconformities are found in the various volcanic units. They can be regional, such as the one between the ancient internal eruptive complex and the intermediate series/Pic d'Antonia formation, or local such as those found in the latter formation.

The steep slopes of the Mesozoic strata on Maio are exceptional. These slopes are usually gentle throughout the Archipelago: 20° to 30° and even less on Sao Tiago.

The generation of these volcanic rocks in the Pic d'Antonia formation, which constitutes a stratified volcanic complex, and the major or minor erosional unconformities have produced intense lithological heterogeneity.

II. GROUND WATER

Organization of the water sector (1984)

As the table below shows, many bodies play a part in the development and management of water resources, but the main responsibility rests by governmental decree with the Ministry of Rural Development. The activities are distributed within this Ministry as follows:

- The Studies and Planning Office co-ordinates the activities of the various technical services and the many foreign aid programmes;

- The Department of Ground-Water Exploitation and Management is responsible for the prospecting, exploitation and control of ground water resources.

- The Department of Agricultural Engineering is responsible for the construction of reservoirs, public water-supply systems, irrigation channels, small dams, etc.;

- The Department of Soil and Water Conservation Services is concerned with works to protect water sources and control erosion, as well as with artificial recharge;

- The Centre for Agronomic Studies carried out physical, chemical and bacteriological analyses of the water;

- The Centre for Equipment and Workshop Maintenance is responsible for the maintenance of all the Ministry's equipment and vehicles, including drilling rigs and pumps.

In 1985 the Government was considering the creation of a single service for water resources, in conjunction with the preparation of new water legislation.

Ground water research

With the exception of some earlier work, the systematic study of ground water really began in Cape Verde in 1971 under a contract between the Portuguese Government and the BURGEAP study company (Paris). A ground-water unit was set up at the same time, equipped with two drilling rigs. The study lasted two years from 1971 to 1973. The first field operations were carried out in 1972 on Sao Tiago: more than 100 exploitation and observation boreholes were drilled. In addition, an inventory of water points was drawn up and a summary general hydrogeological presentation was prepared, together with a preliminary water balance.

When Cape Verde became independent in 1975 these activities were continued under the Department of Ground-Water Exploitation and Management Services (DSEGAS) of the Ministry of Rural Development, as part of several. bilateral, ultilateral and international co-operation projects financed by the European Economic Community (EDF) on Sao Tiago, by the Fonds d'aide et de coopération (France) on Sao Nicolau and Sao Tiago, by the Netherlands on Santo Antao, by UNDP on all the islands (except Sao Nicolau), by UNICEF on Sao Tiago, Maio and Boa Vista, by the United States on Sao Tiago, etc. The prospecting and exploitation operations have been extended in varying degrees to almost all the islands, with the exception of Fogo and Brava. The first detailed studies of hydrographic basins and other zones have been made on Sao Tiago in order to determine the ground-water potential. The digging of the country's first deep gallery has begun on Sao Nicolao.

Distribution of responsibilities in the water sector

Functions	Zone	Ministry
Agro-climatological studies and research	Whole country	
Drilling and tapping - equipment and maintenance	Whole country	Rural Development
Exploitation and distribution of drinking water and irrigation	Rural areas	
Construction of distribution networds and reservoirs	Urban centres	Public Works
Distribution and sale - network maintenance	All population centres	Interior
Desalination of sea water - distribution and administration	Mindelo and Sal	Economy and Finance
Quality control	Whole country	Health
National meteorology	Whole country	Transport and Communications

The following are the main items of research under way at present (1984):

- Feasibility study of the galleries on Sao Tiago and prefeasibility study on Fogo (EDF(BURGEAP);

- Evaluation of ground-water resources in two pilot basins on Sao Tiago (UNDP/TCD).

DESGAS, whose present organigramme is set out below, employs about 250 persons, six of them relatively young and with university education. Most of the technical and site staff have received practical training "on the job" without adequate prior theoretical training.

DSEGAS has the following prospecting equipment:

- Eight drilling units with equipment;
- Seven compressors;
- Six lorries;
- Nine light vehicles;
- Three test pumps (submersible);
- Miscellaneous equipment (piezometric probes, conductimeters, chronometers, etc.).

Department of Ground-Water Exploitation and Management Services

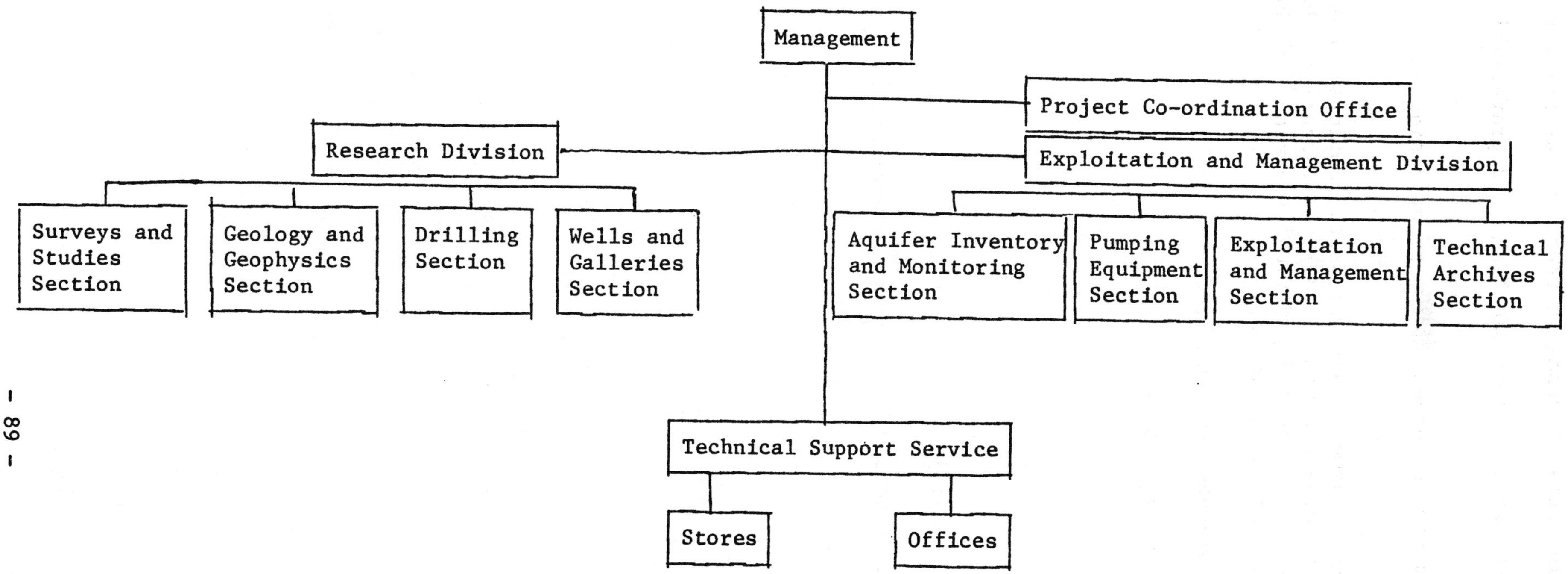

Prospecting Methods

The prospecting methods have always been diverse, but everything depends on the availability of specialized technical staff (hydrogeologists) and the necessary surveying materials (instruments, maps, drilling rigs). As far as possible, the following operations are carried out:

- Preparation or updating of hydrological inventory. Geological field surveys to prepare a geological map and a structural map of the aquifer. It should be noted that the interpretation of aerial photographs helps very little in the determination of the morphological characteristics;

- Preparation of a hydrogeological plan and drilling of boreholes;

- Test drilling with the assistance of a hydrogeologist;

- Exploitation wells;

- Test pumping at constant discharge for 24 hours in the exploitation well;

- Piezometric drilling to monitor the aquifer when the exploitation is fairly intensive or the aquifer is close to the sea.

As indicated above, not all these operations are carried out in all cases. The boreholes are usually drilled without hydrogeological supervision and on the basis of practical criteria (access, proximity of users, etc.).

During the BURGEAP study (1971-1973) a number of electro-geophysical soundings were carried out with fairly satisfactory results, but only recently has this method been used in specific projects. The positive results of the drillings based on this method demonstrate its effectiveness. An experiment with seismic refraction has shown that this method is not suitable, owing both to the geological characteristics and the practical difficulties of implementation.

Description of the main aquifers

Although the study of ground water has been extended to all the islands since Independence, most of the work has been concentrated on São Tiago, and knowledge of the other islands remains very incomplete. Consequently, most of the characteristics described below, in particular the numerical values, relate in fact to São Tiago. However, it is thought that they are equally valid for the other islands, given their geological similarity to São Tiago.

The overall hydrogeological system has not yet been studied in detail.

The general opinion is that there exists a vast single aquifer involving all the formations, with the intermediate series constituting the main reservoir. However, the presence of one or several perched aquifers is not excluded, possibly in part of the intermediate series and, at a greater depth, in the main aquifer or the saturation zone located mainly in the base series.

The aquifers vary greatly in depth, depending on the distance from the edge and the altitude of the soil cover. Difference of more than 100 m can be observed between points only a little distance apart.

It must be noted that the values given in the table below, which summarizes the characteristics of the aquifers, are not entirely representative, for the following reasons:

- They refer only to productive wells;

- In most cases they do not refer to the whole thickness of the aquifer in question;

- The wells have usually been drilled in the valleys. Few are located in level ground ("achadas") where the aquifers are covered by a considerable thickness of volcanic rocks.

Limestones and eruptive limestone complex of Borde (Jurassic-Cretaceous)

These are found only on Maio, surrounding the island's central massif. These fissured limestones, with heavy intrusions of basaltic veins in the eruptive limestone complex constitute the island's main aquifer with stocks evaluated at 55 million m^3 for a saturated depth of 20-30 m below sea level. The great thickness of the formations, their steep slope and their degree of fissuration indicate water circulation at a great depth. The water is fairly saline, with conductivity values of between 1,200 and 3,400 micromhos/cm. The Cl^-/HCO_3^- ratio varies from 1.2 to 1.8.

The geological formations have been grouped into three series according to their hydrogeological behaviour: base series, intermediate series and recent series.

Base series. This is made up of tuffaceous breccia formations of the ancient internal irruptive complex (AC) and the Flamengos (Fl.) and Orgaos (CB) formations. They are fairly compact and impermeable in comparison with more recent formations and act as a substratum. However, they are not totally barren and do constitute a low yield aquifer, with greater porosity in the Flamengos and Orgaos formations. The aquifer studied so far is generally limited to the more fractured upper part, which is some 10 to 30 metres thick. The conditions at greater depth are not known, and it is possible that the islands' true - saturated - aquifer may be found there (see above).

The productivity increases, sometimes attaining very high values, in the veins and especially in those caused by the most recent eruptions. This fact is particularly important in the CA where there are many of these intrusions. The same phenomenon occurs in the Flamengos formation in the few places where pillow lavas are abundant. The recharge in these formations is adversely affected by a thick argillaceous surface alteration, especially in the CA where it can attain five to 10 metres.

Description of the aquifers 1/

Formation and geological age	Island 2/	Production limits (Q) m^3/hour	Transmissivity (T) m^2/day	Storage (S) 3/
Limestones (Jurassic-Cretaceous)	Maio	1-12	70-5,900	0.07
Calco-eruptive complex (Jurassic-Cretaceous)	Maio	4-35	60-190	0.02
Ancient internal eruptive complex (Cretaceous-Eocene)	São Tiago São Nicolau	0-10	1-10	0.03
Flamengos formation (Miocene)	São Tiago São Nicolau	0-45	1-120	6×10^{-4} to 3.6×10^{-3} 0.15
Orgãos formation (Miocene)	São Tiago	0-10	1-4	-
Intermediate series (Miocene-Pliocene)				
- submarine phase	São Tiago	20-100	50-9,000	5×10^{-4} to 10^{-3}
- aerial phase	São Tiago, São Nicolau, Maio, Santo Antao	5-60	5-160	4×10^{-4} 0.05
Alluviums (Holocene)	São Tiago	20-100	800-8,600	-

1/ According to test pumping.

2/ Where the aquifer has been studied and exploited.

3/ Values obtained from very few tests (sometimes only one).

Intermediate series. A number of factors help to make this aquifer the most important:

- Relatively good permeability and porosity owing to the abundance of well-fractured basalts and uncompacted intercalations of alluviums and pyroclastic sandstone;

- Great size in the majority of the islands;

- Favourable recharge conditions in the hilly regions where the rainfall is heaviest;

- Lower section of breccias and tuffs which can act as an impermeable base similar to the CA, giving rise to many springs, as occurs on Santo Antão in the lower complex and on São Tiago in the lower Pic d'Antonia formation.

Apart from the pillow lavas, the most productive zones are usually found in pyroclastic deposits, sandstones and alluviums, with the ancient valleys of erosional unconformity at the base of the series playing an important role.

With their very high permeability and porosity, the pillow lavas of the submarine phase occupy an important position. However, the extreme transmissivity values given in the table above are subject to serious reservation, for they come from tests which recorded downdraws of only a few centimetres and very rapid stabilization. The order of magnitude of the transmissivity coefficient may be 400-1,000 m^2/day.

The irregular distribution of the various products of the subaerial volcanic phase mean a high degree of heterogeneity in the aquifer, with transmissivities ranging up to a factor of 10 over distances of a few hundred metres.

Except in the lower section where they are more abundant and can act as drains, the veins play a secondary role. In the few cases when the boreholes have reached the base of the series, the total depth of the aquifer has attained or exceeded 50 m. An important factor is the position of the base of the series (or even the erosion surface of the base series) in relation to sea level, for this can mean a fairly incurved and probably perched aquifer or indeed a deep aquifer.

At the top of the main aquifer there may be one or several perched aquifers resting on relatively impermeable strata, some of them situated at a high altitude and offering many tapping points and springs.

Recent series. The pyroclastic deposits and the occasional scoriaceous lavas which make up this unit are extremely permeable and porous. Even if they are not aquifers, they do at least constitute excellent recharge zones for the underlying formations. However, this behaviour appears to be due more to the disposition and extent of the outcrops. On Santo Antão where this series is better developed, recent drilling has revealed ground water in the basalts and pyroclastic breccias attributable to this series which have high specific yields (similar to those of the pillow lavas).

Alluviums. Owing to their coarse granulometry they form very permeable aquifers but only in the lower part of valleys where they are sufficiently extensive and thick to permit substantial exploitation. Here they can attain a width of several hundred metres and a depth of 40 metres. A little distance away from the sea, perhaps five kilometres, these values decline rapidly, but the aquifer can still be exploited by dug wells.

Quality of ground water

In order of magnitude the specific hydraulic conductivity indicative of total mineralization ranges between 200 and 5,000 micromhos/cm (± 150 to 4,300 mg/l). It increases naturally in the direction of flow (downstream) but also towards contact with the formations of the base series where it attains values of 1,000 to 3,000 micromhos/cm. The change is quite sudden. The high values are typical of the coastal zones and those above 3,000 micromhos/cm are found on contact with the wedge of salt water. Very high values (9,000-150,000 micromhos/cm) can sometimes be observed in the formations of marine origin as a result of contamination by salt or polluted water. The commonest type is sodium bicarbonate (bicarbonate/chloride according to the most recent analyses). Alkaline-earth ions are also common, especially in the base series, but sulphate is fairly rare.

Few bacteriological analyses have been made so far. They have indicated the contamination of springs and wells by faecal matter, also revealed by the frequent presence of NO_2^-, NO_3^-, and NH_4^+.

Although the salt content is in many cases higher than the drinking-water standards established by WHO, the water can be used for human supply. However, it must be used with care for irrigation, if not by reason of its total salinity, then at least owing to its high sodium content.

Water balance and ground water resources

Of the various headings in the water balance, only the precipitation has been measured directly, but even these measurements have not been made regularly over long periods. This is why various indirect methods of estimation have been used to calculate an approximate value of the balance and/or recharge. In 1974 BURGEAP produced for the recharge the formula $I = 0.25 (P - 300)$ mm, P being the annual precipitation in mm; this formula has been used to estimate the total ground water resources indicated in the table "Ground water resources" (see below).

In 1980 the United Nations project CVI/75/001 established a first provisional balance for the Archipelago, the results of which are given in the table "Trial water balance" (see below).

In 1982 , M. I. Dittrich produced the following values for São Tiago:

	Reception	Interception	Surface runoff	Infiltration	Actual evaporation
mm	360	40	38	50	232
percentages	100	11	11	14	64

The ratio of recharge to average annual rainfall is given by I = 0.21 (P - 142 mm).

For the intermediate series on Santo Antão A. Bosscher (1983) found the recharge/rainfall ratio: I = 0.1 (P - 50 mm).

Lastly, in 1983 BURGEAP established a new balance for the massif of Pico d'Antonia (São Tiago) with a new approach based on infiltration co-efficients ranging between 15 and 50 per cent of rainfall depending on the steepness of the mountain slopes.

In order to give an idea of the order of magnitude, the table "Ground water resources" gives the 1974 BURGEAP estimates and those made by the United Nations in 1980 (project CVI/75/001) (see below).

Trial water balance

Island	Area	Rainfall		Actual evapo-transpiration	Surface runoff		Infiltration	
	km^2	mm/year	hm^3/year	mm/year	mm/year	hm^3/year	mm/year	hm^3/year
São Tiago	991.0	320	317	157	108	108	55	55
Fogo	476.0	450	214	180	182	87	88	42
Brava	67.4	350	24	160	122	8	68	5
Maio	269.0	100	27	75	15	4	10	3
Boa Vista	620.0	75	46	60	10	6	5	3
Sal	216.0	75	16	60	10	2	5	1
Santo Antão	779.0	350	273	157	125	97	69	54
São Nicolau	388.0	150	58	90	37	14	23	9
São Vicente	227.0	75	17	60	10	2	5	1
Total	4,033.4	-	992	-	-	328	-	173
Average	-	246.4	-	122.3	81.2	-	42.9	-
Percentage of precipitation		100		49.6	33.0		17.4	

Data from United Nations project CVI/75/001 (1980).

Ground water resources

m^3/day	United Nations - 1980		BURGEAP 1974
	Available	Exploitable	Total
Boa Vista	8,000	1,600	1,200
Brava	14,000	7,000	4,500
Fogo	120,000	60,000	60,000
Maio	8,000	2,000	3,000
Sal	3,000	300	300
São Tiago	150,000	85,000	60,000
Santo Antão	145,000	71,000	80,000
São Nicolau	25,000	8,000	12,000
São Vicente	3,000	500	800
Total	476,000	235,400	221,800

III. EXPLOITATION OF GROUND WATER

Governmental Bodies

Although many bodies are involved in the exploitation and use of ground water (see above), only the Department of Ground Water Exploitation and Management Services is concerned with the construction of tapping and drawoff installations up to the wellhead. The Drilling Section employs 23 persons, including seven drillers. Their training is mainly practical; some of them have more than 10 years work experience, but none of them has more than four years of primary education.

The Section has eight drilling units with the following equipment:

Equipment model	Type	Capacity
1 Koehring-Speedstar SS-15-III	Normal rotary and air with compressed air hammer (down-the-hole)	300 m - 6" 150 m - 10"
1 Bonne Esperance FBE 2 N GC	Idem	300 m - 6" 150 m - 8.5"
1 Stenuick HS66D (modified)	Air rotary with compressed air hammer	120 m - 8" 20 m - 6"
1 Stenuick HS66D	Idem	90 m - 8"
2 Schott-Dubon	Cable-tool percussion	150 m - 8" 70 m - 17" 1/
2 Longyear	Coring 2/	250 m - 2"

1/ Actual capacity much lower owing to the lack of test tubing and difficulties in hard rocks.

2/ Mainly for geological surveying (galleries, foundations, etc.)

The yield of the rigs is very irregular; it is not unusual for equipment to be out of commission for a year or more, almost always as a result of mechanical difficulties. From 1980 to 1984 a total of 223 boreholes was drilled with a total depth of 17,784 m, i.e. an average of 45 per year or 2,957 m. Of this total, 24 per cent has proved productive, 26 per cent was drilled for observation and/or study purposes, 25 per cent was drilled for the laying of foundations or mineral prospecting, and 25 per cent was abandoned.

There is also a Wells and Galleries Section with a staff of 28, which is concerned with the digging of wells and construction of installations to tap springs or the building of small galleries. This work is generally done by hand, with the help of a compressed-air drill in the case of galleries.

Tapping and use of ground water

Tapping methods. As was stated above, the first drilling for water was carried out in 1972. Up till then the ground water had been taken from springs, filtration galleries and dug wells.

At present the springs are tapped by different types of installation, either small accumulation basins or actual small galleries no more than a few metres in length, or a few dozen metres in exceptional cases. However, there are longer galleries: the one at São Martinho Pequeno (São Tiago) is probably the longest at 710 m. The first long gallery (2,200 m) of recent times was constructed on São Nicolau with technical assistance and supervision by BURGEAP. From 1950 to 1960 many filtration galleries were built, first in the longitudinal and then in the transversal alluviums along underground dams. They still produce large quantities on Santo Antão, but on São Tiago and São Nicolau where they are located in alluviums of little depth or close to the surface, many have dried up or reduced their yield. The system of springs and filtration galleries continues to be the most important means of exploitation, supplying about 50 per cent of all the ground water extracted in the country. On Brava, Fogo and Santo Antão it is virtually the only means of exploiting ground water.

As might be expected, the dug wells are located mainly in alluviums, but many of them also exploit the underlying volcanic rocks which usually belong to the base series. Only rarely do they penetrate more than one or two metres into the aquifer; their productivity is therefore very low. In thick alluviums their yield rarely exceeds 100 m^3/day; they are built with diameters of two to five metres and depths of four to 20 m; their exploitation by motorized surface pumps for irrigation is relatively recent, and drinking water is usually drawn off in buckets. Wells are particularly numerous on São Tiago, accounting for about 50 per cent of the supply.

Drilling for water is usually carried out by means of compressed-air drills with diameters of 200 to 250 mm. The installations are furnished with 150-160 mm PVC tubing. Their depth has been pushed beyond 100 m in recent years, and as far as 170 m. In June 1984 there were 142 exploitation boreholes producing some 20,000 m^3/day. They are very unevenly distributed: most of them are on São Tiago and there are none on Brava, Fogo or Santo Antão. The commonest drawoff method is by vertical-spindle turbine pump driven by a diesel motor. About 50 per cent of the boreholes have this equipment; the next commonest method is wind power - 20 per cent; the remaining 30 per cent is produced by submerged electrical pumps, surface turbine pumps with diesel motors, and a few hand pumps and pumps driven by aerogenerators.

The table below (Exploitation of ground water) gives the daily yields for each island by these various methods. Although the figures relate to different periods, they do give a fairly accurate idea of the present situation.

Exploitation of ground water

(m^3/day)

Island	Springs and galleries	Dug wells	Boreholes No.	1/	Total
Boa Vista	22 1/	480 5/	6	210	712
Brava	-	-	-	-	2,500* 6/
Fogo	-	-	-	-	1,500* 3/
Maio	10 5/	1,160 5/	11	1,730	2,900
Sal	-	270 3/	2	40	310
São Tiago	10,200 2/	19,600 2/	94	15,836	45,636
Santo Antão	-	-	-	-	28,000* 3/
São Nicolau	1,200 4/	200 4/	26	1,532	2,932
São Vicente	-	1,500 2/	3	650	2,150
TOTAL	-	-	142	19,998	86,640

* Almost exclusively from springs and galleries.

Data produced by:

1/ Livramento, D.R. and Pieterse, N., 1984.

2/ Fernandopulle, D., 1979.

3/ United Nations project CVI/75/001, 1980.

4/ BURGEAP, 1981.

5/ Beurskens, H.J.M., 1981.

6/ BURGEAP, 1974.

Utilization. For various reasons, in particular because the water is drawn off for many different uses, it is difficult to obtain reliable values for the amounts of water consumed by each use.

The larger part - if not all - of the water is used for irrigation and as drinking water. The BURGEAP estimates for 1980 are reproduced in the table "Ground-water utilization in 1980 (according to BURGEAP estimates, 1981)" (see below).

The value for drinking water seems to be consistent with the existing delivery systems which supply about 170,000 people, or some 58 per cent of the population, usually in an inadequate manner. The remaining 42 per cent of the population take their supplies themselves from springs and dug wells in a total amount estimated at 2,500 m^3/day. The total production of water for human supply is in the order of 4,600 m^3/day.

Most of the 68,000 m^3/day used for irrigation is consumed on São Tiago and Santo Antão, which have 94 per cent of the irrigated land - a total area of 1,790 ha. According to the Central Corporation for Land and Rural Development, Investment and Resource Management (SCETAGRI) (1981), sugar cane is grown on 1,029 ha (57 per cent), potatoes on 355 ha (20 per cent) and bananas on 181 ha (10 per cent). The remaining 13 per cent of the land is used for the production of vegetables, maize, manioc, etc.

Cost of ground water

The ex-well production costs of ground water 1/ are given below on the basis of BURGEAP data (1981). They do not include the costs related to the distribution networks.

	Tapping	Drawoff	Total
Boreholes			
- São Nicolau	2.5	3 to 27	5.5 to 29.5
- São Tiago (J. Varela)	1	16	17
Private dug wells	3	2 to 5	5 to 8
Springs	2	0	2
Galleries	3.5 to 7.5	0	3.5 to 7.5

1/ Cape Verdian escudos per m^3 (US $1 = 50 Cape Verdian escudos).

These figures have been obtained by different methods of calculation and thus they are not fully comparable; they do however indicate the order of magnitude of the values. It can be seen that the drawoff, which includes the costs of equipment and operation, has a strong influence on the total cost.

Deterioration of the aquifers

The main difficulties encountered in the exploitation of ground water are connected with the intense drought which has affected the country since 1970. It has caused a considerable decline in piezometric levels, and many springs and galleries and many dug wells have dried up or had their yields reduced. In addition, some boreholes drilled since 1970 to combat the effects of the drought have been affected by declines in piezometric levels sometimes in excess of 10 m. The overexploitation of the aquifers has also contributed to these declines, but it is difficult to quantify the relative effects of these various causes.

The most seriously affected aquifers are those located at altitude which are small in size and/or perched, especially alluvial aquifers. Most of the filtration galleries constructed in these aquifers have now dried up. The large aquifers found in favourable recharge conditions, for example the aquifer in the Pic d'Antonia formation, have not suffered so much. On São Tiago, for example, the discharge of the tapped springs of Aguas Verdes which supply Praia has hardly declined at all. In contrast, on São Nicolau where there are hardly any springs in contact with the substratum, the number of springs declined by 60 per cent between 1969 and 1980 and by 80 per cent between 1956 and 1980.

Sea-water intrusion has occurred in some coastal zones, in particular in the alluvial aquifers and pillow lavas. This intrusion has not usually been very extensive. However, in places it has caused the abandonment of a number of dug wells used for irrigation. It has probably been more extensive on the low islands; for example, on Maio in the area of Calheta sea water had penetrated in 1980 up to 2 to 3 km inland, apparently under the combined influence of the decline of the already low piezometric level, exploitation, drought and perhaps reforestation. In order to improve the aquifers by artificial recharge, many small dams have been built in the valleys. These are not really artificial recharge installations but rather installations designed for multiple purposes, such as erosion control, control of the torrential runoffs, and the creation of additional arable land. This last objective involves rapid filling with fine and coarse sediments, which casts doubt on the effectiveness of these installations for artificial recharge.

Ground-water utilization in 1980
(according to BURGEAP estimates, 1981)

Island		Drinking water (m^3/day)	Irrigation (m^3/day)
São Tiago total		1,340	33,800
Town of Praia	900		
Towns of Assomada, Tarafal and Porto de Pedra Badejo	240		
Rest of the island	200		
Santo		145	30,500
São Nicolau		112	1,500
Brava		-	1,400
Maio		14	680
Fogo		300	280
Boa Vista		70	80
São Vicente		100	400
Total		2,081 (+2,500) 1/	68,640

1/ About 2,500 m^3/day should probably be added for personal supplies not taken from the water-supply networks.

Requirements and future development

According to the first National Development Plan, the long-term goals (year 2000) require the following water resources:

Drinking water

Rural population (260,000): 11,200 m^3/day

Urban population (160,000): 5,200 m^3/day

Total: 16,400 m^3/day

Irrigation

(3,600 ha) 137,000 m^3/day

Industry 900 m^3/day

Total (year 2000): 154,300 m^3/day

The available ground-water resources, estimated at 220,000 to 235,000 m^3/day (see above), are generally sufficient to meet these requirements. However, not all the resources are easy to exploit. In the light of the various technical and economic difficulties which must be overcome in mobilizing the resources, a number of estimates have been made of the production increase (m^3):

	BURGEAP 1974	United Nations 1980	BURGEAP 1981
Boa Vista	600 x 2	600	1,200 - 1,700
Brava	1,000 x 2	2,300	1,800 - 2,200
Fogo	5,000 x 2	58,500	15,800 - 27,000
Maio	1,500 x 2	1,500	2,200 - 2,600
Sal	300	-	150 - 250
São Tiago	15,000 x 2	54,000	15,500 - 20,500 (+ 5,000)
Santo Antão	10,000 x 2	43,000	17,000 - 23,000
São Nicolau	3,000 x 2	5,600	3,400 - 4,600 (+ 1,800)
São Vicente	200	-	1,100 - 2,100
Total	72,700	165,500	58,150 - 83,950 (+6,800)

With a very small increase of 65,000 m^3/day, the total exploitation would be approximately 150,000 m^3/day, which would meet the requirement for 2000. However, it must be remembered that this general situation does not apply to all the islands or to each individual sector; in some places the resources/exploitation balance can show a deficit. In any event, very intensive and effective artificial recharge is highly advisable, especially for the solution of local problems. According to BURGEAP (1981), most of the future exploitation should be by means of extensive galleries, in the long if not in the short term.

For future borehole exploitation, account must be taken of the production cost elements described above. Given the relatively high drawoff cost, it would seem preferable to use the cheapest techniques - wind power for example - even though in this case the number of boreholes would have to be increased to compensate for their low discharge.

Since the only existing fresh-water resource is ground water and since little is konwn about the actual available flows, any exploitation will have to be strictly monitored and regulated in the light of the aquifer's behaviour.

III. REFERENCES

Alves Costa, M. (1956): 1° e 2° Relatório de Reconhecimento Hidrogeológico do Arquipélago de Cabo Verde.

Babau, M.C., Alves L.M. and Silva, R.L. (1983): Les conditions générales au Cap-Vert in Desenvolvimiento e Pesquisa no Longo Prazo em Cabo Verde. Instituto Gulbenkian de Ciencia, Lisbon, Portugal.

Babau, M.C. (1983): Evolution de la pluie annuelle de 1885 à 1983 à la station de Praia, Ile de Santiago. World Meteorological Organization and Ministry of Rural Development, Republic of Cape Verde, unpublished.

Beurskens, H.J.M. (1981): Wind Energy for Water Pumping in Cape Verde. SWD, Amersfoort, Netherlands.

Bosscher, A. (1982): Groundwater Availability and its Use in Irrigated Agriculture on Santo Antão, Cabo Verde. Directoraat Generaal Internationale Samenwerking, Gravenhage, Netherlands, unpublished.

Brodbeck, J.F. and Haupt, M. (1982): Description hydrogéologique des ribeiras dos Engenhos et da Boa Entrada, Sta. Catarina, project CVI/79/001, UN/DTCD, Praia, Cape Verde, unpublished.

Brodbeck, J.F. (1984): Géologie d'Achada Baleia et de Tarrafal, project CVI/82/004, UN/DTCD, Praia, Cape Verde, unpublished.

Brodbeck, J.F. (1984): Observations sur la géologie et l'Hydrogéologie de l'Ile de Santiago (République du Cap-Vert), unpublished.

BURGEAP (1974): La Mise en Valeur des Eaux Souterraines dans l'Archipel du Cap-Vert. Rapport de fin de mission. Ministry of International Co-ordination, Lisbon.

BURGEAP (1979): Campagne de Forages d'Eau dans l'Ile de S. Nicolau 1978-1979. Ministry of Co-operation, French Republic and MRD, Republic of Cape Verde.

BURGEAP (1980): Campagne de Forages Galleries et Captages dans l'Ile de S. Nicolau 1979-1980. Ministry of Co-operation, French Republic and Ministry of Rural Development, Republic of Cape Verde.

BURGEAP (1981): Etude pour l'Organisation du Service de l'Eau. Ministry of Co-operation, French Republic and Ministry of Rural Development, Republic of Cape Verde.

BURGEAP (1982): Campagne de Forages Galleries et Captages à S. Nicolau et Santiago 1981-1982. Ministry of Co-operation, French Republic and Ministry of Rural Development, Republic of Cape Verde.

BURGEAP (1982): Programme de Recherche d'Eau pour la Ville de Praia et le Développement des Cultures. Compte Rendu de Mission de Fin de Campagne. Rapport Final. Secretariat of State for Co-operation and Planning, Republic of Cape Verde.

BURGEAP (1983): Etude Générale du Massif du Pico da Antonia, Santiago 1982-1983. Ministry of Co-operation and Development, French Republic and Ministry of Rural Development, Republic of Cape Verde.

COMARO (1960): Estudo hidrogeológico do arquipélago de Cabo Verde, Lisbon.

Consortium for International Development and Utah University (1982): Tarrafal Water Resources, Cape Verde. Contract AID/Afr.-c-1403. Final Project Report.

Dittrich, I. (1982): Une Délimitation des Composants du Bilan Hydrique pour Santiago. Un modèle semi-déterministique. Ministry of Rural Development, Praia, Cape Verde, unpublished.

Fernandopullé, D. (1977): Investigations Hydrogéologiques de l'Ile de Maio. Project CVI/75/001, unpublished.

Fernandopullé, D. (1979): Resources en Eau Souterraine. Iles du Cap-Vert (Rapport Intérimaire). Project CVI/75/001, UNDP/DTCD, Ministry of Rural Development, Praia, Cape Verde, unpublished.

Haupt, M. and Mota Gomes, A. (1982): Projecto João Varela: Pesquisa de Aguas Suberrâneas. Relatório Hidrogeológico. Project CVI/82/004, UN/DTCD and Department of Ground-Water Exploration and Management Services, Ministry of Rural Development, Praia, unpublished.

Junta de Investigação do Ultramar (1969): Carta Geológica. Ilha de Maio. Escala aprox. 1:30,000, Lisbon.

Junta de Investigaçoes do Ultramar (1973-1975): Carta Geológica de Cabo Verde. Ilha de Santiago. Hojas escala 1:25,000, Lisbon.

Livramento D.R. and Pieterse, N. (1984): Revisão de Potencialidades de Bombagem Eólica. Cape Verde/Netherlands bilateral project on renewable energy, Ministry of Rural Development, Praia, Cape Verde, unpublished.

Mitchell-Thomé, R. (1972): Outline of the Geology of the Cape Verde Archipelago. Geologische Rundschau, 61 (3), Stuttgart, Federal Republic of Germany.

Mitchell-Thomé, R. (1976): Geology of the Middle Atlantic Islands. Beiträge zur Regionalen Geologie der Erde, Band 12. Gebr. Borntraeger, Berlin.

Monteiro Alves, L. (1983): Rôle de l'Argométéorologie dans le Développement Socio-Economique de la République du Cap-Vert. Centre Interuniversitaire des sciences de l'Environnement, Arlon, France.

Mota Gomes, A. (1982): A Hidrogeologica de São Tiago. Department of Ground-Water Exploration and Management Services, Praia, Cape Verde.

United Nations (1980): Recherche et Mise en Valeur des Eaux Souterraines. Cap-Vert. Conclusion and recommendations of project CVI/75/001, United Nations, New York.

United Nations (1983): Investigación y Desarrollo de Recursos de Agua. Cabo Verde. Conclusiones y Recomendaciones del Proyecto. United Nations, New York.

SCETAGRI (1981): República de Cabo Verde. Esquisse de Schéma Directeur de Développement Rural des Iles du Cap-Vert. Ministry of Co-operation and Development and Secretariat of State for Co-operation and Planning, Cape Verde.

Secretariat of State for Co-operation and Planning (1983): Primeiro Plano Nacional de Desenvolvimiento 1982/85, Praia.

Serralheiro, A. (1976): A Geologia da Ilha de São Tiago (Cabo Verde). Imprensa Portuguesa, Lisbon.

CENTRAL AFRICAN REPUBLIC

Area: 622,984 km^2

Population: 2.45 million (United Nations estimate, 1983)

I. BACKGROUND

The Central African Republic is a landlocked country in the middle of the African continent.

Natural regions

The country has a low relief marked by erosion, with occasional inselbergs or cuirassed buttes; the average altitude is 400 to 600 m. The soil is of the tropical leached ferralitic or ferruginous type. The northern part of the country has two mountain ranges which cover only two per cent of the territory: in the north-east the Bongo and Dar Djala mountains with Toussoro (1,330 m), Ngaya (1,326 m) and Koubo (1,310 m); in the north-west, an extension of the Adamaoua plateaus of Cameroon, the Yadé mountains (1,309 m), Mt. Ngaoui (1,410 m) and the Bakore mountains (1,242 m).

The conjunction of these two ranges or the "Oubangui ridge" forms a threshold between the upper Chad basin in the north and the Oubangui plains in the south. One-third of the territory lies in the Chad basin and two-thirds in the Congo basin; the valleys of the Ouham and Aouk form a vast plain, while south of Bangui lies the small valley of the Oubangui. Most of the soil in these plains is hydromorphic and alluvial.

Climate

The Central African Republic is subject to the high pressure zones - anticyclones of Libya in the north and of St. Helena in the south - which govern the weather of Central and West Africa. During the northern winter the Libyan high pressure zones push a flow of dry air towards the country with a north-east wind - the harmattan. During the northern summer the St.Helena high pressure zones push moist air across the country from south-west to north-east; this is the monsoon. The line between these two air masses, called the intertropical front, determines the local seasons.

The average annual rainfall is fairly high: 1,700 mm in the south of M'Baiki and 1,800 mm at Rafai; it decreases steadily northwards to 700 mm at the Nzili waterhole on the frontier with Chad and Sudan. The isohyets follow the parallels but veer south-east at the Sudanese frontier. Their regular progress is occasionally interrupted by relatively dry belts (1,400 mm). There are two periods of maximum rainfall, the main one in August-September and the second in May-June. The rainy season varies in length from over 300 days in the south to about 125 days in the north-east.

There is little variation in average annual temperature from one region to another. It is 23.4°C at Bouar and 26.5°C at Birao. The seasonal range is wider in the north (6.8°C at Birao) than in the south (2°C at Berberati).

The measured and calculated evaporation potential for 1969-1971 is:

- 1,500 to 1,600 mm in the equatorial zone south of the 4th parallel N;

- 1,600 to 1,900 mm in the intertropical zone between the 4th and 9th parallels/N;

- Over 1,900 mm in the Sahelian zone north of the 9th parallel N.

Surface water

Two hydrographic systems - Sudanese in the north and equatorial in the south - collect the surface water: the Upper Chad basin with in the east the Chari fed by the Bahr Aouk, Bamingui, Gribangui and Bahr Sara, draining Ouham and Nana Bakassa, and in the west the Logone with the Pendé and Mbéré; the north Congo basin with the Oubangui fed by the Mboumou and Uele and on the right bank by the Ouara, Chinko, Kotto, Ouaka, Mboko and Lobaye, and in the south-west the Sangha fed by the Kadei, Mambéré and Nana. The regime of the watercourses follows the precipitation pattern.

At Bangui the Oubangui has a a mean annual flow of 4,350 m^3/s, ranging from a minimum of 700-1,000 m^3/s to a maximum of 11,000 m^3/s at the end of the rainy season. The water level varies by five to seven metres between these two periods. The Lobaye and the Kotto have similar regimes; their mean annual flows are 750 and 450 m^3/s respectively.

The watercourses of the Upper Chad basin have very irregular patterns. They are basically torrential, but the flow can dry up completely. The Ouham at Bossangoa has an annual flow of 240 m^3/s, with a minimum of 20 m^3/s and a maximum of 900 m^3/s. For the Bamingui at Bamingui the figures are 20 m^3/s, 0.4 m^3/s and 77 m^3/s respectively.

The sediment discharge measured at Batangafo in the Ouham basin (44,700 km^2 - annual flow 11 x 10^9 m^3) is 56 mg/l of dissolved solids. The corresponding specific erosion is 10 t/km^2/year for mechanical erosion and 7.5 t/km^2/year for chemical erosion.

Geology

The bulk of the country consists of crystalline metamorphic Precambrian formations of the African shield. The surface formations occupy a quarter of the country and include:

- Quaternary deposits: Neo-Chadian formations of lacustral origin with thick layers of recent clays and alluviums covering the hydrographic networks, the depressions and the swampy areas of the Sangha basin;

- Tertiary sediments: ferruginous sandstone, sands and cuirasses forming the edge of the Chad basin and, in the south, silicified sandstones, sands and loess of the Congo basin (Banbio sandstone).

Two thick Secondary strata - Carnot and Mouka Ouadda sandstones - rest in angular discontinuity on the basic granitic complex.

Crystalline basement rock occupies most of the country. It is subdivided into:

- Discontinuous eruptive formations including large masses of batholitic granite (Bouar plateau, Mt. Yade) and doloritic veins (upstream of Bangui, Nola region);

- Precambrian terminal with little metamorphism, overlying the base complex and including three units: upper unit with series of schists, quartzites, limestone and sandstone; middle unit consisting mainly of carbonated formations with a glacial conglomerate at their base; and lower unit represented by quartzites alternating with sericitic schists;

- Base complex covering the greatest area and subdivided into two groups: upper group formed of quartzites and schists; and lower group formed of gneiss, micaschists, amphibolites, granulites, migmatites and anatectic granites.

II. WATER RESOURCES

Institutional aspects

The following are the main bodies concerned with water and water supply:

- In urban areas: the National Water Corporation (Ministry of Public Works and Urban Development) for supply, and the National Hygiene and Health Education Service (Ministry of Health) for sanitation;

- In rural areas: the Department of Agricultural Engineering (Ministry of Agriculture and Livestock) for supply.

There is no national service with specific responsibility for the study and evaluation of ground water.

Up to 1982-1983, years of devastating drought which caused the Central African Republic to declare itself a disaster zone, water supply and sanitation were not considered priority matters for development. The launching of the International Drinking Water Supply and Sanitation Decade made the country aware of the problem and prompted it to devise a national policy and strategy with respect to water supply and sanitation, which took the specific form of

- The formation of the National Water Supply and Sanitation Committee (CNEA) in September 1982;

- The organization of the first national seminar on water supply and sanitation in May 1983;

- The establishment of the Secretariat of State for Water Supply in July 1984.

Hydrogeological research

The country still has no hydrogeological reports or maps, even partial ones. The documents on a number of old well and borehole projects contain no geological or hydrogeological data. However, it is still possible in general terms to distinguish three main aquifer systems:

- The Precambrian formations which occupy the greatest area of the country; these rocks are mostly impermeable and form discontinuous aquifers in their fracture and alteration zones;

- The two massifs of continental sandstone, which form vast reservoirs mostly charged by rainfall;

- The upper Chad basin, which is covered by sands and clays of the continental terminal, is the only aquifer of regional size.

The study and evaluation of the water resources of these formations is of vital importance for ground-water development and exploitation.

Studies and projects

Several ground-water study and development projects were planned or in operation in 1984 under the auspices of the following bodies:

- Swedish International Development Authority (SIDA). This project is concerned with 100 developed springs, 140 equipped boreholes and the organization of maintenance teams for these installations. It operates in the prefectures of Sangha, Haute Sangha, Nana-Mambéré and Ouham-Pendé;

- National Ouham-Pendé Development Agency (ACADOP). This project is supported by German technical assistance and the goal is to dig 500 wells in four years; it is the follow-up to a preparatory phase which included hydrogeological studies and 80 reconnaissance boreholes;

- National Agricultural Development Corporation (SOCADA). The project goal is to establish 150 water points in the prefectures of Basse-Kotto, Kémo-Gribangui and Ouaka;

- International Children's Fund (UNICEF). The project goal is to establish in three years 250 equipped boreholes in the prefectures of Ouham and Gribangui-Economique.

The SOCADA and UNICEF projects include an inventory of water resources, village leadership, and hydrogeological and geophysical studies.

- United Nations Development Programme (UNDP). The project goal is to organize a hydrogeological study service with a view to the preparation of a master plan for ground-water resources and their exploitation in the Central African Republic. The executing agency is the United Nations Department of Technical Co-operation for Development.

UNDP has also financed 17 boreholes at Bangui and the establishment of a wells team at Birao.

III. EXPLOITATION OF WATER

Bodies responsible for water resources

The National Water Corporation (SNE), which is responsible for water supplies in urban areas, has the assistance of the Ministry of Public Works and Urban Development in the construction and maintenance of installations. In rural areas the Department of Agricultural Engineering of the Ministry of Agriculture and Livestock is responsible for the planning and maintenance of water points.

The following water-drilling equipment was available in 1984:

- The Secretariat of State for Water Supply had two Mobil Drill B53 rigs passed on by the United States Peace Corps (USAID);

- SIDA had one cable-tool rig, ancient but operational, and one Atlas Copco B80 rig;

- UNICEF was using at Atlas Copco B80 rig.

Use of ground and surface water

Water supplies for Bangui and the five other towns which have distribution networks are taken from surface water which requires full treatment.

At Bangui the water treatment plant has a capacity of 1,500 m^3/h and 4,500 subscribers representing 11 per cent of the town's population. Four other urban centres have treatment plants: Bambari - 300 m^3/h, 10 per cent subscribers; Berbérati - 200 m^3/h, 9 per cent subscribers; Bouar - 110 m^3/h, 10 per cent subscribers; and Bozoum - 40 m^3/h, 5 per cent subscribers. However, most of the urban population takes its supplies mainly from wells, springs, pools and a few standpipes.

In rural areas the water is provided by rivers, pools, springs and the collection of rainwater, as well as by undeveloped and developed wells. There are some modern installations which were financed by external assistance but most of them are unusable owing to the failure to make prior hydrogeological studies, followed by poor planning, or somply because of the lack of any maintenance organization.

This is true of the three big projects described below:

- Project of the European Development Fund (EDF) (1967-1970) for the development of 53 springs and digging of 388 wells. A small number of these installations was still in use in 1984;

- Project of the United Nations Capital Development Fund (UNCDF) and the Ministry of Health for the construction of wells equipped with hand pumps (1979). All these installations have been abandoned;

- Boreholes project of the United States Peace Corps (1977-1980). This project had a number of setbacks owing to the lack of hydrogeological studies.

There are about 100 positive boreholes equipped with hand pumps, but they are out of service.

A number of new foreign assistance projects have been completed, including the SIDA project: the 100 springs and 140 equipped boreholes are operating perfectly thanks to the efficiency of the maintenance service.

Estimate of future needs - problems

Within the framework of the International Drinking Water Supply and Sanitation Decade, the Government of the Central African Republic has established the policy of guaranteeing 100 per cent cover of drinking water requirements in urban areas and 50 per cent in rural areas. In urban areas this very ambitious programme entails the extension of the networks of the five main towns and the complete equipment of 14 additional towns. The rural areas, including 36 suburban centres and 8,000 villages, are to have 2,500 water points by 1990.

Current projects include the equipment of the five additional towns and extension of the Bangui network, the establishment of 400 water points, and the development of about 50 springs, in addition to the SIDA and ACADOP projects.

The support and mass participation of the people and increased training of leaders and national staff are essential to the success of these programmes.

The hydrogeological and geophysical studies, the field reconnaissance work, the drilling operations and their equipment and the technical activities present no major difficulties.

IV. CONCLUSION

The Central African Republic is a wet tropical country and uses surface water and ground water from shallow aquifers to supply its people, especially in rural areas. This solution is economical in the short term, but the poor sanitation conditions are causing water-borne diseases to increase. The great distance from water points and the exhaustion of many wells and springs are causing further deterioration in the living conditions of an already deprived people.

The Government has endorsed the goals of the International Drinking Water Supply and Sanitation Decade and has therefore undertaken a far-reaching programme for urban and rural water supplies.

The exploitation of ground water is only beginning: studies, surveys, and the mobilization of resources are now taking place, especially with the participation of United Nations bodies (UNICEF, UNDP, DTCD), with a view to establishing a master plan for the country's ground water resources and their exploitation.

The needs are very large and require the participation of co-operation organizations working in partnership with the Central African Republic and of humanitarian non-governmental organizations.

V. REFERENCES

Mestraud, J. L. and Bessoles, B. — Géologie et ressources minérales de la République centrafricaine - Etat des connaissances fin 1963. BRGM, 1963.

Calegari, Sergio — Expertise sur la prospection des nappes aquifères en République centrafricaine, 1981. European Development Fund, 1981.

WHO/PNUD — Document de planification du secteur eau et assainissement, 1982.

Boulvert, Y. (ORSTOM) — Carte pédologique de la République centrafricaine, 1983.

CHAD

Area: 1,284,000 km^2

Population: 4.79 million (United Nations estimate, 1983)

I. BACKGROUND

Natural regions

The territory of Chad covers the eastern half of the Chad basin. Viewed as a whole, the Chad basin has two parts of unequal size. The larger, northern part extends over several countries. It lies between the Jos plateau (Nigeria) and the Air (Niger) in the west, the Tibesti in the north, the Ennedi and the Ouaddai in the east, and the Mandata mountains (Cameroon) and the Guéra massif in the south.

The southern part lies almost entirely in Chad between the Mandata mountains and the Pala plateaus in the west, the Oubangui ridge in the south (from the Yade mountains to the Jabal Mela), and the Birao ledge and the Guéra massif in the east and north.

The lowest part of the whole basin is in the north-east sector in the Djourab and Bodelé country (155 to 180 m in altitude). The present Lake Chad occupies a higher position (about 280 m). It is held at this altitude by the dune dam of the Kanem erg. The bed of the Bahr El Ghazal or Soro, which formerly ran from the lake as far as Djourab, no longer contains water in the present climate. The relief in Kanem, Eguey, Toro and Djourab consists mainly of dune formations. The alternating dunes and depressions create a monotonous landscape with differences of altitude of only 30 at the most. The mountain massif of the Tibesti has volcanic peaks which provide the highest points (Emi Koussi - 3,415 m, Toussi - 3,315 m) and the high plateaus of Tarso at 2,000 to 3,000 m.

The eastern edge of the basin rises steadily in gentle glacis to the crystalline massifs of Ouaddai. The depositional glacis in the centre of the basin have varying degrees of induration or are covered with dune formations; further east they merge gradually into erosional glacis, at first thinly covered but then bare at the edge of the massifs. The crystalline formations rise sharply above the glacis in inselbergs. The Guéra massif (central Chad massif), where Mt. Guédi in the Abou-Telfane rises to 1,508 m, presents the same aspect of mountains raised above the surrounding heavily cuirassed glacis and running towards Lake Fitri in the north-west, the Chari plains in the west, and the Salamat plains in the south and south-east. In the south the area of the Chari and Logone basins consists of low sandclay plateaus (the Koros) and depositional flood plains. There is no apparent variation in the relief except in the rainy season when it is easy to distinguish the flooded from the unflooded land.

The Koros rise slowly to the southern edge of the crystalline rock. A wide rock pediment stands out 500 to 600 m higher, with residual relief of the inselberg type. The spurs of the Yade massif (Lam mountains) rise sharply from these erosional glacis.

The southern, central and eastern glacis are often covered with ferruginous crusts which protect the relief and prevent the reestablishment of the vegetation.

The plains of the lower Chari and lower Logone slope imperceptibly down to Lake Chad. The only inhabitable spots are small bluffs on the banks of the watercourses at the top of which are perched villages which escape the annual floods.

The lake itself is held back in the north by the Kanem dunes; it works its way between the dunes and forms a large number of oblong islands at its north-eastern edge.

Climate

Chad has a very varied climate. There are three main climatic zones, from south to north:

Sudanese zone

This zone extends roughly from 7°5' to 10° n; it has heavy rainfall with an annual average of 1,000 to 1,400 mm. The population is sedentary and agriculture well developed, with food and cash crops: millet, sorghum, groundnuts and cotton.

The heavy rainfall produces the following annual variations in temperature:

- In January the maximum temperature is about 30°C, while the minimum varies from 15°C to 20°C, producing an average of about 25°C;

- In March the zone is slightly hotter with an average of around 35°C;

- July the weather becomes very rainy and the air quite humid; the minimum temperature is about 20°C;

- In November, the last month of the rainy season, the temperatures rise to a maximum of 35°C, with a minimum of 18°C, while the daily averages range from 20.5°C to 20°C. Owing to the heavy rainfall the evaporation is less heavy in this zone that in the rest of the country. It is in the order of 1,900 mm/year.

Sudano-Sahelian zone

This zone extends roughly from 10° to 11°50' N; it receives less rainfall that the Sudanese zone. The annual rainfall of 700 to 1,000 mm makes it suitable for certain food crops, mainly millet, sorghum and groundnuts. Livestock is also fairly well developed. The population thus consists of herdsmen and farmers.

The annual temperature variations are roughly the same as in the Sudanese zone.

- In January the maximum temperature is about 30°C and the minimum between 15°C and 20°C, producing an average daily temperature of about 25°C;

- In March the weather becomes hotter. The maximum temperature is about 38°C and the minimum about 20°C, with a daily average of around 30°C.

- In July the weather is fairly mild owing to the atmospheric humidity and the rainfall. The maximum temperature varies between 25°C and 30°C, while the minimum remains above or at 20°C; the average temperature is around 25°C;

- In November the weather is slightly milder than in July. The maximum temperature is about 35°C and the minimum between 15.5°C and 18.5°C, with an average daily temperature of between 20°C and 15°C. The evaporation is about 2,100 mm/year.

Sahelian zone

The Sahelian zone extends roughly from 11°5 to 14° N and the rainfall is only 200-700 mm. In the north, in the Borkou-Ennedi-Tibesti desert (BET), the total annual rainfall is below 500 mm. The Sahelian zone has more herdsmen than farmers. Only early crops are grown here. The people are mainly herdsmen, in constant search of watering places and pasture for their livestock. The annual temperature variations are almost the same as in the Sudano-Sahelian zone. The great heat causes considerable evaporation of surface water and transpiration from plants. The annual evaporation in this zone is around 2,200 to 2,300 mm.

The BET desert region in the far north has very great temperature variations during the year. For example, in January the maximum temperature ranges from 17.5°C to 27°C, while the minimum temperatures are below or at 13°C - 15°C, with daily average of 10°C to 20°C.

In contrast, it is very hot in July. The maximum is above 40°C and the minimum about 25.5°C.

In November the weather is cooler with a maximum temperature of 22.5°C to 27.5°C and a minimum of about 17.5°C; the daily averages range from about 17.5°C to 22.5°C.

The po;ulation of this zone consists mainly of nomadic camel herdsmen.

II. SURFACE WATER

Permanent watercourses

The Republic of Chad is drained by two main rivers, the Chari and the Logone, which are fed mainly by tributaries outside the country.

The Chari, upstream of its confluence with the Logone, receives the following tributaries:

Right Bank	Left Bank
Bahr Salamat	Bahr Illi (Padjouek)
Bahr Keita	Bahr Sara
Bahr Aouk	

The Bahr Sara, just above the Manda bridge (on the Bahr Sara) receives a largeish tributary, the Maudoul, and on its right bank, a few kilometres before it joins the Chari, the Bahr Ko.

The flows, the highest levels and the area of the catchment basin for the various hydrometric stations, from upstream to downstream on the Chari, are as follows:

Station	Catchment basin (km^2)	Annual flow (m^3/s)	Year	Highest level (cm)	Crest flow (m^3/s)
Saha	193,000	326	1961	725	2,090
Heli-Bongo	217,000	243	1962	710	-
Bousso	450,000	945	1961	610	3,980
Mailao	500,000	946	1961	750	4,220
D'Djamene	600,000	1,280	1961	910	5,160

Source: Monographie du Tchad and Annuaire hydrologique de la Republique du Tchad 1977-1978.

Tributaries of the Chari

Station	Watercourse	Catchment basin (km^2)	Year	Highest level (cm)	Crest flow (m^3/s)
Baibokoum	Logone	21,360	1956	785	
Moundou	Logone	34,900	1956	630	3,640
Lai	Logone	56,700	1955	605	3,730
Tarangara	Bahr Salamat	135,000	1961	560	93
Kyabe	Bahr Keita	14,000	1964	740	545
Golonoosso	Bahr Aouk	96,000	1962	580	968
Manda	Bahr Sara	79,600	1955	750	3,670

Source: Annuaire hydrologique de Tchad 1977-1978.

The Logone is the largest tributary of the Chari; above its confluence with the Chari (at N'Djamena) it receives the following tributaries:

Right bank	Left bank
Ba - Illi	Tangilê
Pendė	Nya
Lim	

During the high-water period at Bryor and Ere the water of the Logone is captured by the rivers of the Benoue basin - Mayo Kebi and the Kabia.

Tributaries of the Logone

Station	Watercourse	Catchment basin (km^2)	Year	Highest level (cm)	Crest flow (m^3/s)
Bologo	Tandjilé	-	1969	780	210
Argao	Nya	-	1970	620	415
Moulkou	Ba-Illi	-	1970	545	-
Gore	Pendé	-	1970	670	1,060

Source: Annuaire hydrologique du Tchad 1977-1978.

Lake Chad

Lake Chad is the country's main water resource. Its basin is one of the best known in Africa, for it has been the subject of many studies.

Lake Chad is situated in the Sahelian zone and is fed by three rivers, two of which are not in Chad (Komadougou Yobê in Nigeria and El-Beid in Cameroon) and have much smaller flows than the Chari. In an average year the total recharge is about 50 billion cubic metres. In a normal year the hydrological balance of Lake Chad is as follows:

Recharge from the Chari-Logone system;	43.25×10^9 m^3
Recharge from other rivers;	1.92×10^9 m^3
Rainfall on the lake;	7.25×10^9 m^3
Total	52.42×10^9 m^3
Evaporation from the lake	49.95×10^9 m^3
Losses through infiltration	2.47×10^9 m^3

In the period of mean flow the Chari-Logone system delivers 40×10^9 m^3. After the severe drought in 1972-1973 the total recharge was only about 25×10^9 m^3 (Annuaire hydrologique 1977-1978 .

After the highest water level observed at the N'Djamena measuring station (Public Works station) in 1961, the following flow characteristics were recorded:

- Water level	9.10 m
- Flow	5,160 m^3/s
- Mean annual flow	170 m^3/s
- Flow volume	$53.7x10^9$ m^3
- Total stocks	$10.5x10^9$ m3

The area of the lake under water is 23,500 km^2.

The sediment discharge in the lake varies between 2 and 2.6 million tons a year.

In addition to Lake Chad there are six other less important lakes:

- Lake Fitri (Batha prefecture) which has a catchment basin of 1,300 km^2 and is fed by the River Batha;
- Lake Iro in Moyen-Chari which has a catchment basin of 455 km^2 fed by the crest waters of the Bahr Asoum;
- The lakes (Tikem, Fiangat and Léré) in Mayo Kebbi which are fed by the Kabia (Tikem-Sianga) and Mayo Kebbi (Léré).
- Lake Ounianga in Borkou-Ennedi-Tibesti.

Seasonal watercourses

In the north of the country, especially in Ouaddai, all the watercourses (wadis) flow for only a few hours after heavy rain. The main wadis are the Rimé, Karmé and Enné. The Bertha and its tributary the Nelmele flow only from July to October.

III. GEOLOGY

The most ancient Precambrian rocks, with an average age of about 1,000 million years, are found around the edges of the country; these are the granitic and crystallophyllitic rocks of the Tibesti and the Ouaddai and the Lam mountains in the south. They also emerge as more recent sediments in some places in the middle of the basin. It is thought that the Chad basin is broken up into a series of small parallel basins which can be identified from the general orientation of the crystalline ridges (north-north-east/south-south-west and west/north-west).

Soundings and aeromagnetic prospecting indicate that the basement rock is regular and slopes gently towards the south-west to the western frontier of Chad. The Tibesti sandstones which form the Tarsos in the interior of the massif also belong to the Precambrian period.

The Paleozoic series (Cambrian-Ordovician, Devonian and Carboniferous) appear only in the north of the country in Borkou and Ennedi where they form large sandstone tables. The deposits are contemporary with the first manifestations of life, witness the fossils of animal and vegetable organisms (bilobates and harlania of the Tassilian sandstones, fish of the Carboniferous period).

The Mesozoic period is little represented in Chad. In northern Ennedi the Erdi sandstones are probably contemporary with the Nubian sandstones. In the Mayo Kebbi region the Lame series marks the most eastern extension of the marine transgression through the Benoue trench (Cenomanian).

Apart from the Eocene marine deposits in northern Tibesti, the Tertiary formations in Chad are mainly continental terminal. The continental terminal has been thoroughly studied in southern Chad where it covers a large part of the basins of the Logone and Moyen-Chari; it includes:

- A middle series of red sands;

- An upper series with varying degrees of crust.

The continental terminal series are well preserved only in the regions where they were protected by indurated formations. The interfluvial areas known as Koras became cuirassed during the Quaternary period.

The main layer of this cuirass runs from the southern crystalline rock to the Mayo Kebbi region. It is also found around the central massif of Chad as far as the Ouaddai.

Quaternary formations occupy most of the Chad basin. The studies carried out by ORSTOM and BRGM have dated these formations fairly accurately. The chronology of their deposits is identical with the variations in the level of Lake Chad resulting from climatic changes over the past million years. The wet periods favoured the deposition of diatomites and clays, the periods of alternating seasons produced more sandy deltaic deposits, while the dry periods led to the formation of sand dunes.

The most ancient lacustral extension of the Quaternary period reached Borkou in the north, Ouaddai and the central massif in the east, and in the south the region of the Koros and the crystalline shelf of Mayo Kebbi across which the waters of Paleo-Chad were repeatedly poured towards the Atlantic in the Bènoué. The deposits corresponding to this extension form a clay-sand series with limestone concretions. A shift to a dryer climate caused the lake to retreat and produced the first delta of the Chari and the deposition of sand dunes which is most visible in Cameroon and Nigeria.

The second extension of the lake (wet period) is marked by the line of the lacustral belt at the 320 m contour. During this period the direction of discharge of the Benoue was reestablished through the Toubouri lakes, and the Recent clay series was deposited. The second delta of the Chari was formed after the retreat of the lake.

The last transgression is delimited by the sand belt at an altitude of 278-290 m.

The present delta of the Chari is the result of the retreat of the waters of the lake to their present limits.

IV. HYDROGEOLOGY

The Quaternary formations are of complex and varied composition, and they can range from pure sand to sand with a very marked clay content; they can also be of a clearly kaoline type (for example, the Koro de Lai aquifer, where the water circulates through tiny channels).

These abrupt variations in the clay or sand content are very common, and the Quaternary sediments usually take the form of overlapping lenticular units.

The three main groups of aquifers in Chad are:

- The "sanies" aquifer in northern Bastra;

- The Chari-Baguirmi aquifer;

- The shallow "Oudian" aquifer in Kanem.

In addition to these three large aquifers, there are two other types of aquifers:

- The aquifers of the Koros and the southern region;

- The fragmentary aquifers, usually underflows of the granitic regions.

The "Koros" aquifers in the south

The recharge of the sanies aquifer comes partly from the underflows of the alluvial valleys of the Ouaddai and Batha massifs and their ancient beds. Most of the water probably comes from rainfall, which infiltrates swiftly in these very sandy regions.

Chari-Baguirmi aquifer

This region has a succession of compact and very flat clay-sand formations dotted with small higher areas of dunes or with clay depressions which are flooded in the rainy season and which increase in size further south. The average depth of the Chari-Baguirmi aquifer is 35 to 40 m. In the north-west and west the aquifer rises to a depth of about 20 m as it draws closer to Lake Chad, the Chari and its tributaries (Bahr Ligua).

In the south-west and south lie the flood areas of the Chari and its eastern tributaries. In view of the abundance of surface water in this area, the inhabitants have never had to dig wells.

Kanem aquifer

The recharge of this aquifer is fairly complex: most of the recharge is contributed by Lake Chad through its spreading grounds at the southern edge of Kanem; the rainfall must also supply very large quantities of water, for it infiltrates very easily in these sandy formations.

In the northern part of Kanem a study of the chemical properties of the water seems to indicate a very large recharge from the southern slopes of the Tibesti (rainwater and condensation) and from the north-west slopes of the Ennedi.

A feature of the water of the Kanem aquifer is its heavy concentration of sodium salts (carbonates and sulphates), but these salts do not occur everywhere.

At the edge of the Batha and Kanem regions, in Harr and Guetty, there is an aquifer which lies no more than 10 m deep.

Aquifer of the Koros and southern region

The region of Chad south of the 11th parallel N is well watered by a sufficiently dense hydrographic system and the problem of water does not arise; most of the villages take their supplies directly from the Chari and the Logone or their tributaries. Very productive underflow aquifers are found at shallow depths in the valleys of these watercourses.

On the borders of the Koros the aquifer is clearly larger, for its storage rock consists of beige sands separated by an episodic stratum of lateritic cuirass.

Aquifers of the granitic regions

These are local underflow aquifers in the deep valleys of the granitic massifs; the ground water flows very easily through the coarse granitic sands, from which good yields can be obtained.

To sum up, there exists throughout Chad a number of large aquifers which should make it possible to develop the areas which are at present virtually uninhabited and uncultivated.

V. GROUND-WATER RESOURCES

The main technical services for water resources are the Hydrological Service and the Water Office, both under the authority of the Department of Water Resources and Meteorology (Ministry of State for Agriculture and Rural Development). The Water Office has specific responsibility for the study and evaluation of ground water and it performs the following functions:

- Compilation of ground-water files;

- Direction of ground-water research and exploitation;

- Planning of water resources development;

- Preparation of ground-water research and exploitation projects and market studies for the water-supply works for towns, villages, industries and suburban districts;

- Control of drilling and digging operations for ground-water reconnaissance and exploitation;

- General study of the country's ground-water aquifers to obtain better knowledge of the available resources and their exploitability;

- Participation in urban sanitation studies;

- Briefing of visitors interested in ground-water problems.

From 1950 to 1979 the Water Office carried out several hydrogeological studies financed by FAC and EDF. This produced a better overall picture of the country's situation with respect to ground water and its exploitability. However, these studies were not always taken as far as necessary. They were conducted by expatriates, usually furnished by French technical co-operation as the country had almost no national specialized staff. A policy for training national staff was established to cope with this problem.

Most of the prospecting was by electrical sounding.

The ground-water resources consist essentially of the resources of the large area of shallow groundwater known as the Chad basin and of the stocks of the subsurface aquifers.

The aquifer of the Chad basin is the most heavily exploited. All the wells and 80 per cent of the boreholes tap the water of this aquifer, which has a fairly good recharge and extends over almost all the country, except for the mountain regions of BET, Ouaddai and Guéra and other areas where the basement rock outcrops. There are two hydrogeological units: one in the north, the Lowlands of Chad, drains the waters of the Kanem, the Niger aquifer and the BET aquifer; the other, in the south, is recharged by the waters of Lake Chad, the Chari and the Logone and the Chari-Baguirmi horizon.

The two units are separated by the ground-water watershed which starts from Chittati, passes close to Ati and ends at Abéché.

The southern unit can be subdivided into smaller units, including the Koros and Logone horizons.

The Chad basin has two recharge zones: internal and external.

The external zone consists of the mountainous regions of the Hoggar, Ouaddai, Dargour, etc.

The internal zone consists of the Lake Chad synclines where the rainfall is heavy. It has a good recharge from surface water: lakes, the Chari and the Logone and other watercourses. However, apart from the Bahr Azoum, the watercourses of southern Chad drain the aquifer; this phenomenon ceases below Kim in the case of the Logone which begins to recharge the aquifer, and below Niellim for the Chari. The reason for this in the case of the Chari is that after Bousso the river flows only over clay alluviums; from N'Djamena to the lake the two watercourses recharge the aquifer. The Batha recharges the aquifer as far as Ati. The waters of the lake itself seem to feed the neighbouring regions, in particular the Chihati aquifer and the Batoum-Ammak depression. The water of the basin is mostly calcium bicarbonate or calcium and sodium bicarbonate, sometimes with heavy sodium concentrations in areas of heavy evaporation. The mineral content of the water is 1-3 g/1, sometimes as high at 6 g/1.

The basin has a series of aquifers with different hydrogeological properties formed by the following geological formations, which may contain confined aquifers:

- The Pliocene-Quaternary formation in the central part of the basin where several aquifers can be distinguished: a Pliocene aquifer in the Koro region with alternate strata of fluvial-lacustral sedimentations and dunes, and an Early Quaternary lacustral aquifer in Moji.

The water-bearing rocks of the Pliocene-Quaternary formation are up to 500 m thick and they have good permeability. In the lower levels of this series, which also consist of sands, a confined aquifer known as the "middle pressured aquifer of the Chad formations" offers excellent resources around the lake, where it is artesian.

The depth of the top of the aquifer varies from 250 to 350 m.

The water of this aquifer has a low mineral content; yields of 2 1/2 are obtained from the wells.

- The continental terminal contains argillaceous sandstones and sands covering vast areas:

- In the north from the Lowlands to Batha, where the traditional wells (sanies) sometimes have to be 80 m deep to reach the aquifer;

- In the south of the country, where this formation includes a ferruginous cuirass altered by erosion.

This aquifer is 100 to 200 m deep. The water has a low mineral content which increases from the south towards the Lowlands. The average yield of the wells is 27.7 1/s.

- The third aquifer consists of arkosic sandstones, micro-conglomerates and calcareous micro-breccias of the lagoonal-lacustral Upper Cretaceous period, which are found only in Mayo Kebbi and have poor permeability. There may be a confined aquifer under the thick surface layer of marls and schists.

These aquifers have so far received little attention owing to the availability of surface water and shallow local aquifers which are more easily exploited.

The external recharge zone contains hardly any water. For example, the Carboniferous marine limestones and the Cambrian kaoline sandstones which form a continuous aquifer in BET have poor permeability. A large part of this area (where the basement rock outcrops) consists of Precambrian fissured granitic rocks of poor permeability. The fissures are less than 50 m in depth.

The boreholes which reach this aquifer at Largeau have yields of about 1,000 m^3/day. However, the traditional method of obtaining water is by means of wells which usually tap the resources of the alluvial horizons.

A large amount of fresh water is contained in these alluvial horizons. In the Saharan zone the ground water is usually found in lenses. In addition to the fresh water resources suitable for domestic use, the country's subsoil contains fairly large stocks of thermal and mineral water.

Although these resources are located far from the big centres, they can be exploited as sources of energy for the neighbouring settlements.

The shallow aquifer is subject to fluctuations depending on the region, the rainfall and the flood waters of the rivers. The results of recent deep soundings (Conoco) seem very promising.

However, the data needed for a reliable estimate of deep resources can be provided only by additional deep reconnaissance soundings.

VI. EXPLOITATION OF GROUND WATER

Government bodies

Ground water is of vital importance for the country's economic and social development. Ground water is extensively used both for drinking water (100 per cent of the urban water supply comes from wells and boreholes) and for pastoral water supplies.

In order to ensure rational and fair utilization of water, the Government established by decree the National Pastoral and Village Water Supply Office (ONHPV) under the authority of the Ministry of Livestock and Pastoral Water Supply.

This body is responsible for the maintenance and replacement of rural water supply installations and the construction and equipment of new installations.

The Office is run by national technicians backed up by expatriates and it is administered by an administrative council chaired by the Director-General of the Ministry of Livestock and Pastoral Water Supply.

As it was established only recently, the Office is still at the stage of internal organization and equipment. It is not therefore possible at present to give an accurate picture of its human and material resources, although they do seem clearly insufficient.

The country's largest drilling company was SOTRAHY, but it departed Chad after the events of 1979.

Utilization of ground water

With almost 6,000 waterpoints, including some 2,000 cement wells and more than 400 wells equipped with pumps, Chad is in a good position compared with other African countries with respect to programmes for the utilization of ground water.

The waterpoints give the following yields:

- 20 m^3/day from a traditional well or a cement well in granitic areas;

- 40 m^3/day from a cement well in sedimentary areas;

- 400 m^3/day from a borehole.

Urban uses

In Chad the urban water supply comes almost exclusively from ground water. However, N'Djamena had developed to such an extent before the events of 1979 that it was planned to take part of the water from the Chari and to treat it for distribution to the consumer; it was thought that in the long term ground water alone would not be sufficient to meet the needs.

With respect to the use of ground water for urban supplies, it must be pointed out that there are only 21 boreholes or wells supplying drinking water for 11 urban centres (N'Djamena, Sahr, Moundou, Abéché, Kelo, Koumra, Doba, Pala, Moussoro, Mao and Fianga). In future years about 10 other urban centres ought to be quipped with a water-supply system. In 1974 8.4 million cubic metres of drinking water were supplied for an urban population of 600,000, i.e. 40 litres a day per inhabitant.

Rural uses

The use of ground water in rural areas depends on the region. In the regions which are sufficiently well-watered, the wells supply water for human consumption, as the animals can be watered directly from pools, lakes and waterholes.

In some regions, particularly north of the 13th parallel N where there is almost no surface water, there are many wells of mixed use. The water points can also be used to supply water to small-scale farming operations. It is common to find small irrigated areas around the wells and it is possible to envisage at the national level the establishment of a system of village irrigation using ground water. For example, a borehole discharging 400 m^3/day can easily irrigate a minimum area of 6 ha. The multiplication of such irrigated areas throughout the country could make a substantial contribution to the efforts to achieve food self-sufficiency.

Industrial uses

The biggest consumers of industrial water are the Sahr textile factory, the Banda sugar mill, the Logone breweries at Moundou, and the 22 mills of Cotton Tchad.

The 1975 water requirement was estimated at 13,000 m^3/day, i.e. about five million cubic metres a year.

Agricultural uses

The current annual requirement for irrigation water for rural development is estimated at about 552 million cubic metres, which would be sufficient for all the irrigated land, i.e. 19,090 ha.

The types of crops envisaged are rice, wheat and sugar cane.

Future needs

An estimate of the future needs for 2000 is given below. This forecast takes into account the growth rates of the population, livestock and irrigated areas.

Domestic needs

The World Bank estimates that by 2000 the population will be seven million and the urbanization rate 6.7 per cent a year. The country's population will be distributed as follows:

Population	1983	2000
Urban	900,000	2,700,000
Rural	4,100,000	4,300,000
Total	5,000,000	7,000,000

According to a document prepared by the Inter-African Committee for Hydraulic Studies (CIEH) entitled "Utilization of water resources and land in the savannah regions", the daily consumption of water in Chad in 2000 will be as follows:

- Rural areas: 50 litres a day per inhabitant;
- Urban areas: 200 litres a day per inhabitant.

The total annual consumption will be as follows:

- Rural areas: about 80 million cubic metres;
- Urban areas: about 200 million cubic metres.

The total domestic requirement will therefore be 280 million cubic metres a year.

Livestock needs

The same CIEH document envisages that the cattle population will stabilize at five million head and the sheep and goats at 10 million head, with 600,000 camels. The average daily consumption of water by livestock is estimated as follows:

20 litres a day per head of cattle;

5 litres a day per sheep or goat;

35 litres a day per camel.

The total quantity of water consumed each year by each category of animal is as follows:

- Cattle: about 40 million cubic metres;

- Sheep and goats: about 20 million cubic metres;

- Camels: about 8 million cubic metres.

This makes a total of about 70 million cubic metres.

Irrigation

The areas suitable for irrigation are distributed by region roughly as follows:

Lac	60,000 ha
Kanem	5,000 ha
Assalé	13,000 ha
Ouaddai (geographical)	1,000 ha
Batha	20,000 ha
Bahr Azoum	200,000 ha
Chari	163,000 ha
Logone	200,000 ha
Total	662,000 ha

The 19,090 ha irrigated at present represents barely three per cent of this total.

The CIEH document envisages that 72,000 ha will be irrigated in 2000 in Chad, i.e. barely 11 per cent of the total area suitable for irrigation and less than four times the area irrigated at present. If this figure is accepted and if 24,000 m^3 is taken as the annual amount needed to irrigate a hectare of land, the total amount of water needed will be about 1.7×10^9 m^3.

On the basis of the estimates given above for population, livestock and irrigation, the water requirement for 2000 can be established at more than 2,000 million cubic metres a year.

VII. PROBLEMS

Shallow ground water is virtually unused south of the 13th parallel N, except in large urban settlements. North of this line, larger quantities of water are drawn off owing to the temporary nature of the waterholes created in the rainy season. In contrast, the degree of exploitation is very worrying on the Kanem and Bathagar plateaus where measurements show the aquifer to be remarkably stable. Recharge by infiltration of meteoric water is very problematical, and the exploitable aquifer must be monitored closely.

Surface and ground water is used for the permanent crops which occupy a total area of 1,652,000 ha and which also make direct use of rainwater; in addition to the 19,090 ha officially inventoried, the irrigation system also supplies water to a large number of small plots, with the water taken directly from watercourses, waterholes and lakes.

VIII. CONCLUSIONS

Ground water is of vital importance for the country's economic and social development. Ground water is extensively used both for drinking water (100 per cent of the water requirement of the urban centres comes from wells and boreholes) and for pastoral water supplies.

The problem of water is all the more important as Chad is a Sahelian country. The Government is therefore determined to solve it. In the years to come the implementation of a rational water-supply policy will reduce the cost of water, especially in the urban centres, where it is sold at the following rates:

- 0 to 4 m^3: CFAF 77/m^3;
- 5 to 1,000 m^3: CFAF 168/m^3;
- 1,000 to 1,300 m^3 and above: CFAF 77/m^3.

Number of waterpoints inventoried, September 1977

Prefecture	Boreholes	Soundings	Large-diameter wells	Other traditional	Total
Batha	7	56	340	246	649
BET	4	0	18	14	36
Biltine	47	51	93	660	851
Chari-Baguirmi	139	90	389	519	1,137
Guéra	1	14	65	118	198
Kanem	12	0	410	615	1,037
Lac	49	0	0	54	103
Logone occidental	10	13	45	43	111
Logone oriental	2	17	49	116	184
Mayo Kebbi	11	16	216	89	332
Moyen-Chari	69	16	195	278	558
Ouaddai	1	52	97	831	981

IX. REFERENCES

Abadie, J. — Etudes géologiques et hydrogéologiques dans la région de Largeau (Tchad). BRGM report LAM 64 A 4, Archives of Agricultural Engineering and Agricultural Service. N'Djamena, 1964.

Abadie, J., Cluseau, R. and Schneider J.L. — Contribution à l'hydrogéologie du Nord Kanem. Prospection par sondages électriques à l'est du Bahr-el-Gazal entre les 14ème et 16ème parallèles. BRGM report DS 64 A, Archives of Ministry of Public Works, N'Djamena, 1964.

Abadie, J. and Gagnière, G. — Enquête hydrogéologique et préparation de l'implantation des sondages de reconnaissance dans le sud de l'Ennedi, le Batha, l'Ouaddai, le Kapka, au cours du 1er semestre 1963. BRGM report LAM 64 B 1, 1964.

Abadie, J. and Gagnière, G. — Carte hydrogéologique de reconnaissance au 1/500,000. Notice explicative de la feuille Batha. Rapport de synthèse de la feuille Batha. BRGM report LAM 67 A 2, Archives of Hydraulics Service, N'Djamena, 1964.

Barbeau, J. — Hydrogéologie du basin tchadien. Annual report of AEF Geological Service, 1954, pp. 7-13, 1954.

Barbeau, J. — Hydrogéologie du bassin tchadien: zone sud (Chari-Baguirmi). Bulletin of AEF Department of Mines and Geology, No. 7, 1956, pp. 95-96, 1956.

Bonnet, M. and Schneider, J.L. — Fluctuations des nappes au Kanem, Chari-Baguirmi et Batha (Tchad). Programme 1964-1966. BRGM report 68 YAO 6 LAM, Archives of Ministry of Roads and Transport, Studies and Hydraulics Service, N'Djamena, 1968.

Gagnière, G. and Plote, H. — Programme quadriennal d'aménagement hydraulique de l'Ouaddai. Etude du secteur nord-ouest. Campagne 1964-1965. Etudes géophysiques et sondages de reconnaissance. Exposé et interpretation. BRGM report LAM 66 A 05, Archives of Ministry of Public Works, N'Djamena, 1966.

Gagniere, G. — Programme quadriennal d'aménagement hydraulique de l'Ouaddai. Etude des secteurs nord-est et centre ouest. Etudes géophysiques et sondages de reconnaissance. Campagne 1966. Exposé et interprétation des resultats. BRGM report LAM 67 A 3, Archives of Ministry of Public Works, N'Djamena, 1967.

Philipart, A. — Enquête hydrogéologique dans la région nord-east (Batha, Guéra, Ouaddai) (Tchad). BRGM report LAM 64 A 3, Archives, 1964.

Plote, H. — Programme quadriennal d'aménagement hydraulique de l'Ouaddai. Etude du secteur nord-ouest (4 vol.). BRGM report LAM 64 A 9, Archives of Ministry of Public Works, N'Djamena, 1964.

Plote, H. *Programme quadriennal d'aménagement de l'Ouaddai. Etude du secteur nord-est. Campagne 1964-1965. Rapport général. Inventaire des points d'eau et programme*. BRGM report LAM 66 A 4, Archives of Ministry of Public Works, N'Djamena, 1966.

Plote, H. *Programme quadriennal d'aménagement hydraulique de l'Ouaddai. Etude du secteur sud. Campagne 1965-1966. Rapport général et programme des travaux pour 1966-67. Inventaire des points d'eau sur les feuilles Abougoulem-Adré et Abéché*. BRGM report LAM 67 A 6, Archives of Ministry of Public Works, N'Djamena, 1967.

Plote, H. *Programme quadriennal d'aménagement hydraulique de l'Ouaddai. Etude du secteur centre-ouest. Sondages de reconnaissance dans la région d'Abeche. Campagne 1966-1967. Exposé et interprétation des résultats*. BRGM report LAM 67 A 7, Archives of Ministry of Public Works, N'Djamena, 1967.

Plote, H. *Programme quadriennal d'aménagement hydraulique de l'Ouaddai. Etude du secteur sud. Rapport général et programme des travaux. Inventaire des points d'eau sur les feuilles* Guereda et Am Zoer (4 vol.). BRGM report 68 YAO 4 LAM, Archives of Ministry of Public Works, N'Djamena, 1968.

Plote, H. *Programme quadriennal d'aménagement hydraulique de l'Ouaddai. Etude du secteur sud. Sondages de reconnaissance. Campagne 1967-1968. Exposé et interprétation des résultats*. BRGM report 68 YAO 7 LAM, Archives of Ministry of Public Works, N'Djamena, 1968.

Plote, H. *Programme quadriennal d'aménagement hydraulique de l'Ouaddai. Notice explicative et registre des aménagements et travaux de prospection proposés*. BRGM report 70 RME 034, Archives of Ministry of Public Works, N'Djamena, 1968.

Plote, H. *Implantation de 32 puits FED (lot No. 1) - Ouaddai (Tchad)*. Report BRGM 70 RME 036, Archives of Department of Agricultural Engineering and Water Supply, N'Djamena, 1968.

Plote, H. *Programme quadriennal d'aménagement hydraulique de l'Ouaddai. Hydrogéologie de l'Ouaddai. Rapport de synthèse (Tchad)*. BRGM report 70 YAO 004 LAM, Archives of Ministry of Public Works, N'Djamena, 1968.

Schneider, J.L. *Enquête hydrogéologique dans le Chitati septentrional (Kanem, Tchad)*. BRGM, report, Fort Lamy, Archives of Ministry of Public Works, N'Djamena, 1963.

Schneider, J.L. *Carte hydrogéologique de la République du Tchad au 1/500,000. Feuille Mao. Notice explicative. Rapport de synthèse*. BRGM report LAM 66 A 3, Archives of Ministry of Public Works, N'Djamena, 1966.

Schneider, J.L. — Rapport des fluctuations des nappes du Kanem, Chari-Baguirmi, Batha, année 1963. BRGM report LAM 67 A 1, Archives of Department of Public Works, Hydraulics Service, N'Djamena, 1967.

Schneider, J.L. — Carte hydrogéologique de reconnaissance au 1/500,000. Feuille Fort-Lamy. Notice explicative. Rapport de synthèse. BRGM report LAM 67 A 4, Archives of Department of Public Works, Hydraulics Service, N'Djamena, 1967.

Schneider, J.L. and Teixido, L. — Programme des 153 puits d'hydraulique pastorale dans les préfectures du lac de Kanem et du Borkou-Ennedi-Tibesti (Tchad). Rapport d'implantation (2 vol.). BRGM report LAM 67 A 5, Archives of Ministry of Public Works, N'Djamena, 1967.

Schneider, J.L. — Carte hydrogéologique de recconnaissance du Tchad au 1/500,000. Rapport de synthèse et note explicative de la feuille Pays-Bas. Largeau. BRGM report 68 YAO 005 LAM, Archives of Ministry of Public Works, Studies and Hydraulics Service, N'Djamena, 1968.

Schneider, J.L. — Notice explicative de la carte hydrogéologique au 1/1 500,000 de la République du Tchad (FAC). BRGM report 70 RME 029, 1970.

Torrent, H. — Etudes géologiques et hydrogéologiques dans le Mayo-Kebbi (Tchad). BRGM report LAM 64 A 5, Archives of Ministry of Public Works, N'Djamena, 1964.

Torrent, H. — Données complémentaires sur la géologie du bassin du Logone-Chari résultant de l'étude des coupes des 125 puits FED. BRGM report LAM 64 A 6, Archives of Ministry of Public Works, N'Djamena, 1964.

Torrent, H. — Carte hydrogéologique de la République du Tchad au 1/500,000, feuille Moundou et notice explicative. Rapport de synthèse. BRGM report LAM 64 A 10, Archives of Ministry of Public Works, N'Djamena, 1964.

Torrent, H. and Philipart, A. — Implantation de 165 puits dans les préfectures du Chari-Baguirmi, du Batha, de l'Ouaddai et de Biltine. Etude hydrogéologique et socio-économique avec fiches d'implantation. BRGM report LAM 66 A 2, Archives of Department of Agricultural Engineering and Water Supply, N'Djamena, 1966.

Torrent H. — Carte hydrogéologique du Tchad au 1/500,000 et notice explicative. Feuille Bongor. Rapport de synthèse (FAC). BRGM report LAM 66 A 6, Archives of Ministry of Public Works, N'Djamena, 1966.

UNDP/FAO — Commission du Bassin du Tchad (Cameroun, Niger, Nigeria, Tchad). United Nations Development Programme, Rome, 1973.

Zoubovsky, J.P. — Etude des différents rapports concernant la préfecture de Guera - Géologie, Hydrogéologie, Climatologie, Hydrologie, 1968.

COTE D'IVOIRE

Area: 322,500 km^2

Population: 9.2 million (United Nations estimate, 1983)

I. BACKGROUND

Climate

Côte d'Ivoire is a coastal State on the Gulf of Guinea with a seaboard of about 500 km between Cape Palmas and Cape Three Points. The country is mostly flat with local relief; it is more hilly towards the west, with some points above 500 m, the highest being Mt. Tonkoui (1,189 m) and Mt. Momi.

In the south the terrain is "humpy". It has many low-lying areas and a few isolated or grouped domes, such as Mt. Niénokoué (623 m) in the forest zone and Mt. Niangbo (643 m) in the savannah.

From south to north, at about 600 km from the coast, come the following vegetation zones, with decreasing rainfall: the tropical forest, the wet wooded savannah and the dry wooded savannah. The forest area forms a V in which higher ground towards the west and east encloses an area of savannah in the middle.

The maximum average annual levels of precipitation are in the order of 2,400 mm in the coastal zone, with a minimum of 1,200 mm in the north-east. The rainfall is fairly evenly distributed in time in the south; towards the north it tends to be concentrated in a short rainy season from July to September.

The country has three climatic zones:

- The equatorial type in the coastal zone where the temperature ranges between 21° C and 33° C, with a very high humidity rate of over 80 per cent and abundant rainfall, especially in the west - with 2 m a year at Tabou.

- The tropical type in the forest zone and the south of the savannah zone, with wider temperature ranges (14° C to 39° C), humidity rates of up to 70 per cent and moderate rainfall (1,000 to 1,400 mm). This part of the country has two dry seasons alternating with two wet ones. The "big" rainy season is the longest (beginning in December) and the most intense (with the heaviest rainfall in June); the "small" rainy season lasts about two months, from August to October.

- Lastly, the Sudanese type in the savannah zone - except for the southern part. Here there is a marked contrast between the dry and rainy seasons. The annual rainfall declines from a maximum of 1,000 mm (Odienné, Dimbokro) to 400 mm in the Dabakala region. The monthly temperature range is 7° C in the dry season and 5° C in the rainy season. The north has a single rainy season with the maximum in August and a dry season of six to eight months marked by the Harmattan, a strong wind which reaches its maximum force in January, when the average temperature range is about 18° C.

Monthly rainfall distribution

(mm)

	J	F	M	A	M	J	J	A	S	O	N	D	ANNUAL
Transitional equatorial type:													
ABIDJAN	29	42	115	155	355	595	226	30	55	189	187	109	2,087
TABOU	53	48	88	134	416	533	156	83	221	203	195	145	2,275
SASSANDRA	23	24	74	104	295	511	157	23	48	100	142	87	1,588
Moderated transitional equatorial type:													
DIMBOKRO	16	46	123	151	182	184	95	60	138	130	63	19	1,207
BONDOUKOU	15	43	74	131	169	166	80	64	186	177	51	17	1,173
SOURRE	33	56	139	166	173	245	107	100	220	191	120	54	1,604
Transitional tropical type:													
FERKESSEDOUGOU	5	19	47	91	140	158	186	299	251	112	32	110	1,450
ODIENNE	3	14	39	76	118	176	289	383	292	164	54	10	1,618
Mountainous type:													
DANANE	24	68	127	171	180	305	347	347	408	217	75	35	2,304
MAN	16	57	116	159	159	214	207	260	320	170	56	22	1,597

The total annual evaporation, measured in a Colorado pan, increases from the forest zone in the south (1,000 mm) to the savannah in the north (2,000 mm). It is strongly influenced by altitude and vegetation.

II. HYDROLOGY

River system

Côte d'Ivoire has abundant surface water, with four main rivers: Cavally, Sassandra, Bandama and Comoé; several small coastal rivers: Tabou, San-Pédro, Nionniourou, Bonbo, Agnéby, Mé and Bia; and in the north some tributaries of the Niger: Baouié, Bagoé, Kouroukellé, Dégou, Kankélaba; and tributaries of the Black Volta: Koulda and Banlou.

The Cavally rises in Guinea and is almost 700 km long. The Taté station, 60 km above the river's mouth, is downstream of a basin of 28,000 km^2. The Sassandra also comes from Guinea; it is 650 km long and has a basin of 75,000 km^2. The Bandama rises at Boundiali in Côte d'Ivoire; it is 1,050 km long, with a basin of 95,000 km^2.

The Comoé has its source in Burkina Faso. It is 1,160 km long and its basin has an area of 78,000 km^2.

The main rivers flow north to south into the Atlantic Ocean. The coastal rivers do likewise and, with exception of the Bia, they have their sources in Côte d'Ivoire, but some of them discharge into lagoons just before the coast. The tributaries of the Niger, on the other hand, rise in Côte d'Ivoire but flow northwards towards Mali.

The rate of flow of the country's watercourses is determined mainly by the rainfall, the influence of the individual characteristics of the basins, such as relief, vegetation and soil type, being fairly limited.

Hydrological systems

The country has four types of hydrological system:

- The "transitional tropical" system

The influence of this system is felt mainly in the northern part of the country beyond the Ferkessedougou-Touba line, i.e. in a region of savannah which is fairly well wooded - depending on the annual rainfall. In the important watercourses of the Bagoé and the Haut-Bandama this system produces a single crest in August, September and October, followed by a rapid drop in the water level in November and December.

- The "transitional equatorial" system

This system is found in the southern part of the country beyond a line through Abengourou, Toumodi and Soubré, i.e. the forest zone. The double crests of the Agnéby and the Bia are typical, with:

- A first high-water period in June-July;

- A second high-water period in October-November;

- A low-water period in August-September followed by an even lower one from December to March.

- The "moderated/transitional equatorial" system

This system is found in the central part of the country, which is covered by savannah except for a few places in Baoule where the forest predominates.

The Nzi at Ziénoa and the Kan at Tiebissou are fairly typical of this system with average water levels in May and November and two moderate crests a year. This type thus lies between the two previous ones.

- The "mountainous" system

This system is found in the mountainous region in the west of the country (Man, Danané, Toulépleu, Taï). The rainfall is abundant, ranging from 1,600 to 2,500 mm a year. The vegetation is of the forest type.

The high waters occur from April to October with the maximum in September; the low-water period is in January-February. The specific rate of flow is fairly high at this time.

It must finally be noted that most of the big watercourses in Côte d'Ivoire have mixed hydrological systems owing to the large area of their catchment basins.

Hydroelectric potential

Since the average fall of the various watercourses is fairly gentle (less than 1m/km), their hydroelectric potential is generally moderate; but it is of some interest because of the waterfalls which occur. Some sites are fairly suitable for hydroelectric development and a few of them have been exploited: the dams on the Bia (Ayamé I and Ayamé II), on the Bandama (Kossou and Taabo), and on the Sassandra at Buyo.

An important hydroelectric project is to be built on the Sassandra at Soubré. Some other sites are of interest: the section of rapids on the Cavally between Taï and Grabo where the fall is 1m/km; on the Comoé above Malamalasso where there are several kilometres of rapids with a fall of 30 to 40 mm; and between Attakro and Aniassué.

It must be noted that with the construction of the Soubré dam Côte d'Ivoire's electricity requirements will be fully satisfied.

Description of dams in Côte d'Ivoire

	Elevation	Capacity (billions of m^3)	Output (GWH/year)
AYAME I	90.5	1,000	80
AYAME II	69	70	120
KOSSOU	206	28,000	80
TAABO	124	630	960
BUYO	200	8,000	900

III. GEOLOGY

Most of the country lies in the zone of Precambrian crystalline basement rock called the "West African shield". The remaining part is the coastal sedimentary basin which covers about 9,500 km^2, of which 1,700 km^2 is lakes and lagoons.

The oldest basement formations include gneiss, granites and migmatites. This is rock of "Prebirrimian" Lower Precambrian age. These formations of crystalline rocks are most common in the west of the country (migmatites are found everywhere). The metamorphic "Birrimian" formations of Middle and Upper Precambrian age are mostly of sedimentary origin. They are generally schistic rocks with sandstones and conglomerates in places. These metamorphic formations with granite intrusions are accompanied in places by volcano-sedimentary formations of tuffs, breccias, green rocks and schists (Proterozoic). They form bands running north-north-east/south-south-west, heavily folded and sometimes raised to the vertical.

The coastal sediments form a wedge which narrows towards the north where it terminates at the edge of the basement rock less than 35 km from the coast.

The coastal sedimentary basin is crossed from west to east by a fault with a displacement of several thousand metres separating two distinct zones: one to the north where the sedimentary layer is not usually more than 300 m thick, and to the south a deep basin where the bedrock sinks to 4,000 or 5,000 m, vertical to the coast, as has been demonstrated by oil boreholes which penetrated Tertiary and Cretaceous formations (Upper and Middle) of marine facies.

The coastal basin has the following morphological units:

- The lagoons on the site of an ancient hydrographic system, now submerged, with very thick mud deposits on their beds.

- The "high plateaus" north of the lagoons with altitudes of 40 to 100 m. They are composed of detrital continental formations of Miocene-Pliocene and probably Upper Cretaceous age ("terminal continental"). The terimanl continental formation rests directly on the basement rock. In places it overlies a thin stratum of Paleocene (mostly limestone) or Maestrichtian (sand-sandstone) formations. It consists mainly of coarse sands in its lower and sand-sandstone clays in its upper part.

- The "low plateaus" south of the lagoons with altitudes of 8 to 12 m. They consist of Quaternary formations with a series of marine clays interbedded with sands (10 m) overlain by a sand-clay continental formation (20 m).

- Sand bars run along the shore, 2 to 6 m high and a few dozen metres thick.

IV. HYDROGEOLOGY

Aquifers

- Crystalline basement rock

The relatively detailed study of the hydrogeological characteristics of the zone of crystalline basement rock in Côte d'Ivoire began with the launching in 1975 of a programme to install 9,000 village water points (by 1985 there were 11,500). Since the end of 1977 about 2,000 water points have been installed, leading to the following conclusions:

Granitogneiss aquifers

In the quartz-grained granitogneiss rocks the ground water is located in the sands of the alteration zone, 20 to 30m thick on average but sometimes 60 m, and in the fractures of the bedrock. The transmissivity of the sands in the water-bearing part overlying the bedrock is in the order of 10^{-4} to 10^{-5} m^2/s.

Generally speaking, the most sandy, even gravelly, strata, at the level of which are found the largest volumes of water, are located at the base of the alteration, and this depth must be reached by the wells and boreholes.

Sand-clay layers are usually found at the surface. The crystalline bedrock suboutcrops in only a few places: Séguéla, Mankono, Touba.

The annual fluctuations in the piezometric level or "aquifer downdraw" varies from a few metres to 6 to 10 m. The lowest values are found close to depressed areas.

The fissuration of the bedrock is discernible from aerial photographs and by examination of the geological features determined in the field: lines of backwaters and watercourses, configuration of laterite cuirasses, and distribution and type of vegetation. In the schistic rocks of the Birrimian system the alteration zone is mainly argillaceous and barren. The ground water is found in the quartz seams or seams of volcanic matter.

Birrimian aquifers

The Birrimian rocks have been identified mainly from data collected on the sites of the 2,000 boreholes in the "cocoa belt". In these formations there is a danger that the installations may become blocked by fine particles from the alteration stratum, which is mostly argillaceous. There is little ground water available at depths beyond 30 m in the substratum. The incidence of water flows is inversely proportional to the thickness of the argillaceous alteration stratum. They decline rapidly when this thickness exceeds 40 m and are virtually non-existent beyond 60 m.

The thickness of the alteration stratum itself depends on the nature of the underlying Birrimian rocks. It attains an average of 45 m in schists but does not exceed 30 m when the schists are interbedded with sandstone or quartz. These hard strata, which seem to make the formation more resistant to erosion, are the most suitable for ground-water exploitation. Finally, the presence of detrital or volcano-detrital intercalations usually coincides with the most productive sections.

It is agreed that the fissures in the substratum contain a widespread aquifer.

In most cases, wells drilled in sands are between 15-20 m and 35-40 m deep, and the water is usually found at an average depth of 10-20 m below the level of the soil. The yields are usually in the order of 1 to 3 m^3/h per metre of downdraw.

- Coastal basin

This basin covers 6,000 km^2 of terminal continental formations, of which 500 km^2 on the edge of the bedrock in the north is barren ("dry wedge") and 1,700 km^2 consists of coastal Quaternary formations (low plateaus and sand bars). In their upper part these latter formations contain an aquifer which can supply fresh water in amounts ranging from a few to several dozen m^3/h, with very good discharges in places in coarse sands: for example, a borehole at Treichville discharges 310 m^3/h with a downdraw of 3.1 m (the aquifer overlies a clay stratum which protects it from intrusion by the underlying salt water). The salt wedge has been identified at depth in the sands at Vridi.

- Widespread aquifers

The only widespread aquifers are the one in the terminal continental formation which covers no more than three per cent of the territory (most it is divided into small discontinuous aquifers) and the one in the coastal

Quaternary formations (plateaus and sand bars). This ground water is stored in the sands of the upper part and exploited by wells and small boreholes which can deliver considerable amounts of fresh water. The horizons of coarse, very permeable sand found in some places are extremely productive; one of the Treichville boreholes, for example (southern quarter of Abidjan - low plateaus), delivers 210 m^3/h with a drawdown of 3.1 m. At this point the lower clay stratum protects the aquifer against upward intrusion by the underlying salt water.

In contrast, in another borehole in the Vridi peninsula (sand bar) a salt wedge has been identified at depth, with salinity ranging from 100 mg/l to 25 g/l in a homogeneous sand series. This aquifer is recharged by direct infiltration of rainwater and rests in density equilibrium on the salt water or on an impermeable substratum; it is vulnerable to domestic, industrial and saline pollution. This aquifer is exploited at Vridi at present to supplement the supply to certain industries. The terminal continental formation has been more fully studied in the central part of the basin than at its two extremities.

A study of the water resources available to supply the town of Abidjan has provided information about the hydraulic characteristics of the central part of the aquifer. The transmissivity is low to average in the north of the basin: $T = 1$ to 4×10^{-3} m^2/s. It is high to very high in the Abidjan area: $T = 10^{-2}$ to 3×10^{-1} m^2/s. The storage co-efficient 'S' has been determined at two points in the area; the values are S = 14 per cent and S = 9 per cent.

Water quality

The terminal continental water has a low mineral content, with the dry residue usually between 40 and 150 mg/l; it is always very aggressive, with pH close to 4.5, a CO_2 content of 90 mg/l, 1 to 4 mg of Ca + Mg, 3 to 6 mg/l of HCO_3, 7 to 14 mg/l of Cl and 10 mg/l of SiO_2.

The water of the bedrock has a moderate mineral content, with the dry residue of a number of samples analysed ranging between 250 and 500 mg/l, or even up to 650 mg/l; the pH of the fissure water is generally close to 7 (neutral pH).

The recharge conditions are excellent, with effective infiltration of 50 per cent of the rainfall, i.e. 20 m^3/s or a water layer of 0.35 metres or 640 million m^3, an amount close to the drawoff, i.e. 10 million m^3/year.

The boreholes in the catchment area exploited for water supplies have produced the following basic data:

Transmissivity: low in the north: 1 to 4.10^{-3} m^2/s

high at Abidjan: 10^{-2} to 3.10^{-1} m^2/s

Storage co-efficient: 10 to 15 per cent.

V. GROUND-WATER EXPLOITATION

The Village Water Supply Service of the Water Department of the Ministry of Public Works, Construction, Posts and Telecommunications is the government body responsible for drawing up the national water programme. It plans, implements and supervises this programme.

To implement the vast drinking water supply programme adopted by the Government, the technical services concerned have opted for the exploitation of ground water by means of wells and boreholes. The construction of these works has produced a picture of the water-bearing capacities of the main geological units.

The various aquifers of the basement rock are hydrogeologically similar. They are composed of fractures in the granitic or metamorphic rocks which are fed with water contained in the overlying alterites. The fracturation is similar in the crystalline rocks and there is very little variation in the hydraulic behaviour.

In granitogneiss formations 77 per cent of the boreholes have proved positive, with an average thickness of alterites of about 20 m. In the schistic formations, where the average thickness of the alterites is about 30 m, the success rate has been 85 per cent. The air-lift discharge is 2 m^3/h in the granitogneiss formations and 3 m^3/h in the schistic formations. It must be pointed out that the sinking of large-diameter productive wells is difficult in the schistic formations although the alterations are thicker; this is due to the large quantities of clay.

The coastal sedimentary basin has the best aquifer system; the success rate is 100 per cent, except when technical difficulties are encountered. It must be noted, however, that the aquifers are very deep on the Bongua plateau (piezometric level between 70 and 80 m).

Thus more than 12,000 water points were installed between 1975 and 1984. These are mainly wells drilled either by down-the-hole hammer or by rotary drilling, sometimes by auger. The works are undertaken by specialist enterprises after international tender. As stated above, the works are carried out under the supervision of the Water Department, which is also responsible for drilling studies.

Some 11,500 water points were installed between 1975 and 1984 in rural areas. Côte d'Ivoire has a national drilling company - FOREXI. Small-diameter wells are drilled in hard rocks and thick formations and large-diameter (1 m) in unconsolidated formations less than 25 m deep. In this case, excavator rigs of the Calweld type are used.

In the cocoa belt and other "Birrimian" areas small-diameter wells are usually drilled by the down-the-hole hammer method powered by compressed air or by the tricone rotary method. It is accepted that at least three exploration holes must be drilled to produce two exploitable wells; of a series of almost 500 drillings throughout the country, nearly 85 per cent have proved positive.

The terminal continental formation is exploited by many water points in the central part of the country; but in some places the water points are far apart. The thickness of this formation is irregular owing to the fairly contrasted topography of the top of the crystalline substratum.

The water of the terminal continental formation is not very mineralized, with a dry residue ranging from 50 to 150 mg/l. It is aggressive (pH = 4 to 5). The exploitation at Abidjan draws off about 50 million m^3/year (130,000 m^3/day). Almost as much is drawn off from the rest of the basin. As an example, according to some estimates the requirements of the metropolitan area will be in the order of 250 to 300 million m^3 in 2000.

The exploitable reserves are about seven billion m^3. This figure is close to that of the renewable resources and shows that the present rate of drawoff is far lower than the aquifer's capacity; and the aquifer also has a large discharge through springs, especially in the Banco ravine west of Abidjan.

The Abidjan boreholes discharge 150 to 250 m^3/h with a drawdown of 5 to 15 m, the level of the aquifer when stationary being 30 to 50 m, with a well depth of 35 to 65 m.

Some wells in the terminal continental formation are more than 100-400 m deep.

VI. CONCLUSIONS

The drilling of so many wells and boreholes in Côte d'Ivoire since 1974 has provided a better picture of the ground-water potential both in the coastal sedimentary basin and in the areas of crystalline basement rock, where the depth of the base has also been determined, together with the thickness of the alterations in some places and the probable discharges; but much still remains to be done with respect to the hydrology of the ground water. This is particularly true of the recharge of the aquifers in the crystalline bedrock, i.e. the renewal of the resources which determines the life of the installations in terms of the rate of exploitation. Progress must also be made in the determination of the slope of the fractures detected by geophysical methods, with a view to better exploitation of their water.

However, it can be concluded from the present state of knowledge that Côte d'Ivoire as a whole has sufficient ground water, in both quantity and quality, to meet the needs of its rural population and much of the urban population, as well as those of industry.

VII. REFERENCES

Barraud, J.P. — Reconnaissances hydrogéologiques en arènes granitiques par sondanges électriques en moyenne Côte d'Ivoire. CIEH, 1961.

BNETD/ERH — Catalogued archives of water points in Côte d'Ivoire of the National Office for Technical Studies and Development; Branch: Evaluation of Water Resources, Abidjan (various dates).

Brancart, R.Y. — Note sur le bassin sédimentaire de Côte d'Ivoire. Unpublished document of the Hydrocarbons Department, Ministry of Mines of Cote d'Ivoire, 1977.

BRGM/CIEH — Water resources planning map for Côte d'Ivoire, Ghana, Togo and Benin (Explanatory note by M. Bourgeois), 1979.

Camerlo, J., Fahy, J.C. and others — Premiers résultats de la campagne destinée à l'alimentation en eau des villages de la "Boucle du Cacao" en Côte d'Ivoire. BL CIEH No. 26, August 1976.

Camerlo, J., Fahy, J.C. and Haubert, M. — Premiers résultats des forages destinés à l'alimentation en eau des villages réalisés dans le Birrimien de la Côte d'Ivoire. United Nations, 1976. United Nations Water Conference, Mar del Plata, Argentina, 14 to 25 March 1977.

Degallier, R. — Interprétation des variations naturelles du niveau des nappes souterraines. Application aux données provenant du bassin versant de Korhogo - Côte d'Ivoire. BRGM report 75 SGN 383 AME, 1975.

Delany, F. — Mission hydrogéologique des régions à substratum schisteux (Côte d'Ivoire). Rapport de fin de mission. BRGM report DAK 66 A2, 1966.

De Spengler, A. and Delteil, J. — Le bassin sédimentaire-tertiaire de Côte d'Ivoire. Société de développement minier de Côte d'Ivoire (SODEMI), 1983.

Fahy, J.C. — Le programme d'hydraulique villageoise en Côte d'Ivoire: premiers résultats acquis après 18 mois de travaux. Unpublished note DCH, Ministry of Public Works, 1977.

Geomines Ltee — Inventaire hydrogéologique appliqué à l'hydraulique villageoise en Côte d'Ivoire, 1982.

SCET-IVOIRE, SODECI, HOLFEDER and SCET-INTERNATIONAL Group — Etude hydrogéologique de la nappe d'Abidjan pour l'alimentation en eau potable d'Abidjan. (Report, 1982)

Guérin-Villeaubreuil, G. Hydrogéologie en Côte d'Ivoire. BRGM document No. 20, 1962.

Maillary, J.C. Carte hydrogéologique de reconnaissance au 1/1 000 0000 de la Côte d'Ivoire. Paris, BRGM/SODEMI, 1964.

Martin, L. and Tastet, J.P. Le quaternaire du littoral et du plateau continental de Côte d'Ivoire. Rôle des mouvements tectoniques et eustatiques. Assoc. Sénégal No. 33-34, 1972.

Pitaud, G. Etude hydrogéologique du continental terminal dans la région d'Adiaké. Localisation d'une zone favorable à l'implantation d'une forage d'exploitation d'une eau de table d'excellente qualité. SODEMI Report No. 248, 1970.

Sasif (various) Coupes techniques et géologiques de forages de la Société africaine de sondages, injections, forages de Côte d'Ivoire - Abidjan.

SCET-IVOIRE and others Carte piézometrique a 1/50 000 de la nappe du continental terminal de la région d'Abidjan s.l. Unpublished document of SCET-IVOIRE, 1977.

SERI Programme national d'hydraulique villageoise en Côte d'Ivoire. (Village water supply report - Water Department), 1984.

SOGREAH and others Ville d'Abidjan - Etudes préliminaires à l'établissement des projets d'alimentation en eau et d'assainissement d'Abidjan. Rapport No. 10: Etudes des ressources en eau. UNDP/WHO report - Ministry of Public Works of Côte d'Ivoire, prepared by SOCREAH, ref. R 10912-10, 1972.

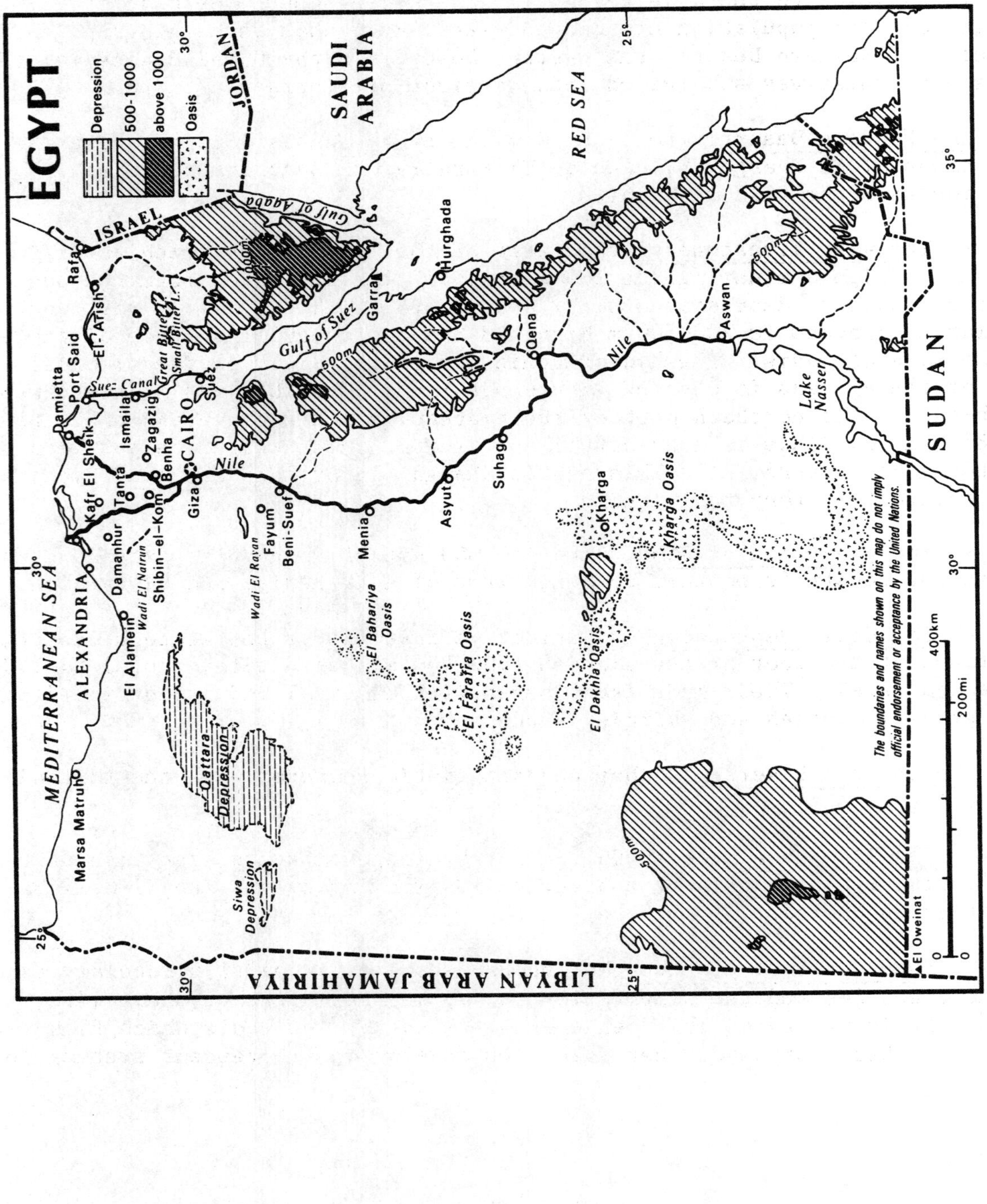
EGYPT
Depression
500-1000
above 1000
Oasis
MEDITERRANEAN SEA
RED SEA
SAUDI ARABIA
JORDAN
ISRAEL
SUDAN
LIBYAN ARAB JAMAHIRIYA
Gulf of Aqaba
Gulf of Suez
Suez Canal
Great Bitter L.
Small Bitter L.
Nile
Lake Nasser
ALEXANDRIA
CAIRO
Marsa Matruh
El Alamein
Damanhur
Kafr El Sheik
Tanta
Damietta
Port Said
Ismaila
Zagazig
Benha
Shibin-el-Kom
Giza
Suez
El Arish
Rafa
Hurghada
Garra
Qena
Aswan
Suhag
Asyut
Menia
Beni-Suef
Fayum
Wadi El Natrun
Wadi El Rayan
Qattara Depression
Siwa Depression
El Bahariya Oasis
El Farafra Oasis
El Dakhla Oasis
Kharga
Kharga Oasis
El Oweinat
500m
1000m
25°
30°
35°
0
200mi
400km
The boundaries and names shown on this map do not imply official endorsement or acceptance by the United Nations.
3145.7x

EGYPT

Area: 1,001,000 km^2

Population: 44.53 million (United Nations estimate, 1983)

I. BACKGROUND

Physiography

In addition to the Nile Valley and its Delta, where the majority of the arable land and population are concentrated, Egypt has three geographical areas: the Western Desert, the Eastern Desert and the Sinai Peninsula; these areas are themselves subdivided into distinct physiographic units.

i) The Western Desert, where the New Valley region is located, with the oases of El Kharga, El Dakhla, El Farafra, El Bahariya and Siwa, includes the following units:

- The vast limestone plateau west of the Nile Valley, with a cliff some 150 m high dominating the low sandstone plateaus which form the base of the topographical depressions of El Dakhla and El Kharga. This vast and undulating plateau is 200-500 m high and has an irregular surface dissected by the beds of wadis. The hydrographic system is mainly of the arborescent type. Part of the plateau is covered with dunes running north-south or north-west/south-east. The northern part of the plateau includes the areas of El Bahariya and El Faiyum. This is a desert of the "serir" type, while the southern part, including El Farafra, El Dakhla and El Kharga, is of the "hammada" type, with vast horizontal stony tablelands.

- The great sea of sand, with 150,000 km^2 of sand dunes extending as far as the country's western borders, west of El Bahariya and El Farafra.

- The oasis depressions, distinctive features of the Western Desert, are found at the foot of the plateau (100-200 m), at altitudes of 1 to 150 m above sea level. Their beds are covered with alluvial and lacustrine deposits, with gravel, cobbles and shifting sands.

ii) The Eastern Desert also has characteristic features with the following units:

- The limestone plateau in the north along the south-east edge of the Nile Delta, the Nile's lower course, the Gulf of Suez, and the El Mellaha mountains in the east.

- The sandstone plateau (Nubian sandstone) of typical morphology extends from the Red Sea and the Nile Valley as far as the southern border with Sudan. It is marked by the deep narrow valleys of the wadis which flow into the Nile. Secondary wadis can sometimes develop in arborescent systems, but their direction is clearly determined in most cases by the fractures - and the joints of the Nubian sandstone formation. South of Wadi El Allaqi the Nubian sandstone is much more deeply eroded, producing a complex system of mesas and flat-topped alignments.

- The Red Sea Mountains rise to 1,500 m. They have a contrasted, deeply eroded topography. Most of the wadis draining the South-Eastern Desert come from this area, which forms the region's main hydrographic basin. A similar landscape is found in the region between the eastern mountains and the Nubian sandstones. Vast granodiotitic systems underlie the great sand plains interrupted by the trachyte ridges typical of this region. Parallel arrangements of low peaks and valleys along the lines of dykes are also found. The coastal zone forms a narrow strip of land between the basement rock and the shore of the Red Sea. It is covered by sediments of Middle Cenozoic and Quaternary age, with the exception of Wadi Araba and the plateau of North Galala, where Paleozoic and Upper Mesozoic sediments predominate. The coastline is low between Suez and El Zaafarana and it widens from that point as far as Safaga and again from the vicinity of Ras Benas as far as the Sudanese border in the south. Copious deposits of gravel are found along a gently sloping terrace ending abruptly at the sea which has been eroded by the wadis draining the mountain region beyond the plain.

III) The Sinai Peninsula includes:

- The triangular massifs in the south, constituting the most mountainous part of the peninsula. Mt. Musa and Mt. Catherine are the highest peaks, at 600 m above sea level.

- The central limestone plateau of central Sinai, 1,000 m above sea level, composed mainly of Mesozoic and Tertiary sediments. It is deeply eroded by valleys bounded by abrupt escarpments. The beds of these valleys fall sharply, producing cascades during flood crests.

- The Rangdom Hills, running north-east/south-west and bounded by the sand-dune plain in the north and the plateaus of El Tih and Egmä in the south. They have steep to moderate slopes.

- The sand-dune plain, which extends from the Suez Canal in the west as far as Rafah in the east, and from the Mediterranean in the north to the Rangdom Hills in the south. These wind-formed dunes rapidly change their shape and position.

- Aqaba - the coastal plains are wider along the Mediterranean and narrower along the Gulf of Suez and Gulf of Aqaba, except for the El Gaa plain whose maximum width is 20 km.

The depressions are of considerable importance in the desert areas, for ground water is often found in them at moderate depths; they are moreover protected from the wind and therefore from evaporation. These are the most suitable areas for human settlements and agriculture. Egypt's main depressions are, from south to north:

Wadi El Natrun	(-20 m)
Qattara	(-134 m)
Siwa oasis	(-18 m)
El Faiyum	(-60 m)
El Bahariya oasis	(+100 m)
El Farafra oasis	(+100 m)
El Dakhla oasis	(+100 m)
El Kharga oasis	(+50 m)

Climate

With the exception of the Valley and a few oases, 93 per cent of the country is desert. It rains in Egypt only from October to May, the summer months (June to September) being "dry". It has rained on occasion in June or September, but there are hardly any valid data on any light rain that may have fallen in July or August.

The heaviest rainfall, although still moderate (about 200 mm/year), is found in the coastal zone where it rains on 40 days a year, mainly from December to February - an effect of the "Cyprus low-pressure areas".

The rainfall generally shows a marked and rapid decline from the coasts towards the interior. In the middle of the Delta it is only 80 mm/year, falling on 20 days. Rain is even rarer in Upper Egypt. At Assiout, 350 km south of Cairo, the annual average is only 7 mm and the maximum recorded in a season is 25 mm, falling in a single storm.

The rainfall is not only meagre, it is also very irregular from year to year. At Alexandria, for example, the season's maximum can vary from 50 to 400 mm from one year to the next. At Marsa Matrouh in the Western Desert the annual average is 160 mm. The rain can be torrential, with one storm bringing 100 mm in a single day.

In the Eastern Desert and the Sinai torrential runoffs can occur in the wadis after a storm, usually in October-December. Such storms may be rare but they can be very violent at times. At Quseir on the Red Sea coast, for example, the average annual rainfall is only 4 mm, but more than 30 mm has been recorded in a single storm.

The maximum rainfall and river crests in the Sinai and the Eastern Desert occur about every 20 and 11 years respectively, and every seven years in the coastal zone of the Western Desert.

The following evapotranspiration occurs:

Western Desert:	from 770 mm (Matrouh) to 2,600 mm (Dakhla oasis)
Eastern Desert:	1,540 mm (Quseir)
Sinai:	1,440 mm (El Tor)

Geology

Egypt is part of the foreland structure on the northern edge of the Arabo-Nubian Shield. In southern and south-eastern Egypt Precambrian and Lower Cambrian basement rocks outcrop or suboutcrop (southern Sinai, anticlinal of the Red Sea, Aswan and El Uweinat). These are igneous and metamorphic rocks offering good prospects for mining.

The basement rock is overlain by a mainly sandstone formation produced by epicontinental sedimentation, the thickness of which increases from the south (500 m) to the north where, at the sites of a number of boreholes, it attains 2,500 m and marine facies are found in places. This formation is of Paleozoic or Lower Mesozoic age. The main aquifers in Egypt are associated with this system, generally known as "Nubian sandstone".

Towards the north-west, after the Paleozoic-Mesozoic sandstones, lies a formation of carbonate rocks interbedded with clay about 2,500 m thick. This formation, which dates from the Upper Mesozoic (Jurassic and Cretaceous) and Lower Cenozoic (Paleocene) periods, offers typical examples of karstic hydrology and occupies a large part of the plateau zone.

Egypt's geological landscape acquired most of its present characteristics in the era following the Paleocene. The main geological events in that era were as follows:

- Widespread lifting of the continental surface, retreat of the sea, continental deposits, with volcanic and hydrothermal activity during the Oligocene and Lower Miocene periods;

- Formation of the Red Sea graben;

- Local north-south intrusions of the sea during the Oligocene, Miocene and Pliocene eras (caused by oscillations in the level of the Mediterranean);

- Formation of the Nile basin, first in the form of a marine gulf in the Pliocene era, then as a body of fresh water flowing from south to north during the Pleistocene;

- Successive climate changes producing desertification, the present features being the accumulation of sand dunes, degradation of the soil and vegetation over most of the zone, increasing salination, and lack of rainfall.

II. AQUIFER SYSTEMS: BACKGROUND

Egypt has three main aquifer complexes corresponding to the sedimentary systems which overlie the Precambrian crystalline basement rock.

i) The lower detrital formation or "Nubian sandstone", consisting mainly of sandstone in thicknesses ranging from 100 to 2,500 m. This aquifer is artesian. Within Egypt's borders the piezometric level declines from an altitude of 500 m in the south-west to approximately sea level in the north.

These Nubian sandstones also extend into Libya, Chad and Sudan to form the artesian basin of north-eastern Africa. The chemistry of the ground water of this vast system varies from place to place, with a dry residue generally below 500 ppm.

Carbon-14 dating and other radio-isotope techniques indicate that most of the water infiltrated to deep levels during one or several rainy periods. This fossil water is between 20,000 and 40,000 years old. Hydrogeologists generally agree that the aquifer might be recharging at present from small recharge zones outside Egyptian territory by infiltration of equatorial rainfall and the waters of the Nile in northern Sudan.

The underground runoff is towards several natural or artificial outlets in the vast area west of the Nile, in particular towards the depressions of Qattara and Siwa, the oases of El Kharga and El Dakhla, and the oases of El Farafra and El Bahariya.

ii) The formation of "middle limestones" of Canomanian to Upper Eocene age, composed of fissured limestone which is well karstified in many cases and common throughout Egypt. Mention may be made here of several rich sources of water fed from these limestones, which so far have received hardly any study.

- West of the Nile, almost all the natural springs of El Farafra, El Kharga, Sitra, Arag, El Bahrein, Garra, Siwa, Wadi Rayan, etc., and almost all the wells in the limestone plateau north of the Qattara depression;

- East of the Nile, including the Sinai, the many natural springs lying along the Red Sea anticlinal, at Helouan and in adjacent areas as far as the east of Cairo (heavily mineralized water), in the fault region of western Sinai (sulphur springs of Hamman Faraun; hot salt springs), and bordering the calcareous plateau of eastern Sinai.

iii) Lastly, the formation of Upper Oligocene to recent age, which is mainly detrital but does have some organic limestones such as the Mediterranean calcarenites of the coastal plain which form undulating terrain. These deposits of marine or continental origin form an aquifer along the whole length of the one thousand kilometres of coastline at an altitude close to sea level. This aquifer, in which the salinity is about 1,000 ppm, rests on a mass of salt water produced by sea-water intrusion. However, subartesian conditions are found in a few places, as are perched aquifers, attributed both to the change of facies and to the geological structure of the terrain. Recharge is by direct infiltration of local rainfall.

One of Egypt's very important aquifers is found in the sand and gravel formations deposited in the depression of the Nile and the Delta - the Nilotic deposits of the Pleistocene era. This aquifer is about 350 m thick and rests on Pliocene clays which can be considered an aquiclude. It is practically saturated with water of about 500 ppm salinity and is recharged from the waters of the Nile and the network of irrigation channels. In this part of Egypt the ground-water aquifer is continuous and sinks gradually from south to north. After the construction of the Aswan Dam the ground water in the Nilotic deposits began to rise, apparently as a result of the marked difference of head between the water in the dammed lake (+178 m) and the surface of the water in the river to the north (maximum level of about +90 m). As the lithology of the deposits in the Nile Valley is not homogeneous, isolated lenticular formations of clay are found - explaining the subartesian conditions which occur at certain points. The volume of renewable water stored in this aquifer is probably about 600 million m^3 and its discharge zone includes the following outlets:

- Wadi El Natrun (-20 m) and perhaps the Qattara depression (-134 m) in the west as well;
- Wadi Tumeilat, the Bitter Lakes and the Suez Canal in the east;
- The lagoons along the Mediterranean coast in the north;
- The many wells in the coastal strips.

At the northern edge of the Delta there are heavy intrusions of sea water, caused perhaps by the excessive pumping of ground water.

Recent hydrogeological studies tend to indicate that these aquifer systems are in contact with each other at many points.

III. GROUND WATER IN THE WESTERN DESERT

Nubian sandstone

South of the 29° 30' parallel N the Nubian sandstones stored great amounts of infiltration water during the rainy periods at the end of the Quaternary epoch and they have retained this water since then. In ancient times this water was exploited in the oasis depressions of El Kharga, El Dakhla and El Bahariya. These same regions have been exploited by 400 boreholes of various depths for about thirty years. The table below gives data for the El Uweinat area east of the Gilf Kebir plateau in the south of the Western Desert.

Aquifers in El Uweinat

Type of well (by yield)	Type 1	Type 3	Type 4	Type 5
Yield (m^3/h)	6	40	18	25
Transmissivity (m^2/day)	1,688	2,437	1,987	44
Permeability (m/day)	28	50	40	0.60
Storage	17.5×10^{-2}	5.3×10^{-3}	11.7×10^{-2}	5.3×10^{-4}
Drawdown (m) ($Q = 50\ m^3/h$)	1.36	1.36	1.18	56
Salinity (ppm)	810	1,260	1,280	1,230

The depth of the piezometric level varies from a few to almost 50 metres.

It should be noted that the average permeability at the El Bahariya oasis is 6.2 m a day and the transmissivity 8,000 m^2 a day.

It has been established by the radio-isotope method that the Paleozoic-Lower Mesozoic aquifer system varies according to its geographical location, as shown by the following table.

	El Kharga	El Dakhla	El Farafra	El Bahariya	Nile waters
Deuterium H_2* (percentages)	-83.4	-83.9	-79.1	-79.7	+25
0_{18}* (percentages)	-10.9	-11.0	-10.5	-10.4	+ 3

* Standard Mean Ocean Water.

These results tend to indicate that there is a separation between the oasis water of the north and south and the Nile, which thus does not feed the Nubian sandstones.

It has been estimated that the daily recharge by rainfall of the Nubian sandstones, which are more than 3,000 m thick west of the El Farafra ocasis, is in excess of four million cubic metres, exclusively between the 29° 30' parallel N and the Egyptian-Sudanese border. During the period 1963-1970 the annual drawoff was about 80 million m^3, increasing between 1970 and 1981 to 300 million and 450 million respectively, and to an estimated one billion cubic metres in 1985-1990.

The drawoffs of ground water at the four oases in the southern part of the New Valley have been calculated by mathematical model as follows:

Ground-water drawoffs - New valley

(Southern oases)

(millions of m^3/year)

	Wells	Evapotranspiration	Total
El Kharga	50	55	105
El Dakhla	115	35	150
El Farafra	1	6	7
El Bahariya	25	30	55

In the vicinity of the High Dam Lake the thickness of the Nubian sandstone varies from 200 to 600 m and the depth of the aquifer from 20 to 60 m over a strip of land 10-kilometres wide around the shores of the lake. Beyond 10 kilometres the depth of the water varies from 150 to 200 m. The recharge of the Nubian sandstone by the waters of the Nile amounted to 5 million m^3 a year before the construction of the High Dam. The lake began to form in August 1964, and the recharge of the Nubian sandstone from the lake has increased to 1.5 billion m^3 a year, with roughly equivalent distribution for the eastern and western parts. In this vicinity the salinity of the ground water in the Nubian sandstone is about 200 ppm.

It has been determined by means of deep boreholes that in the Siwa oasis, to the south of the El Qattara depression, the fresh water overlies deep salt water, while in the north of the Western Desert only salt water is found.

Depressions have been dug in the aquifer by the drawoff of surface water in the oases over more than a thousand years, and vertical hydraulic gradients have been created, drawing deep water up towards the surface. Where deep aquifers have been exploited, the gradient is reversed and water is drawn down from the higher levels.

It is clear that the Qattara depression and the El Kharga oasis are the main outlets and that El Dakhla and the El Bahariya oasis make only small contributions. Before 1960 there was a rough equilibrium between the recharge from the south and the discharge through the natural springs and traditional wells of the oases and the flow towards the Qattara depression. With the exploitation of deep wells since 1960 this balance has been upset, and the water table at the oases has been observed to decline, particularly at El Kharga. Modelling of the water balance in the New Valley and the Qattara depression has indicated ground-water flow and escape (discharges of the springs and wells and evapotranspiration losses) as follows:

	Millions of m^3/year
El Farafra area	1
El Bahariya area	100
El Bahrein area (south of Qattara)	50
South of Siwa	140
Qattara depression	90
Total	381

The effects of various pumping schedules on the water levels have been simulated for a period of 100 years for an aquifer lying 1,000 m below the soil.

The discharge safety levels, up to which pumping does not threaten the water resources, are as follows:

	Millions of m^3/year
El Farafra (Nubian sandstone)	365
El Bahariya "	182
Siwa (Middle Miocene)	140
Total	686

Source: General Petroleum Co., 1981.

Aquifers of the Upper Mesozoic to Quaternary periods

These aquifers are fed from the underlying Nubian sandstones through faults and grabens. Moreover, there are no impermeable strata blocking the upward migration of the water. The Upper Mesozoic-Lower Cenozoic formations, which consist mainly of carbonate rocks, have not been extensively explored along the northern edge of the Western Desert, except for the fractured chalk of the El Farafra oasis and the fractured limestone of Middle Eocene age along the natural lake zones south of the Qattara depression.

Minor aquifer systems of small extent and potential can still be of considerable economic and social importance in areas where there is no other source of water.

The Oligocene aquifers consist of sands, clays, gypsums and calcarenites. They are found in the eastern part of the Western Desert north of the 28th parallel N. They extend as far as the banks of Wadi El Natrun in the east and the El Moghra oasis (north-east of the Qattara depression) in the west. These aquifers are recharged by direct infiltration of rainfall and surface runoff.

Miocene-Pliocene and Quaternary aquifers are found in interbeds of argillaceous calcarenites and limestones with some sandy strata. The water has an acceptable salt content in the Siwa oasis (Middle Miocene), at Moghra (Oligocene-Miocene), in the Qara oasis (Miocene) and in the area of Wadi El Natrun (Quaternary Pliocene). The aquifer is recharged exclusively by direct infiltration of rainwater and transfers from the Oligocene aquifer in the north-east of the Western Desert (from El Moghra to Wadi El Natrun). In the case of Wadi El Natrun and the El Rallat depression to the north, the flow towards the Miocene-Quaternary aquifer is mainly from the western part of the Nile Delta. The annual recharge from Rayah El Beheiri and the Nubariya canal can attain 9 million m^3. About 18.5 million m^3 a year can be taken from Wadi El Natrun to irrigate 5,000 feddans (2,000 ha), while according to the mathematical model the total drawdown after 100 years will be 28 m. The salinity of the water varies from 200 to 5,000 ppm from east to west of Wadi El Natrun and it can attain 20,000 ppm in the north-west in the areas west of the Nubariya canal along the northern part of the El Rallat depression.

Oligocene-Miocene to Quaternary basalts outcrop west of the High Dam lake, at the El Bahariya oasis, north of the El Faiyum depression (at Begel Qatram), west of the area of the Gizeh pyramid, and in the vicinity of Abu Roash. Boreholes drilled in the basalts of the El Bahariya oasis have revealed a considerable volume of good-quality ground water.

Pleistocene and Holocene sediments outcrop along the Mediterranean coast in the form of oolitic calcareous sandstones (in isolated hills), sand in layers and dunes, and alluviums of sabkhas and wadis. The hills, coastal dunes and wadi deposits constitute the main aquifers of the coastal region. The ground water is exploited by wells equipped with pumps, windmills and infiltration galleries. At Fuka, for example, the wells draw off 50,000 m^3 a year and the windmills 24,000 m^3; at Baggush the galleries produce 200 m^3 a day; at Hekma the windmills pump 400 m^3 a day. At Negila 100,000 m^3 is drawn off per year, with 100,000 m^3 a year at Wadi Maguid. At El Qasr the pumps and windmills draw off 75,000 m^3 a year and the galleries - with a total length of 11 km - 360,000 m^3 a year.

IV. GROUND WATER IN THE EASTERN DESERT

Ground-water research in the Nubian sandstones began only recently and was to conclude in 1986. It has been carried out in conjunction with mining operations. Ground water is present in the fractures of the crystalline basement rock, including metavolcanic rocks, gneiss, schists, and granitic and serpentinite intrusions. The presence of veins assists the flow of ground water along the main wadis. The fracturation of these rocks depends on the tectonic conditions, while the deep alteration is determined by age and paleo-climate.

In Wadi Lahmi, west of Ras Benas, the aquifer is 26 m deep. The wells yield 100 m^3/day. At Wadi Karim (east of Quseir), the depth is 30 m and the yield 10 m^3/day. At Wadi Basha south-west of Hurghada the aquifer is at 16 m and the drawoff is 60 m^3/day. The water has a fairly high salt content: from 2,000 to 3,000 ppm.

The storage of the ground water in the fractured rocks is poor, as is the recharge. By contrast, the alteration zones which have a good storage co-efficient are not very permeable. The best location for wells is in places where the fractures are overlain by water-saturated altered rocks (wadi alluviums). The ancient volcanic rocks are very moderate aquifers except in fault zones. On the other hand, the more recent volcanic rocks give good yields in areas with sufficient rainfall.

The Nubian sandstones of the Eastern Desert satisfy local requirements, particularly in the main wadis of the central section of the Desert and near the eastern shore of the High Dam lake and the shore of the Red Sea. This aquifer system extends from the northern part of the Gelala plateau as far as Egypt's southern borders and into Sudan. It has hardly been explored at all, except at El Laquita (drilling depth: 450 m) and near the shores of the High Dam lake (drilling depth: 600 m). Promising aquifers have been found along Wadi Araba west of the Gulf of Suez. The salinity at El Laquita is up to 400 ppm, with heavy concentrations of iron and hydrogen sulphide. The recharge is estimated at 300 million m^3 a year.

In the Ras Gharib plain (at Shagar for example) the Nubian sandstone has been reached at a depth of 900 m. The wells at Shagar yield 3,000 m^3 of drinking water a day. The salt content is as high as 2,500-3,500 ppm at Wadi Araba and in the mining boreholes at Quseir - an indication of very poor recharge by rainfall.

The consumption of water in future years (m^3/day) in the three zones of the Eastern Desert is estimated as follows:

	1985	1990	1995	2000	2010
Zaafarana-Gemsa Bay (south of Gulf of Suez)	10,500	19,000	25,000	28,000	34,000
Ghardaka-Safaza-Quseir (central part)	28,000	45,000	90,000	132,000	144,000
Mersa-Alam-Berenia (southern part)	200	3,500	7,000	11,000	15,000
Total	38,700	67,000	122,000	171,000	193,000

Six wells have been drilled in the Shagar area. The aquifer was reached at 70 m. The drawoff is 5,000 m^3/day. Four other wells have been drilled south of Shagar (Ras Shukein) with a capacity of 4,500 m^3/day. The salinity of the water ranges between 2,600 and 3,400 ppm.

Water has also been drawn from the phosphate mines between Shagar and Quseir: 500 to 1,000 m^3/day from the Hamzawein mine, with a salt content of 2,400 to 2,800 ppm.

There are good wells along the old Qena-Safaga railway line; the two wells at Bir Merkh yield 115 m^3/day, with salinity between 400 and 800 ppm.

Most of the now abandoned mines in the Red Sea Hills have deep wells in hard rocks yielding up to 40 m^3/day of water of varying salinity.

The production cost of the ground water varies from LE 0.07 1/ per cubic metre for the wells in the Nubian sandstone to the west of the Western Desert mountains and the shallow wells in the coastal zone to LE 0.12 per cubic metre for the deep wells in the coastal zone (Shagar).

It has not yet been possible to evaluate the ground-water resources of the limestone plateau in northern Galala owing to its uneven permeability, which depends on the local fracture conditions. The southern piedmont of the Galala plateau (the area of the St. Paul monastery) warrants exploration, for here the Nubian sandstone is not very deep. Between the two Galala plateaus the plain of Wadi Araba, overlying a thick aquifer of Nubian sandstone, constitutes a very substantial reservoir of slightly brackish water.

V. GROUND WATER IN THE SINAI PENINSULA

Ground water accounts for more than 90 per cent of the peninsula's consumption. The volume of water stored in the aquifer system (Nubian sandstone - Quaternary) is estimated at about 75 billion m^3. Moreover, the Nubian sandstone receives a recharge of 100 million m^3 a year, i.e. 67 per cent of the total recharge of the aquifer systems. The aquifers include

---

1/ LE: Egyptian pound.

interbedded sandstones and clays and some layers of sandy limestone. The salinity of the water varies from 1,000 to 2,000 ppm. The fractured limestones have hardly been explored or studied at all; their water has varying concentrations of minerals (800 to 10,000 ppm). There are other smaller aquifers (fractured crystalline basement rock, fractured volcanic rocks, Quaternary deposits) which can be of local importance when there are no other sources of exploitable water. Quaternary aquifers are found only in the coastal zone or in the wadi beds. The water quality varies from good to brackish (700 to 3,000 ppm). The water requirements in the Sinai are estimated at the following amounts (hm^3/year):

Year	Domestic	Tourism	Industry	Agriculture
1985	10	-	-	35
2000	100	2	80	100

At present, almost 50,000 m^3/day is taken from the Quaternary aquifers of the Sinai and about 2,000 m^3/day from all the other aquifers together.

VI. GROUND WATER IN THE NILE VALLEY AND DELTA

The Nile Valley and Delta constitute one of the world's biggest ground-water reservoirs, with a capacity of 400 billion m^3: 120 million in Upper Egypt and 280 in the Delta. The annual recharge is estimated at 8.6 billion m^3: 5.6 billion in the Delta and 3 billion in Upper Egypt.

In Upper Egypt there are three characteristic levels in the alluviums of the Valley. The upper level consists of clays, silts and fine sands and is 4-20 m thick; it overlies the main aquifer, which includes sands, gravels and cobbles, with shallow argillaceous lenses. The thick level of 15 to 250 m is of Pleistocene age. It overlies layers of clay and neogenic limestone. The transmissivity is in the order of 8,000 m^2/day; the permeability is 100 m/day. The storage co-efficient ranges from 7×10^{-6} to 8×10^{-3}. The vertical permeability has been estimated at 0.04 m/day.

The thickness of the ground-water reservoir in the Delta zone increases towards the north, from Cairo (100 m) to the Mediterranean coast (900 m). The transmissivity is around 100 m^2/day, the storage co-efficient from 10^{-4} to 10^{-3}, the permeability from 0.02 to 0.05 m/day. The drawoffs from the aquifer for domestic, agricultural and other needs, and the losses, have been estimated as follows:

Nile Delta	hm^3
- Drainage towards the branches of the Nile	212
- Vertical drainage through clay strata	97
- Flow towards the El Natrun aquifer	50
- Yield of the pumping stations (irrigation and domestic uses)	1,600 (1980)
TOTAL	approx. 2 billion m^3

Upper Egypt Valley	
- Drainage towards the river	3,500
- Pumping	1,300
TOTAL	almost 5 billion m^3

It is estimated that the following additional amounts could be drawn off without risk:

2.5 billion m^3 in the Delta

1.5 billion m^3 in Upper Egypt

In the Delta the quality of the ground water varies considerably from year to year. In Tahrir province the salinity ranges from 480 to 1,920 ppm and it can reach 16,000 ppm in the north of the province. West of the Cairo-Alexandria road the salinity varies from 7,500 to 16,000 ppm. In other parts of the Delta it ranges from 500 to 4,000 ppm and even higher.

In general, the water quality in Upper Egypt is satisfactory for all uses (120 to 1,200 ppm). Since the completion of the High Dam the water quality of the aquifer has been deteriorating owing to the rise of the piezometric level and the exposure of the ground water to increased evaporation, with a consequent increase in its salinity.

Owing to the development of irrigation in the Nile Valley and Delta following the construction of the High Dam, the piezometric level has risen by an average of two metres in Upper Egypt and by one and a half metres in the Delta. In some areas bordering the western part of the Delta the water table has risen by 28 m in the north and 20 m in the south, with an increase of 50 million m^3/year in the volume of water stored in the aquifer.

VII. CONCLUSION

It is necessary to stress the considerable importance of ground water in Egypt, its close connection with the waters of the Nile, and the very pronounced changes brought about in the ground-water system by the building of the High Dam.

Egypt's Desert Institute is responsible for the study, evaluation and exploitation of ground-water resources. Detailed studies of the various aquifers are being carried out under the Institute's auspices at present; both traditional methods and modern techniques are used: various analogical and mathematical models, radio-isotopes and remote sensing. All this activity will eventually result in an evaluation of the potential of these aquifers, the possibilities for exploitation of which are of interest to many public bodies.

GAMBIA

Area: 11,295 km^2

Population: 620,000 (United Nations estimate, 1983)

I. BACKGROUND

Gambia is a flat low country surrounded on three sides by Senegal and crossed from east to west in its centre by the lower course of the River Gambia.

Climate

The climate is of the "Sahelian"type with a short rainy season from July to September followed by a long dry season lasting nine months. The annual rainfall declines gradually from 1,200 mm in the south and south-east to 900 mm in the north and north-east. Gambia has endured the periods of drought which have afflicted the Sahel since 1973 with a pronounced reduction in the annual rainfall.

The average monthly temperature at Banjul ranges from 24.7°C in January to 28.5°C in October. The daily range is 6.5°C in the rainy season (August) and 11.7°C in the dry season (January). In the interior the average figures are three to four degrees higher than at Banjul.

The relative humidity is about 80 per cent in the rainy season (August) but it falls to 46 per cent in the dry season (January).

The potential evaporation is around 2,000-2,500 mm/year and the actual evapotranspiration is 750-900 mm a year.

Surface water

The River Gambia, which rises in Guinea in the Fouta Djalon mountains, crosses Gambia from east to west. The area of the river's basin is about 70,000 km^2, of which only 10,500 km^2 is in Gambia.

At Gouloumbo in Senegal, near the eastern border, the average annual flow is 9,000 hm^3, i.e. 285 m^3/s. During the high-water months (August to September) the average monthly flow ranges from 300 to 1,000 m^3/s, with daily crest flows of up to 1,500 m^3/s. During the low-water period, however, the monthly minimum for April or May is about 2-3 m^3/s, and the daily figures can sometimes fall to only 1 m^3/s in very dry years.

The basin's total flow at Banjul is estimated at 12,000 hm^3 in an average year, i.e. 380 m^3/s.

The river begins to fall in November and the decline continues until the first rains in July. Study of the low-water levels shows that the aquifers upstream of the Gouloumbo station (44,000 km^2) feed into the river a volume of water equal to 30 per cent of its total annual flow - even more in dry years.

For the whole of the River Gambia basin, the runoff co-efficient is 0.25 in an average year, and the infiltration figure is about 0.12; most of the rainfall is therefore lost through evaporation.

The river's poor flow rate in the low-water period is the reason why salt water can penetrate some 250 km upstream from the river mouth every year. In the rainy season the fresh-water front moves west to a point 105 km from the sea. In recent years, following the decline in the river's flow rate caused by the drought, the situation has grown worse, with more rapid penetration of salt water along the river.

Geology

The geological formations found in Gambia belong to the system of sedimentary continental and precontinental series of the western edge of the ancient massif of Fouta Djalon. These formations, which can be more than 1,000 m thick, were probably thrown up by a succession of faults due to subsidence in the Upper Cretaceous period which is gradually lowering them.

The most important upper formation, apart from recent sediments, is the Quaternary and Pliocene series known as the "terminal continental". This formation has been shaped by the river and it is not very thick in Gambia, ranging from 10 m to a maximum of 30 m; it includes sands, clays and silts and is often topped by a thick laterite cuirasse. The substratum of the series consists of a layer of yellow clay with occasional lenses of fine sand and silt. The sand is the location of an unconfined aquifer which is exploited at present by large-diameter wells for rural drinking water supplies.

The heterogeneous Quaternary-Pliocene series rests on a more homogeneous formation of coarse sands with occasional silts and clays dating from the Miocene era, which can be up to 30-40 m thick. This water-bearing formation is exploited by a few wells of moderate depth and it is the largest recharging aquifer known in Gambia at present.

The formation's substratum consists of a thick layer of grey and green clays and marls interbedded with thin layers of limestone, all dating from the Eocene era.

No information is available about the formations underlying the Eocene series in Gambia. However, according to data obtained in Senegal, in Casamance in particular, it can be assumed that the Eocene series overlies fairly homogeneous clays and marls of Paleocene to Upper Cretaceous age. This thick formation has not yet been extensively explored, but it does not seem likely to contain any widespread aquifers until the Maestrichtian level is reached. This latter formation consists of a thick layer of sand and sandstone and it is a very productive artesian aquifer widely exploited in West Africa. It has been reached by an axploration borehole at Banjul at the mouth of the River Gambia at a depth of about 400 m. In the light of data obtained in Senegal, it is assumed that the Maestrichtian aquifer is present in Upper Gambia at an exploitable depth of around 200-250 m. It is therefore of interest for the future.

Few studies have been made so far of the structures determining the great thickness of the sedimentary formations. The existing information seems nevertheless to indicate that a number of large structures cross the region from north-east to south-west and that they could affect the hydrogeological behaviour of the aquifers. The country can thus be divided into three main zones: the horst in the west, the central depression, and the upper basin. This division is probably due to shifts in the sedimentary series at great depth. Recent studies in Casamance seem to indicate the presence of east-west faults which displace the Miocene aquifer to a greater depth in the sedimentary basin of Casamance. In Gambia, however, this same formation is found at a moderate depth and almost at the surface in the horst in the west, especially in the Kombo peninsula.

II. GROUND WATER

Works carried out

Since 1978 the Organization for the Development of the River Gambia, which includes three West African States, has been considering the exploitation of the river's surface water. But to date no extensive study of the ground water has been undertaken. Partial studies for drinking water supplies have of course been made both in Senegal and in Gambia.

Following the creation of the Department of Water Resources in 1974 and more recently the Ministry of Water Resources and the Environment in 1981, greater attention has been given to water problems. Towards the end of 1981 a small hydrogeological unit composed of a United Nations expert, a United Nations volunteer, a Gambian hydrogeologist and two technical assistants was set up in the Department of Water Resources and attached to the Rural Water Supply Branch; it was given the task of gathering together the existing data and directing the first investigations.

A summary of existing documents was prepared. In addition to evaluating and cataloguing the water points, the unit collected and interpreted the field data. More recently an electro-seismic prospecting campaign was prepared, launched and supervised by the unit, backed up by an associate geophysicist.

The means available to the hydrogeological unit are still modest: some cross-country vehicles, equipment for measuring the water levels and for simple hydrochemical analysis, a geophysical logging device, etc. No drilling rig is yet available for the studies.

Hydrogeological systems

Two aquifer systems are at present exploited in Gambia: the unconfined Quaternary-Pliocene aquifer and the recharging Miocene aquifer. The deep-lying Maestrichtian aquifer, which is not yet exploited in Gambia, might be worth exploring in the upper part of the country. The main features of these hydrogeological formations are described below.

Unconfined Quaternary-Pliocene aquifer

The unconfined aquifer is fed by rainwater and outcrops at between 5 and 25 m of altitude, approaching zero near the banks of the River Gambia in its lower course. The piezometric variations are in the order of 2 m, with the lowest levels in April-May before the rains in July.

The hydraulic characteristics of this aquifer are variable and of moderate quality. The transmissivity co-efficient is 50-300 m^2/day and the yield per installation 5-10 m^3/h, with a specific yield generally below 5 l/s/m. The hydraulic conductivity is between 10 and 30 m/day. There is no information about the storage co-efficient, but it can be estimated at one to three per cent by analysis of the data. The effective porosity is 10 to 20 per cent depending on the granulometry of the sands.

The physical and chemical quality of the water tested in the laboratory and in the field is as follows: pH generally acid (5.5 to 6.5); conductivity 300-1,500 millimhos/cm at 25°C; chlorides 5-200 mg/l; CO_2 5-50 mg/l; O_2 1-10 mg/l. Most of the water has a low mineral content (with less than 250 mg/l of salt in all), definitely acid but usually of good chemical quality although bacteriological pollution of open wells does occur.

The water balance of the unconfined aquifer shows that the average annual infiltration can be estimated at 123,000 m^3/km^2, i.e. 1,300 mm^3/year or a flow of 41 m^3/s. The present exploitation amounts to only about 10 hm^3/year, from wells and boreholes.

Recharging Miocene aquifer

The Miocene aquifer is recharged between two clay strata. It has probably been fed over a long period from distant areas covering a large part of West Africa. It is also possible that the recharge is vertical from the unconfined aquifer through the less argillaceous layers found at its top. The piezometric levels of the Miocene aquifer appear to be about 2-3 m higher than those of the unconfined aquifer; this is more apparent in the upper basin in the east and the horst in the west, while in the central depression the piezometric levels of both aquifers come close to zero. In the Kombo peninsula, which is part of the horst, the piezometric levels of the Miocene aquifer are very different from those of the Quaternary and they are 4-5 m higher in places.

The Miocene aquifer generally flows towards the west.

The hydraulic characteristics of the Miocene aquifer are much better than those of the unconfined aquifer: the transmissivity ranges from 300 to 1,500 m^2/day, and the flow from 10 to 40 m^3/day approximately. The flow and the transmissivity increase in step with the thickness of the aquifer in the direction of Casamance. The specific yield varies from 10 to 30 l/s/m, and the hydraulic conductivity is generally below 0.2 m/day. The storage co-efficient has been determined at about 3×10^{-4} in a few places.

The physical and chemical characteristics of the water are similar to those of the unconfined aquifer. But there is a gradual increase in the salt content towards the west. At some points near the coast, especially in the horst, the aquifer is contaminated by salt water.

No study has yet been carried out to determine the storage capacity of this important aquifer. It is certainly several times greater than that of the unconfined aquifer; it has even been estimated at about 13,000 to 20,000 hm^3/year. The present drawoff through boreholes is only about 5 million m^3/year.

Maestrichtian aquifer

The only water point in Gambia reaching the Maestrichtian aquifer is at Banjul, with a depth of 400 m. This water point is dry at present but it has delivered heavily mineralized water in the past. In Haute-Cassance near the Gambian border the Maestrichtian aquifer is found at a depth of 200-250 m and it has a transmissivity co-efficient of 500 to 2,500 m^2/day.

The chemical quality of the water in this area is good, but the mineral content probably increases downstream.

A table summarizing the hydraulic characteristics of Gambia's exploited and exploitable aquifers appears below.

Aquifer type	Geological era	Specific yield (l/s/m)	Hydraulic characteristics		
			Hydraulic conductivity K (m/day)	Transmissivity T (m^2/day)	Storage co-efficient S
Unconfined	Quaternary-Pliocene (terminal continental)	$\leqslant 5$	5 to 20	50 to 300	2×10^{-2}
Recharging	Miocene	10 to 30	5 to 20	300 to 1,500	3×10^{-4}
Artesian	Maestrichtian	$\geqslant 50$	0.2	500 to 2,500 and higher	-

Hydrogeological research is still in its early stages in Gambia. The knowledge acquired so far is based on data collected from a number of boreholes drilled during prospecting for water to supply people and livestock.

A geophysical prospecting campaign using the electro-seismic method has been taking place in various parts of the country, with a view to determining the relative location of the producing aquifers, i.e. the unconfined and the Miocene aquifers; it has been accompanied by test drilling.

However, much remains to be done with respect to study of the structure, the hydraulic characteristics and the water balance of the economically exploitable ground-water aquifers in Gambia. Surveys of this kind will have to be undertaken as the exploitation of the aquifers develops.

III. GROUND-WATER EXPLOITATION

Uses

Ground water is exploited in Gambia mainly to supply drinking water to the towns and villages and for livestock.

In 1978 the exploitation of ground water to meet the requirements of Banjul and its vicinity and the small towns in the country's interior was entrusted to the Gambia Utilities Corporation (GUC). Studies were made, especially in a section of the Kombo peninsula, with a view to sinking boreholes for Banjul and its environs. The boreholes in this interior part of the country were drilled by a foreign company.

More recently, in 1979-1980, the United Nations Sudano-Sahelian Office financed the drilling of 28 boreholes to supply water for livestock. The pumping stations at these boreholes are run by the Department of Water Resources.

In addition to these boreholes, the Department of Water Resources has been undertaking since 1976 the construction of large-diameter wells to supply drinking water in rural areas. These installations, which numbered 300 in 1984, are distributed throughout the country. Provision has also been made for about 40 boreholes, mainly to supply rural collectivities.

Ground-water requirements

A partial survey of the demand for and utilization of water has revealed the following water requirements for the various sectors:

Sector of use	m³/day		hm³/year	
	Requirement	Production	Requirement	Production
a) Drinking water				
- Rural	14,000	5,600 (min)	5.1	2.0 (min)
- Urban and tourism	10,000 (min)	17,400 (max)		4.8 (max)
	18,000 (max)	11,500	1.9 3.2	3.9
b) Livestock				
- Cattle	9,700	5,500	2.3	1.5
- Other domestic animals	7,000	7,000	2.6	2.6
c) Irrigation				
- Watering of gardens	16,000	8,000	3.8	1.9
- Commercial irrigation	2,500	1,000	0.6	0.3
TOTAL	59,200 (min) 67,200 (max)	38,600 (min) 50,400 (max)	16.3 (min) 17.6 (max)	12.2 (min) 15.0 (max)

This table shows that the demand is only partially met. The country's population will probably be in excess of 1.2 million in 2000, including about 300,000 in urban areas. It can be assumed that the number of head of livestock will equal that of the population and that the irrigation of cash crops, a major water use, will be carried out over an area of about 500 ha. It is also agreed that 10,000 m³ a day will be needed by the processing and foodstuffs industries. Given these assumptions, the water requirements for the various sectors can be estimated as follows:

Sector of use	m^3/day	hm^3/year
a) Drinking water		
- Rural	36,000	13.1
- Urban and tourism	30,000 (min) 48,000 (max)	5.6 8.6
b) Livestock		
- Cattle	20,000	4.8
- Other domestic animals	12,000	2.9
c) Irrigation		
- Watering of gardens (650 ha)	26,000	6.2
- Cash crops (500 ha)	20,000	4.8
d) Industries	10,000	3.7
TOTAL	154,000 (min) 172,000 (max)	41.1 (min) 44.1 (max)

These estimates indicate that the demand will more than double between now and 2000. A considerable effort therefore remains to be made with respect both to investment and to research and exploitation.

As the aquifers are still little used, problems of overexploitation of the resources do not arise, except in the Kombo area where the increased demand for drinking water for the town of Banjul and its environs and for the irrigation of cash crops may place at risk the delicate equilibrium of the area's aquifers.

The rural areas could be supplied without any deterioration in water quality. But the strong demand for livestock and irrigation purposes will entail careful planning of the resources in the areas to be brought into exploitation.

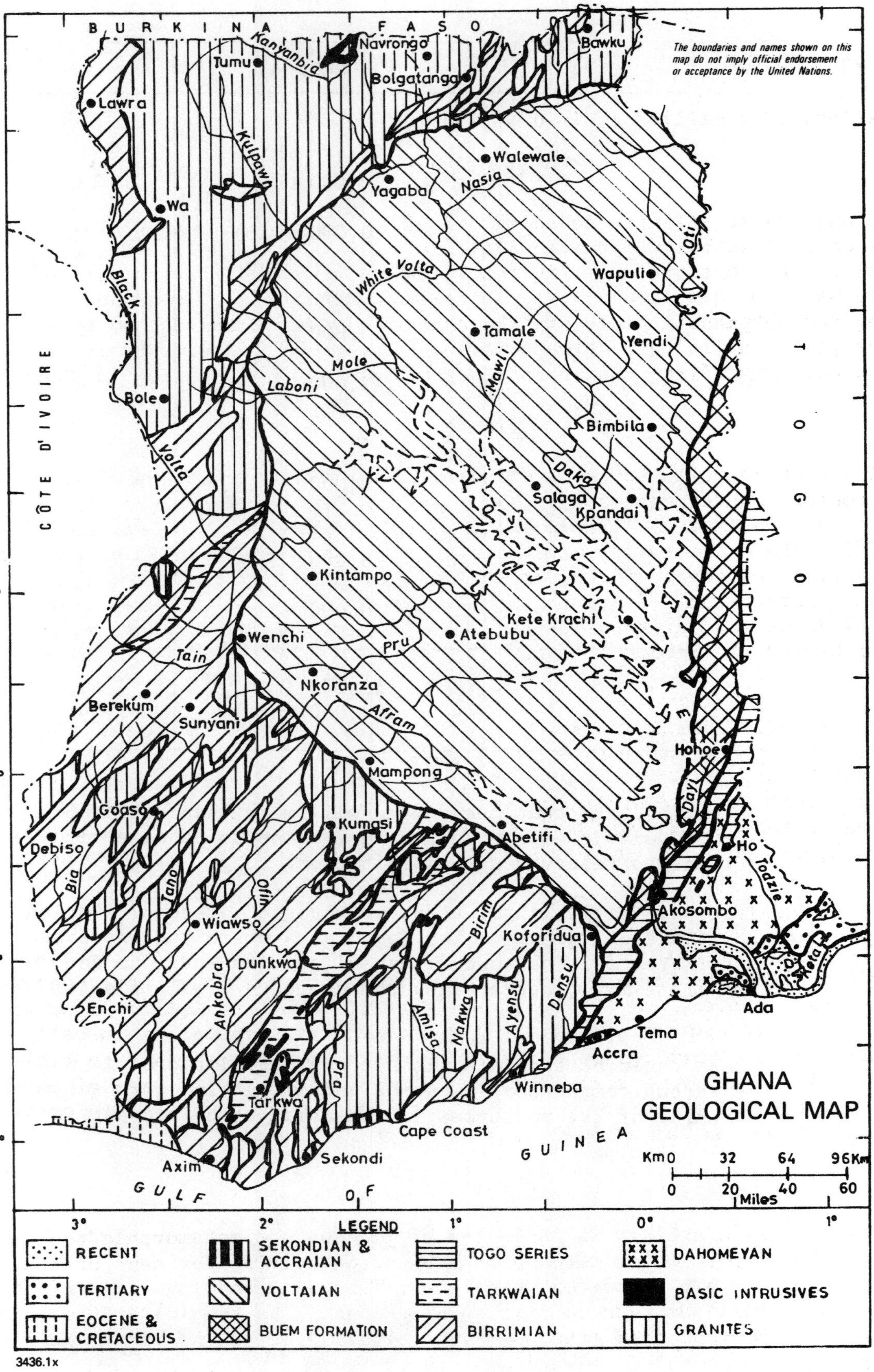

GHANA
GEOLOGICAL MAP
The boundaries and names shown on this map do not imply official endorsement or acceptance by the United Nations.
BURKINA FASO
CÔTE D'IVOIRE
TOGO
GULF OF GUINEA
Navrongo
Bawku
Tumu
Bolgatanga
Lawra
Walewale
Yagaba
Wa
Wapuli
Tamale
Yendi
Bole
Bimbila
Salaga
Kpandai
Kintampo
Kete Krachi
Wenchi
Atebubu
Nkoranza
Berekum
Sunyani
Hohoe
Mampong
Goaso
Kumasi
Abetifi
Debiso
Ho
Akosombo
Wiawso
Koforidua
Dunkwa
Enchi
Ada
Tema
Accra
Winneba
Tarkwa
Cape Coast
Axim
Sekondi
Kanyanbia
Kulpawn
Nasia
Oti
White Volta
Black
Volta
Mole
Laboni
Mawli
Daka
Tain
Pru
Afram
Dayi
Todzie
Keta
Bia
Tano
Ofin
Birim
Ankobra
Amisa
Nakwa
Ayensu
Densu
Pra
LAKE
Km 0 32 64 96 Km
0 20 40 60 Miles
11°
10°
9°
8°
7°
6°
5°
3°
2°
1°
0°
1°
LEGEND
RECENT
TERTIARY
EOCENE & CRETACEOUS
SEKONDIAN & ACCRAIAN
VOLTAIAN
BUEM FORMATION
TOGO SERIES
TARKWAIAN
BIRRIMIAN
DAHOMEYAN
BASIC INTRUSIVES
GRANITES
3436.1x

GHANA

Area: 238,000 km^2

Population: 12.7 million (United Nations estimate, 1983)

I. BACKGROUND

Ghana has an Atlantic seaboard of 560 km on the Gulf of Guinea. Most of the country is below 320 m in altitude. The highest point (Mt. Afadzato) (900 m) is in the south-east, in the Volta region. The eastern part of the country has some high ground: Mt. Akwapim and the Kwahu plateau and the Mampong and Nakpanduri hills, in the Ashanti and northern regions respectively; the hills vary between 85 and 800 m in altitude. The main lowlands are the Accra and Afram Plains in the south and the vast plains in the north.

Climate

The tropical climate is governed by the north-south shifts of the intertropical convergence zone of West Africa. Temperatures are high with small annual variations. The average monthly temperature is between 24°C and 28°C. There are two distinct seasons: the wet season which usually lasts from April to October, and the dry season from November to March. The meximum rainfall is in September in the north, with 800-1,250 mm. There are two periods of maximum rainfall in the south in June and October, with October having the lesser amount. The heaviest rainfall is in the south-east.

The relative humidity is high in the south, with 80 to 90 per cent recorded throughout the year except in January and February when the harmattan blows. The relative humidity is low in the north and can drop to three per cent during the harmattan months. In the south evaporation measured in the pan is usually between 1,000 and 1,780 mm and in the north between 2,000 and 2,800 mm. The evaporation is lowest in the south-west.

Surface water

The country's main river is the Volta which has three main tributaries: the White Volta, the Black Volta and the Oti. The other main watercourses are the Pra, Ankobra, Tano, Bia, Densu, Ayensu, Ochi-Nakwa, Ochi-Amisa and Todze. The main courses are perennial but their tributaries are usually seasonal. Basic hydrological data on the watercourses are given in tables 1 and 2. Lake Bosumtwi is the largest natural lake, with an area of about 48 km^2 and a depth of up to 72 m. The artificial lake on the Volta created by the Akosombo dam has a total area of 8,480 km^2.

Geology

The country's geology is dominated by igneous and metamorphic rocks of the Precambrian base which outcrop or suboutcrop over 54 per cent of the territory. These are Dahomeyan, Birrimian, Tarkwaian and Togo formations crossed by granitic and granodioritic intrusions. The remaining 46 per cent is made up of consolidated (the Buem and Voltaian formations being the largest) and non-consolidated sedimentary rock.

The oldest formation is the Dahomeyan, outcropping in the south-east along a north-east south-west line. It forms the basement of the Accra Plains and of the south of the Volta region. This formation covers four per cent of the country and consists mainly of acid and basic gneiss and migmatite with large amounts of garnet, some quartz and a little marble. In the Accra Plains the formation rises sharply in some places to form inselbergs with altitudes of 30 to 40 m above sea level. The fractures are well developed with separate blocks of garnet-bearing gneiss.

The Birrimian formation is of Middle Precambrian age and covers 27 per cent of the country; it is subdivided into a lower series of phyllites, schists, grauwackes and tuffs and an upper series of lavas with phyllites and grauwackes. This formation usually contains carbonated, ferrous, manganiferous or calcareous elements. It has granitic or granodioritic intrusions.

The Tarkwaian formation is of Upper Precambrian age and covers two per cent of the country. It consists of sediments formed in shallow water: sandstone, tillite, phyllite and conglomerates derived from the Birrimian. The conglomerates are found at the base of the system. They are overlain by:

i) A series of quartzites, sandstones and conglomerates known as the "Banket Series";

ii) An argillaceous formation;

iii) A sandstone formation.

The Tarkwaian formation is more than 2,000 m thick. Gold is found in the conglomerates of the Banket Series.

The Togo series is of Late Precambrian age and covers one per cent of the country. It has a cataclastic and metamorphized base, with quartzites, sandstones, schists, phyllites, metamorphic schists and siliceous limestones. In places there are small intrusions of gabbros, peridotites, pyroxenites and serpentines, with pockets of chromite and asbestos.

Intrustive granites and granodiorites cover about 20 per cent of the country. They are of Middle Precambrian or Post-Birrimian age. They occur mainly in the north, west and south of the country. The "granites" form a heterogeneous body of splites, diorites, granodiorites, syenites, granitogneiss and pegmatites and they are intrusive mainly in the Birrimian and Tarkwaian formations. The Buem and Voltaian are the most important of the sedimentary rocks. The Buem formation (three per cent of the country) is Precambrian and is made up of shales and limestones, sandstones, quartzites and lavas. The Voltaian rocks cover 42 per cent of the country; they are of Paleozoic age and consist of an upper series of limestone and a lower series of schists and conglomerates with limestone in places. A feature of the Voltaian formation is its horizontal disposition and the almost total lack of igneous intrusions. Its limits are market in the north, west and south by vertical cliffs which dominate the lower basement formations.

The other sedimentary formations include Devonian rocks of the Accraian and Sekondian eras and non-consolidated Eocene, Cretaceous, Tertiary and Recent sediments. Accraian and Sekondian rocks occur in small isolated units along the coast. The Accraian formation is usually folded and consists of sandstone, conglomerates, schists and fine-grained sandstone. The Sekondian formation also includes isolated outcrops of these rocks, with occasional bands of cherts and nodules of impure limestone and siderite.

The non-consolidated formations - marly fluvial sands and lacustrine sands, gravels and clays with some volcanic brecchias and tuffs - cover about one per cent of the total area of the country.

II. GROUND-WATER RESOURCES

Institutional aspects

The Government of Ghana has decided to develop the country's water resources for irrigation, watering of livestock, and industrial and domestic uses, including drinking water. This policy is implemented mainly by governmental bodies and institutions, in particular the Ghana Water and Sewerage Corporation (GWSC), the Hydrology Division of the Architectural and Engineering Services Corporation (AESC), the Irrigation Development Authority (IDA), the Meteorological Department and the Water Resources Research Institute (WRRI).

The official text establishing the GWSC states that this body is responsible for the production, distribution and conservation of water for public, domestic and industrial use in urban and rural areas.

The Hydrology Division of the AESC makes hydrogeological studies, including the collection and analysis of hydrological data and publication of hydrometeorological data. It also carries out flood-water regulation and drainage projects and makes studies and implements projects for the protection of the coastal zones. It has four main sections:

- Hydrological and water resources studies;

- Drainage and sanitation;

- Coastal engineering;

- Training.

Summary of hydrogeological research

Up to the end of the First World War the Government gave no attention to ground-water research and development. Private individuals, mining companies, trading companies, missionaries and small communities dug their own wells for their own needs. However, following a severe drought from 1918 to 1920 the Government of the Gold Coast (Ghana's former name) initiated ground-water research and exploitation operations. The Geological Service was then responsible for finding potentially favourable sites, especially in the semi-arid parts of the north and in the coastal plains. In 1937 a water supply division was created in the Geological Service to develop well-digging techniques; it had to cease operations in 1942 for lack of staff but not before it had established a number of wells, waterholes and small dams. A small number of well-diggers had also received job training: a water supply department was formed in 1943 and finally established in 1944 as the Department of Rural Water Development (DRWD). This body operated for 20 years. However, it was the geologists of the Geological Service who made the necessary hydrogeological studies in the towns and villages; electrical geophysical surface prospecting using resistivity techniques was the main research tool. The GWSC was created in 1965 with the specific remit of supplying the country's people with water and exploiting ground water to meet the needs of urban and rural communities. The WRRI was reorganized in 1966 to carry out the tasks mentioned in the preceding paragraph.

The IDA is responsible for carrying out the Government's irrigation policy and in particular for furnishing technical assistance in the preparation of projects. It is also required to make full studies of the possible impact of irrigations projects on the environment and to advise the Government accordingly.

The Meteorological Department is responsible for the collection and processing of hydrometeorological data to be used in the preparation of economic and social development projects. It has five main scientific divisions: climatology, agrometeorology, hydrometeorology, observation networks, and instrument and equipment workshop.

The WRRI is responsible for evaluating the country's water resources through the collection and study of hydrological data, determining research needs with respect to water resources in terms of the priority to be given to the various research programmes, carrying out and co-ordinating water resources research activities, and advising the Government on the adoption of a water policy to produce the maximum social and economic benefits for the country.

The WRRI should have a total of six research divisions: surface water, ground water, water quality, solid content of water, water planning and hydraulics and equipment laboratory. The first four divisions had been established and were operational by 1985.

Ground-water services

At present there are only two services responsible for ground-water exploration and evaluation: the drilling unit of the GWSC and the WRRI. The first body has the main responsibility for ensuring ground-water supplies for remote villages in rural areas and for some urban areas in which uncontaminated ground water is available in sufficient quantities. This drilling unit operates from its main base at Kumasi, the capital of the Ashanti region, from which its teams are sent throughout the country on the Government's instructions and in response to requests from education institutions, missions, industrial companies, farmers and groups of farmers. In many cases this unit collaborates with technical services or foreign bodies in the implementation of drilling programmes financed by external loans and/or grants. The projects are usually concerned with the drilling of shallow tubewells equipped with hand pumps, the organization of repair and maintenance programmes and the training of the villagers in the necessary techniques. Two large projects of this kind have recently been successfully concluded: one in the north (2,500 wells) financed by a loan from the Canadian Government (Canadian International Development Agency) and the other in the south (3,000 wells) financed by the Federal Republic of Germany.

WRRI's ground-water division is mainly concerned with the various aspects of ground-water evaluation: availability, quantity, quality, flow rate, perenniality. It also studies aspects of exploration techniques and equipment, including natural phenomena and surface indicators (such as the presence of certain plants) and geophysical methods.

Prospecting methods

Most of the traditional methods of prospecting for ground water are used in Ghana with local adaptations, especially with respect to the choice of drilling sites. These methods include study of geological and hydrogeological documents, field surveys, interpretation of aerial photographs and natural indicators, geophysical prospecting and test drilling. A brief description of the methods now follows.

Field surveys. The aim is to identify and map the hydrogeological units and their structure and possible sources of pollution: dumps of household refuse, mineral spoil and industrial wastes, cesspools, cemetries, etc. The local inhabitants are questioned about water-resources developments in the past: water levels in the wells in periods of drought, etc.

Interpretation of aerial photographs. This is the most important aspect of hydrogeological prospecting in Ghana. Aerial photographs give information about lithological limits, fracture zones, type of flow, fold axes and the ancient beds of watercourses; this information helps to determine the location of the wells. Aerial photographs are also used to determine the location of inhabited points, access routes to the sites and the location of the work sites.

Geophysical prospecting. This method cannot be taken up blindly, for the operations are costly and require highly qualified personnel which is not always available. Moreover, this technique is not always effective. The resistivity method with a Wenner electrode array is preferred to seismic methods. The very low frequency technique has also been tried but it is still too early to assess its effectiveness as there has so far been no feedback.

Natural indicators. These indicators are useful mainly in the Accra Plains, a triangular area of 6,000 km^2 between the Akwapim hills in the west, the River Volta in the north and east and the Gulf of Guinea in the south. These plains consist of Precambrian rock overlain by alteration zones. The indicators include termiteries and baobabs, whose presence can point to potential aquifers.

Main aquifers

The presence, distribution, size and hydraulic behaviour of the aquifers is determined by geological factors. The aquifers can be divided into two main categories: those of the crystalline basement rock and those of the sedimentary formations.

The aquifers of the crystalline basement rock are linked to the fracture zones and zones of alteration and decomposition and to zones with veins of quartz, aplite and pegmatite. Other characteristics are low transmissivity coefficients, small size and irregular configuration. They may equally well consist of veins of fractured quartz and of decomposed formations up to 30 m thick. Sometimes they form isolated pockets in which the ground-water flow is determined by the surface topography. The storage coefficient is usually between 0.3 and 0.8 per cent, and the transmissivity values ranges from 9×10^{-6} to 3×10^{-4} m^2/day. A brief description of these aquifers is given below.

Precambrian

Dahomeyan: mainly in the Accra Plains and in some parts of the Volta region. The aquifers are found at depths of 25 to 75 m, and their configuration and size are determined mainly by the veins and alterations. They take the form of bands of altered fractures in the bedrock consisting mainly of biotite gneiss with garnet and horablende, schists and migmatites. The transmissivity values determined by valve testing range from 15×10^{-2} to 62.5 m^2/day. The hydraulic conductivity values range from 10^{-3} to 20 m/day. The storage coefficient is 10 per cent in unconfined aquifers and 10^{-4} in semi-artesian conditions.

Birrimian: represented by decomposed metamorphic schists, phyllites with abrupt fractures, schists and quartzitic veins. These aquifers have been intensively exploited in all the regions in which they are found, at depths sometimes in excess of 100 m. The Birrimian phyllites usually contain larger amounts of exploitable ground water than the schists.

Granites of various ages: in the north, west and south of the country. The aquifers are found in the decomposed parts and the pegmatite veins. They feed some 2,500 wells with yields of up to 450 l/m in the pegmatites. The average depth of the aquifers is 50 to 80 m.

Tarkwaian: the aquifers are found in fractured quartzites, phyllites and conglomerates. Most of them have yields of about 80 l/m. The water level in the majority of the wells is usually between three and ten metres below the surface.

Togo series: the aquifers are found in zones of contact with the Buem and Dahomeyan formations. In these zones the aquifers consist of fractured quartzites, schists, phyllites and sandstones. There are two types of aquifer:

1. Shallow aquifers in altered and decomposed formations at depths of up to 25 m;

2. Deep aquifers in fault or fractured zones in the substratum of bedrock and compact rock.

The yields are usually below 80 l/m but can be as high as 750 l/m in some places.

Sedimentary formations

The Buem formation is the most productive, with yields of 200-400 l/m. The Voltaian is the largest formation, covering 42 per cent of the country; it has been little explored for it occurs in regions of very low population density. The main water-bearing zones of the Voltaian formation are found on the periphery of the outcrops in contact with the Buem formation where there is intense fracturing. The available yields in these zones are substantial and can be as high as 350 l/m.

Elsewhere the aquifers have poor yields of below 8 l/m. They are of Cretaceous-Eocene age and found in limestone close to the coast in the south-east and west of the country. These aquifers have the highest yields, particularly in the west where they can deliver up to 500 l/m. Table 3 gives a summary of ground-water research data. With the exception of the Accra Plains, most of the study wells were drilled as part of a village water-supply project for installation of shallow wells. Accordingly, the data refer mainly to depths of 20 to 40 m, the aquifer horizon of the majority of the geological formations.

Water quality

The main purpose of water analysis has been to determine the water's suitability for drinking according to the standards of the World Health Organization. Table 5 summarizes the chemical characteristics of the water of several geological formations. Generally speaking, the undesirable or harmful elements are iron, manganese and chlorine. When the content of these elements is low, the water is usually suitable for consumption. Very high iron content (8 g/l) is sometimes found in isolated spots in the south of the country. Chloride concentrations can be heavy on the coast owing to the penetration of sea water at depth. High concentrations can also be found in the aquifers in the interior, particularly in the Tamale area.

III. GROUND-WATER STUDIES: GAPS AND FUTURE NEEDS

The purpose of most of the current ground-water studies is to determine the extent to which the water can be used now and might be used in the long term for human supplies and crop irrigation. For example, WRRI is studying the crystalline basement rock of the Accra Plains to obtain data on the quality of the water, the flow rate and the fluctuation of levels. WRRI is also studying water quality and level fluctuation in other regions of the country. Studies are also to be made of the recharge of the aquifers and the contamination of ground water by fertilizers, herbicides and pesticides; in view of the considerable increase in the number of wells in low-yield aquifers, it is important to know whether the pumped water comes from renewable or from fossil resources, in order to determine whether it is necessary to control their exploitation. The age of the water has to be assessed in these exercises and a laboratory for study of the isotopes in the environment will be needed. At a later stage it will be necessary to analyze the reaction of the resources to climatic changes; this means that the present network of aquifer monitoring stations will have to be extended to other regions where ground water is exploited. It would also be useful to drill some deeper wells to provide an overall picture of the ground-water system; this cannot be provided at present as most of the wells are shallow. Lastly, when more data are available it will be possible to study the aquifers by mathematical modelling.

IV. EXPLOITATION OF GROUND WATER

Governmental bodies

The Ground-Water Division of the GWSC, which is responsible for water drilling in rural areas, has the following staff: nine hydrogeologists, two chief drillers, nine drillers and 15 assistant drillers. It uses the services of the GWSC laboratory for the chemical and bacteriological analysis of samples of ground water. This laboratory has ten chemists/bacteriologists and 15 technicians.

The drilling equipment includes two rotary rigs (Cyclone TH 60 and Mayhe) and five cable-tool rigs (three 22 New Bucyrus Erie and two 6 ORL). The laboratories are suitably equipped to make the standard chemical and bacterioriological analyses. The hydrogeologists have geological and geophysical instruments for carrying out field surveys. They receive their initial training in Ghana and then continue their studies, including their higher specialized studies, abroad.

WRRI's Ground-Water Division has two research officers aided by three assistants and four technicians. It has a drilling unit with one drilling engineer and four drillers and it has one Cyclone TH 60 rotary rig and two Bucyrus Erie cable-tool units. The prospecting equipment includes a resistivity unit (Soil Test R 60) and a seismic unit (Soil Test MD 5).

The drilling operations are limited by the lack of resources. The foreign exchange regulations make it difficult to buy spare parts for the drilling rigs, vehicles and other equipment. Accordingly, the annual drilling rate is about 2,300 m for GWSC and 1,300 m for WRRI, i.e. 35 and 20 boreholes respectively, with depths of 50 to 70 m. These figures do not take account of the programmes financed by CIDA and the Federal Republic of Germany (see above).

Ghana has only one private drilling company: ANICOBB Company Limited which has only one drilling rig. It has been operating only since February 1985.

Use of ground water

Ghana's ground water is used mainly for human supplies (especially drinking water) and watering of animals. With few exceptions, the low yields and high costs militate against the use of ground water for agricultural, mainly irrigation, purposes. However, the relative drought which affected the country in 1981-1983 encouraged the use of ground water in fish-farming, bee-keeping and small-scale industry. Ground-water aquifers supply most of the water used in the game reserves, national parks and other tourist centres, especially in the north. The main and secondary towns which take a large part of their water supplies from ground-water aquifers are listed in table 4.

V. FUTURE NEEDS

The development of ground water entails the import of drilling equipment and supplies and spare parts for repair of the existing equipment; this requires foreign exchange. It is also necessary to create a national industry to produce hand pumps from local resources. Some 5,000 foreign pumps have so far been installed. Although the choice of pump type was based on fairly extensive tests, experience has shown that the imported pumps are not always well suited for use in Ghana, for the spare parts cannot be manufactured in the country and must therefore be bought abroad. It will also be necessary to organize training centres for the villagers responsible for maintenance of the hand pumps. Foreign courses will be needed to improve the training of the drillers and other technical staff in the operation. maintenance and repair of modern drilling machinery. There is a plan for a project to construct a plant to cleanse the water of its iron and manganese content, for 20 per cent of the country's tubewells deliver water with high concentrations of these elements.

Lastly, remote-sensing methods will have to be used in all the regional ground-water studies; this entails the establishment of a remote-sensing centre furnished with the necessary equipment.

VI. PROBLEMS

The main obstacle to the development of ground water is the lack of foreign exchange for purchase of drilling equipment, scientific instruments and supplies. There is also a shortage of workshops for maintenance and repair of the equipment and pumps.

Furthermore, there is an urgent need to organize the rational planning and utilization of the resources. For example, the terms of the drilling contracts are so restrictive that it is impossible to push bore holes beyond the shallow depths established in the contracts. This leads to failures, whereas deeper drilling would make considerable water resources available.

Another problem already mentioned is the high iron and manganese content in 20 per cent of the wells, especially in the south of the country.

The use of ground water for irrigation is not a common practice owing to the low yields and high costs; however, there are some exceptions to this rule; for example, the small cultivated areas in the Accra Plains and the gardens attached to some houses.

VII. CONCLUSIONS

Ghana's ground-water aquifers furnish the most economical and reliable water supplies in rural areas. The water is usually suitable for rural consumption, except when it has high concentrations of iron and manganese which give it an unpleasant taste; it is harmful in a few cases owing to the high concentrations of nitrates and fluorides.

Some 25 per cent of Ghana's population is supplied with ground water. The average cost of the water in the well is estimated at about US$ 0.30 per cubic metre when hand pumps are used.

Since most of the country is covered by igneous and metamorphic rocks, ground-water exploration is especially difficult and requires fairly advanced geophysical methods, study of the location of termiteries and baobabs and test drilling operations. In the past ground-water research has been seen mainly as a means of meeting the needs of small villages with shallow low-yield wells close to the dwellings. However, large yields could be obtained a little further away and at greater depth. The prospects for development of ground water are promising on the whole, although limited by the relatively modest resources at present available.

VIII. TABLES

Table 1. Flow rates

Station	River	Basin (km^2)	Mean/annual flow (m^3/s) */	Maximum flow (m^3/s)	Date of maximum flow	Flow co-efficient	Sediment discharge
Yapei	White Volta	106,000	261	2,994	9-9-63	-	-
Bamboi	Black Volta	127,000	240	3,192	17-9-63	-	-
Saboba	Oti	54,000	289	2,869	1-10-57	-	-
Dadieso	Bia	6,040	27	143	18-8-68	-	-
Jomuro	Tano	10,300	55	778	25-9-68	0.20	70 mg/1 1966
Alenda	Tano	15,800	143	552	30-6-71	0.20	44 mg/1 1966
Prestea	Ankobra	4,220	49	681	26-9-68	0.25	59 mg/1 1966
Daboasi	Pru	22,600	229	1,557	31-8-68	0.21	450 mg/1 1975
Okyereko	Ayensu	16,400	11	161	4-7-68	0.17	604 mg/1 1975
Manhia	Densu	2,100	12	128	26-6-74	0.17	685 mg/1 1974
Todzenu	Todze	2,750	9	136	3-8-68	-	1,590 mg/1 1975

*/ The mean monthly flow is given in table 2.

Table 2. Mean/monthly flow rates

(m^3/s)

Station	River	March	April	May	June	July	August	September	October	November	December	January	February
Bamboi	Black Volta	13.62	12.15	28.09	72.41	180.88	392.18	960.30	844.96	216.14	86.01	42.96	21.41
Saboba	Oti	3.06	5.01	18.69	57.99	203.14	626.52	1445.62	910.09	148.96	30.81	10.62	4.16
Dadieso	Bia	6.51	17.08	24.86	52.53	60.32	31.44	36.05	49.02	32.74	8.67	3.91	2.63
Jomuro	Tano	11.53	27.39	40.30	102.52	126.67	63.83	59.98	102.21	71.31	28.29	10.45	8.89
Alenda	Tano	-	-	-	-	-	-	-	-	-	-	-	-
Prestea	Ankobra	16.39	28.32	48.12	115.57	105.92	47.10	59.95	82.95	44.41	19.68	7.14	8.50
Daboasi	Pru	81.11	127.67	194.39	399.74	447.37	259.35	268.79	396.76	292.77	148.76	73.12	49.79
Okyereko	Ayensu	3.85	4.87	7.25	26.28	21.95	15.24	17.50	14.73	9.40	5.49	2.61	2.41
Manhia	Densu	1.90	2.41	7.70	27.36	28.29	16.94	22.88	17.93	13.42	3.54	2.21	1.16
Todzenu	Todzu	1.08	3.23	5.38	17.81	20.19	15.26	16.00	14.41	6.23	1.81	0.93	0.74

Table 3. Overall results of ground water studies

Geological formation*/	Percentage of total area	Age	Depth of aquifer (m)	Depth of well (m)	T (m^2/day)	K (m/day)	Q/s (1/s/m)
Dahomeyan	4	Lower Precambrian	25-50 65-75	60-100	14.9×10^{-2}-62.5	6.1×10^{-1}	2.75×10^{-1}
Birrimian	27	Middle Precambrian	25-35	30-80	79	4.9×10^{-2}	1.18×10^{-2}
Granites	20	Middle-Upper Precambrian	20-65		7-120	1.19×10^{-1}	2.53×10^{-2}
Tarkwaian	2	Upper Precambrian	15-35	25-60	331	2.68×10^{-1}	6.38×10^{-2}
Togo	1	Late Precambrian	20-65	20-60	4	3.72×10^{-1}	3.59×10^{-2}
Buem	3	Cambrian	20-35	20-80	50	4.15	2.73×10^{-1}
Voltaian	42	Paleozoic	20-35	20-80	160		6.87×10^{-2}
Sekondian and Accraian	less than 1	Devonian	-	-	-	-	-
Eocene-Cretaceous	less than 1	Eocene-Cretaceous	5-30 250-300**	25-90 300**	870	72	1×10^{-2}
Recent Tertiary	less than 1	Tertiary/Recent	3-25	20-40	-	-	-

*/ See map 2 for geographical location.

**/ Only for the limestone formations in the east (Keta zone).

Table 4. Urban consumption

Region and town		Consumption (m^3/day)
Volta		
Ho		850
Keta		300
Jasikan		150
Kete-Krachi		165
Agbosome	less than	100
Kadjebi		170
Aflao		700
Denu	"	100
Dzodze		325
Anloga		430
Eastern		
Abetifi	less than	100
Nkawkaw		775
Akim Oda		550
Akim Swedru		150
Kade-Akwatia		195
Asuom		185
Akoase		115
Adeiso		135
Suhum		500
Central		
Ayanfuri	less than	100
Dunkwa-Ofin		385
Abora-Dunkwa		120
Western		
Half-Assini		200
Enchi		140
Asankrangwa		145
Ayinase	less than	100
Brong-Ahafo		
Berekum		500
Techiman		575
Wenchi		420
Nsakaw	less than	100
Japekrom	"	100
Badu		150
Sampa		145
Amanten		120
Upper East		
Bole		140
Bawku		770
Zebilla	less than	100
Navrongo	"	100
Paga	"	100
Chiana	"	100
Sandema		100
Upper West		
Wa		820
Lawra	less than	100
Nandom	"	100
Hamale	"	100
Tumu		135
Gwollu	"	100
Jirapa		125
Ashanti		
Chichiwere	less than	100
Tepa		185
Juaben		200
Agogo		425
Akrofuom	"	100
Obuasi		1,350
Northern		
Damongo		300

Table 5. Chemical analysis of the water of the aquifers: overall results 1/

Aquifer and no. of wells studied	Na mg/1	Mg mg/1	Ca mg/1	Cl mg/1	Fe mg/1	Mn mg/1	F mg/1	Ph mg/1	CO^3H mg/1	Dry residue mg/1	Total hardness mg/1	Nitrate mg/1
Dahomeyan Accra Plains (50)	40-1400 200-400	10-5600 200-400	3-700 100-400	29-7000 400-1000	up to 11 0.1-0.5	-	-	6.3-10 6.8-7.3	-	-	40-6500	rare traces
Birrimian Southern Ghana (40)	-	-	-	50-300 10-30	up to 2 under 0.2	up to 0.6 under 0.1	-	5-7 6-6.4	-	-	100-200	-
Granites Northern Ghana (30)	5-55 15	5-55 7-20	15-70 25-35	5-25 7-15	under 0.5 under 0.2	under 0.3 under 0.1	0.2- 3.5 0.3	5.7-7.2 6-6.5	75-300 100-200	-	55-350 70-150	up to 75 under 10
Tarkwaian Southern Ghana (8)	-	-	-	0.2-8 0.5-3.5	0.05	0.1- 0.2	-	5.4- 6.9 6	-	-	60-80	0.2-8 1-3.5
Togo Southern Ghana (30)	10-70 15-40	1-300 10-40	10-275 20-60	5-300 10-50	under 10 1.5-2	under 1 under 3	under 0.6 0.2	6-8 6.5-7	30-250 40-150	50-600 100-300	50-600 100-300	under 0.15 under 0.2
Buem Southern Ghana (45)	5-125 10-40	up to 35 5-25	up to 200 10-50	up to 110 10-50	up to 4.5 0.1	up to 4.5 0.1-1	up to 0.7 0.1- 0.2	6.2- 7.2 6.5-7	up to 300 50-150	up to 400 200-300	100-500 100-300	under 1.2 0.05 - 0.2
Voltaian Southern Ghana (13)	-	-	-	5-35 5-15	maximum 10 0.1-1	under 8 0.1- 0.2	-	4.4- 7.2 5-5.5	-	-	8-40 10-20	maximum 3.6 0.3-2
Eocene- Cretaceous Volta Region (55)	up to 1000 50-200	up to 400 10-50	up to 300 50-150	up to 1000 50-250	6.5 0.5-0.5	0.8 0.2- 0.5	20 0.2- 0.7	6.3- 8.5 7-8	250 100-200	8000 300-700	500 200-300	13 0.01- 0.2
Tertiary Southern Ghana (22)	-	-	-	90 10-40	0.15 0.2- 0.06	0.35 0.05- 0.2	-	4-6.3 4.3-5	-	-	15-80 20-40	5.2 0.5-2

1/ The figures in the first line are maximums and in the second line averages.

X. REFERENCES

Akiti, T.T. — Groundwater Flow in Weathered Granite in the Tamne Basin, North-East Ghana. WRRI publication, 16 pages, 1977.

Amuzu, A.T. — A Study of the Drinking Water Quality of Boreholes in the Rural Areas of Ghana. WRRI publication, 1974.

Amuzu, A.T. — The Quality of Groundwater in the Accra Plains of Ghana. WRRI publication, 7 pages, 1978.

Ayibotele, Nii Boi — A Historical Sketch of Water Resources Development in Ghana - Water Resources Research Institute. WRRI publication, 45 pages.

Ghana Government — Ghana 1977 - an Official Handbook, 1-8, 225-233, 1977.

Gill, H.E. — A Groundwater Reconnaissance of the Republic of Ghana, with a Description of Geohydrologic Provinces. US Geological Water-Supply Paper 1757-K, 28 pages, 1969.

Kesse, G.O. — Prospects of Petroleum and Other Fossil Energy Resources Discovery in Ghana. Ghana Geological Survey Report, No. 75/9, 1975.

Mallari, B.M. — Preliminary Study of Geology and Groundwater Resources of Ghana, 1962.

Nathan Consortium for Sector Studies — Water Resources Development in Ghana Report, 1970.

Quist, L.G. — A Preliminary Report on Groundwater Assessment of the Accra Plains, WRRI publication, 45 pages, 1976.

UNESCO — World Catalogue of Very Large Floods, p.91.

GUINEA

Area: 245,857 km^2

Population: 5.18 million (United Nations estimate, 1983)

I. BACKGROUND

Natural regions

Guinea is a tropical country with four main natural regions.

Coastal Guinea, or Lower Guinea, in the west of the country. The climate is of the maritime tropical type with a rainy and a dry season each lasting six months. The average rainfall is very high (about four metres a year) owing to the proximity of the sandstone escarpments of the Fouta Djallon which face in the direction of the monsoon. The humidity is very high even in the dry season (63 per cent as against 95 per cent in the rainy season), the average temperature is 27°C, and the average annual evaporation is 1,800 mm.

Upper Guinea, in the north-east, is the largest of the four natural regions. The climate is of the Sudanese type with sharply contrasted seasons. The total rainfall is still very high but it does not exceed 1,500 mm/year. The dry season is heavily influenced by the harmattan (a hot dry wind). The highest temperatures occur towards the end of the dry season, with maximums in excess of 40°C and a humidity rate ranging between 39 and 96 per cent. The average annual evaporation is 1,900 mm.

The Forest Region, in the south-east, is an area of higher land where the ancient Precambrian basement rock can reach altitudes in excess of 500 m or even over 1,500 m in the Nimba and Simandou mountains. The equatorial climate and heavy rainfall (1,800-2,500 mm), with no separate dry season, maintain a dense forest and produce very high humidity levels in places (98 per cent). The average temperature is 26°C and the evaporation about 1,600 mm/year.

Middle Guinea, in the north and centre of the country, contains the Fouta Djallon massif which consists mainly of sedimentary formations. This massif has an average altitude of over 700 m and it receives 1,500-2,500 mm of rainfall a year. The climate is very gentle and there are two distinct seasons with maximum temperatures of 32°C. The relative humidity is very high and can reach 97 per cent in the rainy season. The average annual evaporation is 1,500-1,800 mm/year.

Surface water

Guinea's hydrographic network has its source in two mountainous regions, the Fouta Djallon and the Forest Region, which are higher than the rest of the country and most of West Africa.

The Fouta Djallon is the "water tower" of this part of Africa, for a large number of watercourses rise there. The principal hydrographic basins are the Konkouré with its main tributaries, the Kakrima, Kokoulo, Cogon, Koliba, Gambia (tributary: the Koulountou), Bafing (tributary: the Téné), Kolenté and Kaba.

Only the Tinguilinta rises in Lower Guinea.

The mean annual specific flow rates are 35 to 45 $l/s/km^2$ in the Fouta Djallon.

In the Forest Region the hypsometric map shows an important watershed along the line of the 9th parallel N. The Niger and all its right-bank tributaries rise to the north of this line and flow northwards (the Niger, Milo, Sankarani and Niandan). The only left-bank tributary (the Tinkisso) rises in the Fouta Djallon. South of this line the Diani and Makona and their main tributaries flow directly southwards into the Gulf of Guinea.

The mean annual specific flow rates in the Forest Region are also very high (30 to 40 $l/s/km^2$).

Geology

Most of the basement rock in Guinea is metamorphic Archean or granitized and folded Birrimian with a complex sedimentary cover ranging in age from Upper Proterozoic to Quaternary.

The Archean era is represented by granitogneiss, gneiss and amphibolites in the east and south of the country. The middle Precambrian era corresponding to Birrimian) in the north-east and south-west is represented by schists, quartzites and arkoses.

Magamatic ultra-basic, basic and acid intrusions occurred during the Archean and Proterozoic eras, causing extensive contact metamorphism.

Upper Proterozoic formations, which are found mostly in the Fouta Djallon, form the base of the sedimentary cover; these formations consist mainly of alerolites, argillites, conglomerates, etc.

The Paleozoic rocks found in Lower Guinea and the central-west part of the Fouta Djallon include dolomitic limestones, sandstones and schists.

Magamatic ultra-basic, basic and neutral intrusions also occurred in the basement formations and the cover during the Mesozoic era. These include gabbros, dolerites, diorites, kimberlites, dunites, syenites and nephelinics, which are found throughout the country.

Marine intrusions during the Tertiary and Quaternary periods led to the formation of Paleogenic and Neogenic sediments in the coastal plain. The reshaping of the deposits under the influence of climatic changes in these two periods contributed to the formation of widespread alteration crusts.

II. GROUND WATER

Ground-water resources are found throughout the country. The experience of neighbouring countries demonstrates that the various geological formations can give yields which, without being large, are sufficient to meet the needs not only of the villages but also of most of the secondary urban centres.

Ground-water organs

Responsibilities in the area of ground water are at present shared among several government services: In the Ministry of Agriculture (Secretariat of State for Waters and Forests), the National Water Points Service (SNAPE) is responsible for rural water-supply activities.

In the ministry of Energy and the Konkouré (MINEK), the Studies Office is responsible for urban water supplies and supervision of the IBRD/ADB water and sanitation project, which is concerned with water supplies for Conakry and the main interior urban centres.

The management of the main existing or future water-supply systems is the responsibility of the National Water Office (DEG), but some secondary systems are still operated directly by the regional authorities.

The National Water Service collects all the surface-water data and is responsible for management of the corresponding resources; it also supervises the UNDP/IBRD project on the general water development plan of Guinea.

The General Office of the Ministry of Mining and Geology has a hydrogeological section which carries out hydrogeological studies.

SNAPE is the Government service responsible for ground-water prospecting and evaluation and it has taken over the former wells project (co-operation with the EEC) and the former rural water-supply project (co-operation with UNICEF).

Summary of hydrogeological research

Long before Guinea became independent (1958) hydrogeological research had been carried out with a view to meeting the country's water needs at that time. These studies were concentrated in Lower Guinea and managed by French companies and hydrogeologists, including Messrs. Archambault and Degallier, the Compagnie Générale de Géophysique, etc. (1950 to 1957).

After independence the research was extended to other areas:

- In 1960 Mr. Aisensten carried out a hydrogeological reconnaissance mission in the Forest Region and in Upper Guinea;

- In 1960 the General Office of Mining continued the ground-water research work at Conakry;

- In 1961 a Hungarian mission made a number of hydrogeological investigations in several villages;

- In 1976 the mission responsible for the water-sanitation project prepared a summary hydrogeological map and a report entitled "Hydrogeology of the Conakry region";

- In 1979-1982 the Hydrogeological Section of the General Office of Mining, in collaboration with the BRGM, carried out hydrogeological research with a view to supplying drinking water to 18 regional chief towns;

- Lastly, since 1979 SNAPE, as part of an anti-desertification programme, has been sinking wells and boreholes and tapping springs to supply drinking water to the rural population affected by the drought.

The following table sets out some fundamental data summarizing the state of knowledge of ground water.

Description of aquifers and results of research

Geological era	Water-bearing formation	Geographical situation	Probability of obtaining 1 m^3/h (percentage)	Average depth (m)	Type of inst.	Hydrogeological characteristics Coefficient K (m/s)	T (m^2/s)	S (%)
Archean	Granitogneiss	NE to SE	70-80	40-50	P;F	-	-	-
Birrimian	Schists	NE	80	50-60	F	-	2to2.5x10^{-3}	1x10^{-3}
Upper Proterozoic	Schist-sandstone and dolerite series	N	40-60	60	F	-	-	-
Ordovician	Sandstone	N and NW	90	40	P;F	-	2x10^{-4}	5x10^{-6}
Devonian	Sandstone	NW	30-70	50	F	-	8x10^{-5}	1.5x10^{-5}
Mesozoic	Dunites	Conakry	90	50-60	F	10^{-5}	3x10^{-4}	-

P - Well
F - Borehole

K - Hydraulic conductivity
T - Transmissivity
S - Storage

All these formations can be exploited by borehole at depths from 50 to 150 m with as yet unknown probability of success. The Ordovician sandstones and the Conakry dunites are certainly the country's best aquifers.

Ground-water research and exploitation is characterized by:

- Geological conditions which vary greatly from area to area:

- Prospecting focussed mainly on the location of fractures (photogeology, electrical prospecting);

- The tapping of springs at depths greater than 50 m, often in fissured horizons of a bedrock stratum under varying thicknesses of alteration formation.

Hydrogeological research is not yet very developed in Guinea. The present studies are focussed mainly on the identification of basin zones and basement fractures which may contain water resources for immediate exploitation. Accordingly, the characteristics of the water, especially its physical and chemical properties, have not yet been studied in depth. The small amount of existing data is not necessarily representative and should not therefore be given too much weight.

III. EXPLOITATION OF GROUND WATER

The National Water Points Service (SNAPE) is the body responsible for rural water-supply activities. It was established in 1980 and now has:

- A General Office which determines the water-points policy and evaluates the results of the activities;

- An Administrative Office responsible for management of loans, personnel and supplies;

- A Studies and Monitoring Office responsible for collection of the data needed for the establishment of the files, and for hydrogeological studies and the monitoring of the works (public bodies and private enterprises);

- A Works Office which carries out in the field the programmes drawn up by the General Office;

- A Maintenance Service responsible for supervision of the works, maintenance of the installations, and control and sanitation measures;

- A Rural Leadership Service responsible for the transfer of technology to the users.

In 1984 SNAPE had 16 senior staff (managers, engineers, sociologists), 11 assistant engineers, 10 accountants and storekeepers, 12 crew or garage chiefs (site foremen) and 309 temporary staff and specialist workers.

Technical assistance is provided by four BURGEAP staff members (chief of mission, hydrologist, driller, mechanic), two United Nations volunteers and two French national service volunteers, making a total of 336 staff members.

SNAPE has offices and depots at Conakry and two operations bases at Labé and Kankan. The drilling equipment includes a Halco Mast 625 D rig (combined rotary/down-the-hole hammer) and well-digging machinery: compressors, pneumatic tools and several small items of equipment.

Most of the towns in Guinea take their supplies from surface water:

Urban water supplies

(surface water)

Town	Total supply (litres/person/day)	Population served (percentage)
Conakry	50	52
Kindia	27	70
Mamou	24	30
Kankan	30	69
Faranah	100	100
Guékédou	100	100
Nzérékoré	100	100
Fria	-	27
Boké	-	15

It must be stressed that in most cases the distribution system does not reach all the urban population. Accordingly, the total supply, i.e. the amount of water divided by the number of inhabitants, is less than the figures indicated.

Furthermore, a per inhabitant supply of 100 l/day must be seen as a long-term objective; existing financial resources are sufficient, in the short- and medium-terms, for provision of only part of this supply. The following table summarizes the present situation with regard to the use of ground water to supply towns and large centres.

Urban water supplies

(ground water)

Prefecture	Type of installation Wells	Boreholes	Springs	Number of installations Wells	Boreholes	Springs	Average yield Wells (m^3/h)	Boreholes ($m/^3/h$)	Springs (m^3/day)	Comments
Labé	x	x	x	56	13	138	8	13	41.12	P - 6% tested F - 100% tested S - 100% tested
Pita	x	x	x	15	7	56	12	1	19.5	P - 93% tested F - 100% tested S - 100% tested
Gaoual	x	-	x	63	-	3	-	-	-	P - not yet tested F - no boreholes S - not tested
Koundara	x	-	x	51	-	2	17	-	62.4	P - 6% tested F - no boreholes S - 100% tested
Lélouma	-	-	x	-	-	84	-	-	46	P - no wells F - no boreholes S - 100% tested
Dalaba	x	-	x	19	-	5	5	-	51.2	P - only 10% tested F - no boreholes S - 100% tested
Mali	x	-	x	38	-	25	-	-	25	P - not yet tested F - no boreholes S - 100% tested
Tougué	-	-	x	-	-	2	-	-	54	P - not yet tested F - no boreholes S - 100% tested
Koubia	-	-	x	-	-	8	-	-	52.2	P - no wells F - no boreholes S - 100% tested
Kankan	-	x	-	-	52	-	-	8.13	-	P - no wells F - no boreholes S - 10% tested
Mandiana	-	x	-	-	40	-	-	13	-	P - no wells F - boreholes drilled in the enterp. S - no springs
Siguiri	-	x	-	-	25	-	-	-	-	P - no wells F - tested S - no springs
Macenta	-	-	x	-	-	1	-	-	8	P - no wells F - no boreholes S - 100% tested

P = wells
F = boreholes
S = springs
x = installation present
- = nil

In addition to the three types of installation (wells, boreholes and springs) listed in this table, SNAPE furnishes supplies for plots within small irrigated areas (size: 0.2-100 ha) equipped with a network of concrete channels which feed storage basins from which the peasants take water in containers to irrigate crops such as maize, cabbage, onions, beans, tomatoes, potatoes, manioc, etc.

Estimate of future needs

In view of the size of the requirement, SNAPE has had to establish an order of priorities, the first being to establish 2,600 water points in the villages where the situation is most critical.

A medium- and long-term programme in three phases has been established on the basis of increasingly high standards of service.

Phase 1: 10 litres of drinking water per day per inhabitant at the end of the dry season within 500 m for villages of more than 300 inhabitants and within 1,000 m for villages of 100-300 inhabitants. (This phase includes the installation of the 2,600 priority water points mentioned above).

Phase 2: 10 litres of drinking water per day per inhabitant with the distance standards described above and partial cover of villages with fewer than 100 inhabitants.

Phase 3: 20 litres of drinking water per day per inhabitant within 500 m of the users.

The following table gives the number of water points needed for attainment of these objectives on the basis of the standards described above and taking into account future population increases:

Number of water points needed

	Phase 1	Phase 2	Phase 3
1980	4,400	7,900	15,800
1985	4,900	8,900	17,800
1995	6,100	11,100	22,000

In the light of these data, SNAPE has set the target of implementing Phase 1 before the end of 1995, i.e. 6,100 water points including 3,590 before the end of 1991.

The abundance of surface water in Guinea clearly argues in favour of the use of this resource to supply the towns. However, in some cases, Conakry for example, the exploitation of ground water can be envisaged.

Accordingly, a preliminary study followed by sounding and drilling has been carried out under the Conakry water and sanitation project financed by the World Bank and the African Development Bank, with a view to supplying water to seven interior centres.

At the same time, as part of the technical co-operation with France, the BRGM has drilled 27 boreholes for 12 interior centres.

IV. REFERENCES

(unpublished documents and reports in the SNAPE archives)

Evaluation mission for the establishment of a water drilling unit in Guinea. Final report, January 1980.

Drinking water and sanitation planning study in Guinea (WHO/World Bank 1980 co-operation programme).

Water supply project for seven towns.

Final report: town of Koundara, 1982; town of Siguiri, March 1983

Ground-water research in the Labé urban area, June 1983.

Review Mines et Développement, No. 10, April 1982.

General water supply plan for Guinea.

- Middle Guinea - Volume hydrologie, Polytechna, 1981.
- Upper Guinea - Volume hydrologie, Sir Alexander Gibb and others, 1983.
- Forest Region - Volume hydrologie, Sir Alexander Gibb and others, 1983.
- Coastal Guinea - Volume hydrologie, Sir Alexander Gibb and others, 1983.
- Drilling records for Upper Guinea and Middle Guinea:
- Wells
- Boreholes
- Springs

GUINEA-BISSAU

Area: 36,125 km^2

Population: 860,000 (United Nations estimate, 1983)

I. BACKGROUND

Guinea-Bissau is situated on the west coast of Africa between Guinea and Senegal. In addition to its mainland territory it has the Bijagos archipelago which lies just offshore. The archipelago consists of 40 islands of which only 20 are inhabited. The country has 160 km of coastline and 705 km of land frontier. Its maximum length is 330 km and its maximum width 193 km. Not counting the tidal mangrove zones, the permanent *terra firma* is estimated at 28,000 km^2.

Natural regions

Most of the country is flat and low-lying with a wealth of marshes and mangrove swamps along the coast; the maximum altitude is 40 metres. To the south-east near the frontier with Guinea there are some small hills, a prolongation of the Fouta Djallon massif. Lateritic cuirasses and clay-sand formations cover much of the surface. From the morphological standpoint the country can be divided into five main zones: the Boe uplands, the Bafata plateau, the Gabu peneplain, the transition zones (Oio and Forrea) and the coastal plains.

The Boe uplands are flat-topped hills forming small plateaus separated by valleys which are mostly open and covered with ancient lateritic, ferralitic or aluminous cuirasses which produce disintegrated and conglomeratic laterites in the valleys. The escarpment of the Bafata plateau dominates the alluvial plains which are gouged by the meandering lower courses of the Geba and the Corubal.

The Gabu peneplain in the north-east has a slightly undulating relief with gentle slopes and a loose and poorly defined hydrographic system.

The coastal plains are broad in the west and south of the country; they correspond to a band of Mesozoic-Cenozoic formations. They are mostly very low but do have a large number of small elongated ridges.

The transition zones in the regions in which the tidal influence is very small or non-existent form a link between the coastal plains and the Boe uplands, the Bafata plateau and the Gabu peneplain.

Climate

Guinea-Bissau has a wet tropical climate with two distinct seasons: a rainy season from April to October and a dry season from November to March. The rainy season is caused by the moist Atlantic winds which blow from the south-west and west towards the Saharan and sub-Saharan zones. The dry season is caused by the dry winds blowing mostly from the mainland in the opposite direction.

The country can be divided into two climatic regions: one is of the tropical Sudanese type with very low rainfall, wide thermal ranges and high humidity in the rainy season and low humidity in the dry season; the other is of the sub-Guinean or coastal type with higher rainfall, smaller thermal ranges and high humidity throughout the year. The first region corresponds to the western part of the country, the second to the wide coastal band.

There is little vegetation but it is very varied: forest, rain forest in the wet areas, scrub savannah and mangroves.

Surface water

The country is crossed by large rivers which rise in Senegal or Guinea. It also has a number of coastal rivers; these are, from north to south, the Cacheu, Mansoa, Geba-Corubal, Grande de Buba, Tombali, Cumbija and Cacine. Even the smallest of these rivers have wide deep estuaries. They are navigable and provide means of communication with the interior of the country; the main urban areas are situated on their banks. They are tidal as far as 100 km inland. Even the Geba has a tidal wave.

There are 43 climatological stations and three stations measuring the flow rate: two on the Curubal and one on the Cumbija.

The mean annual flow rate on the Corubal over a nine-year period was 14,410 hm^3(maximum 23,710 hm^3; minimum 6,110 hm^3).

In 1979/80 the minimum flow was 6 m^3/s and the maximum 1,100 m^3/s; these figures indicate the wide regional variation of flow rates in the dry and rainy seasons. The water quality is equally varied.

II. GEOLOGY

Guinea-Bissau lies between the Fouta Djallon massif, of uncertain Paleozoic age, and the Mesozoic-Cenozoic Senegal basin. It can be roughly divided into two geological units by a line starting in the north in the Pirada region and passing through Contuboel, Bambadinca and Buba before deviating towards Gadamel and the frontier with Guinea. Paleozoic formations predominate to the east of this line.

To the west of the line most of the terrain consists of Mesozoic-Cenozoic sedimentary formations. These are the most important formations from the hydrogeological stanpoint; all the deep boreholes are found in these formations.

The ancient formations have very few outcrops, for they lie below fairly thick surface formations and are not very fossiliferous. They have been dated by analogy with similar and better-known formations in neighbouring countries and from samples obtained during oil-prospecting drilling operations.

The same can be said of the Mesozoic sedimentary series, even though fewer data have been obtained from the oil-drilling sites.

The oldest formations are Cambrian and are found in the north-east. They lie on microgranites of the metamorphic complex, then the multi-coloured argilites of Pirada and Canquelifa overlain by the schist-sandstones of Cantari, which are themselves overlain by purplish feldspathic sandstones (Caium sandstones) of Upper Cambrian age. As a rule, the Cambrian formations constitute very poor aquifers. Small local aquifers can be found in the alteration zones and in particular in the more permeable sandstone zones. Drilling is difficult and the productivity low owing to the great lateral and vertical variations in permeability. The Cambrian formation is followed by the mainly sandstone Ordovician series (Gabu sandstone) identified at Canjufa, Sonaco, Gabu, Canjadude and Cabuca. Most of this series is thick and steeply inclined (up to 48°). At the base of the series the sandstone is white and coarse-grained with some feldspar and muscovite crystals. It is 170 m thick. These very compact sandstone strata are overlain by less compact and sandier strata of considerable thickness covered by fine-grained sandstone, quartzitic in places, which forms the top of the series, with thicknesses of 10 to 30 m.

Little is known about the ground water in this formation except for the wells in the Canjadude region, where the compactness of the sandstone makes drilling difficult and unproductive. In this region the thick alluviums of the Nharasse river give the best results, although the recharge is uneven and the yield intermittent.

The Silurian formation is represented by the Buba sandstones of Gothlandian age. It has been reached by boreholes at Buba, Guilege, Gadamael and Sangonha and by several oil drillings. It is found in the north-east in the same areas as the Ordovician formations and in the south-east as well. The drillings carried out in this area have penetrated very compact black or carbonaceous schists with fine sandstone intercalations at the top and some doloritic layers. At Có and Safim the formations are mostly metamorphic with micaceous sericitic and pyritic schists interbedded with dolorite. The Silurian formations in the north-east are of similar facies, with major or minor indications of metamorphism and with fine clay sandstones at the top forming a transition zone with the Lower Devonian formation. These formations are very poor aquifers. The positive results of the drilling at Buba and Sangonha are no doubt due to the collection of water in the sand-sandstone strata of the Maestrichtian base which overlie the Silurian sandstone in discontinuity.

The Devonian formation occupies a vast area in the east close to the frontier with Guinea. It forms a large north-west/south-east syncline. The base consists of a sandstone series (Cusselinta sandstone) of Lower Devonian age overlain by a thick schist series (Bafatá schist) of Middle-Upper Devonian age. The Cusselinta sandstone identified in outcrops in certain water courses (Geba, Curubal. etc.) is mostly compact feldspathic and micaceous with pink and brown tints. The Bafatá schists are argillaceous and pale or pink in colour with intercalations of fine-grained quartz sandstone. The Devonian sandstone has been reached at a depth of 20 m during water-prospecting drilling operations in the Nhabijocs plain (Bombadinca); it is much darker and more compact than the neighbouring altered rocks. The alteration zone in the schists has proved completely barren.

The post-Paleozoic series, i.e. the Mesozoic sediments of the Senegal subsidence basin, rests on the Paleozoic formation, whereas in Senegal it is also found above the metamorphic basement rock. Little is known in Guinea-Bissau about the "intercalated continental" of Senegal at the end of which the Mesozoic transgression occurred, but it has been identified at S. Domingos, Có and Safim. In the north-east it consists mainly of schists with some sandstone intercalations and limestone layers in places. In the centre of the country it occurs in clearly continental series which also consist of schists with intercalations of fine dandstone. The series is 120 m thick at Safim and 330 m at S. Domingos.

Nothing is known of the hydrogeology of these formations owing to their great depth (800-1,800 m). It is assumed that the sandstones are important aquifers, as they are in Senegal.

The Middle and Upper Cretaceous formation is of lagoonal or coastal facies, as in Senegal, and of great thickness (1,360 m at S. Domingos and 600 m at Safim). The base consists of schists with some limestone-dolomitic intercalations; it is overlain by dark shales streaked with sand and terminates in a thick sandy layer of Maestrichtian age.

The Maestrichtian formation is certainly an aquifer of major importance in Guinea-Bissau. At the edge of the Paleozoic formations it forms a wide band of sandstone outcrops with varied granulometry and containing layers of schist at the base with very fine argillaceous intercalations throughout the series. In the south, where it rests in discontinuity on the Silurian formation, the sandy layers lead to compact argillaceous sandstones of average or coarse granulometry. It is up to 490 m thick in the north-west (S. Domingos) and 540 m in the west (Cangongue).

The Eocene-Paleocene formation of more marine facies is represented by sandier marl-limestone formations with dolomitic intercalations. It forms a central strip running north to south, bounded in the south by the right bank of the Geba. There are occasional outcrops close to Bissau and the formation has been penetrated by a number of boreholes for groundwater reconnaissance and exploitation. It is an important aquifer and is exploited by deep boreholes.

The continental terminal formation is clearly lagoonal at the base and continental at the top; it includes the post-Eocene and Quaternary series. The lower formations of Oligocene age consist, from bottom to top, of fine-grained sand-sandstone layers, mostly argillaceous with clays at the top. Above this is found a small limestone-marl bank, sandy in places, of marine Miocene age, which is covered by more recent clays and argillaceous sands.

The main aquifers in this series are provided by the Oligocene sand and Miocene limestone strata.

The most recent series are represented by alluviums, clays, agrillaceous sands, coastal dunes and cars, lateritic formations, etc.

Tectonics

The topography is unvaried, and the thickness and regularity of the sedimentary formations indicate little tectonic activity. Series of folds running east/west and fault systems running north-north-east/south-south-west have been identified on the north slope of the Bafatá syncline which runs mainly north-west/south/east and occupies a large part of the east of the province. On the southern slope, however, the tectonics is much less marked.

III. MAINLAND HYDROGEOLOGY

As is the case in neighbouring countries and as the situation outlined above indicates, the mainland territory contains only a small number of geological formations providing constant and fairly extensive productive aquifers with relatively uniform hydrogeological characteristics. Such aquifers are usually found in soft sandstones and sandy or fissured limestones. Continuous aquifers are rarely found in other formations. The granitic and quartzitic rocks are very compact and contain only low-yield intermittent and local aquifers of very variable permeability in the alteration layers. The schists are usually clogged with clay deposits. In these formations water is found only in the porous lateritic cuirasses which may overlie the rock. The physical and chemical quality of the water is better at depth and close to the recharge zones.

Recharge is effected by infiltration of water migrating from the higher Paleozoic zones in the Mesozoic formations at their edge. Most of the tapped water is thought to be of fossil origin. The present recharge compensates for the natural drainage of the aquifer systems.

The isohyet map shows that the coastal zone receives the highest rainfall, owing to the proximity of the ocean. Along this zone the rainfall increases from north to south (2,600 mm/year) with the maximum at Catió-Cacine-Bedanda. In the interior the rainfall declines to 1,500 mm/year.

The distribution of the annual rainfall also varies according to place. Rain falls on 130 days in the north, on 150-160 days at the coast in the west, and on 170 days in the south. The average annual temperatures are fairly high, in the order of 26-27°C; the thermal ranges are highest in the interior: 13°C (maximum 32.8°C; minimum 19.7°C); they are lower on the coast: Bissau 9.8°C (maximum 31.6°C;). At Catió in the Bijagos archipelago the figures are even lower: Bubaque 7.7°C (maximum 30°C; minimum 22.2°C).

The evapotranspiration is very high and prevents any great infiltration of rainwater to recharge the aquifers. The transpiration is higher than the evaporation in the coastal zones where the rainfall is heavy and the forest and rain forest are found. The opposite is the case in the interior where savannah type vegetation predominates.

The infiltration potential in the coastal zone is not really very great.

The empirical formula of L. Turc, which gives a calculated value for real evapotranspiration (Re) in terms of rainfall (R) and temperature (T), has been applied to five stations and gives the following values (mm):

	R	Re	Rainfall infiltration
Nova Lamego	1,500	1,400	100
Catió (south)	2,600	1,600	1,000

In the country's interior, where most of the formations have low permeability and the aquifers are small and discontinuous, the actual stocks are smaller than those indicated in the table above. Other factors come into play, such as the drainage of the aquifers towards the low-lying areas and the intense exploitation in the dry season. In fact, the Gabu plain is the part of the country least well provided with ground water. In contrast, in the southern part of the coastal zones the effective infiltration can reach levels which offset the decline in the stocks of the surface aquifers in the dry season. However, this infiltration is probably smaller than the calculated value.

Furthermore, the sunshine rate is very high (62 per cent), which indicates that the evaporation potential of the atmosphere is independent of the relative humidity; it is therefore very high.

In conclusion, the recharge of aquifers in Guinea-Bissau is favoured by the rainfall pattern but impeded by the topographical and geological conditions (stratigraphy and lithology) and the amount of evapotranspiration. The recharge is generally insufficient.

However, there are enormous stocks of fossil water in the Mesozoic-Cenozoic formations of the Senegal basin which would not be affected in the short term by intense exploitation.

In contrast, excessive exploitation of the small resources of the surface aquifers could lead to their complete exhaustion in the short term.

Description of the aquifers

The continuity of the Mesozoic-Cenozoic aquifers means that the results obtained at a few sites can generally be extrapolated to the full extent of these formations. Where ground-water resources are concerned, the country can be divided into two zones separated by the eastern limit of the Mesozoic-Cenozoic Senegal basin. Continuous and extensive aquifers are found in the western sedimentary zone. The eastern zone, with its Paleozoic and Ancient series, has only local aquifers of relatively small extent and yield. It includes the highest areas with least rainfall and greatest evaporation. Accordingly, the surface aquifers dry up in the dry season. Boreholes must be drilled with great care and preceded by geophysical prospecting; otherwise the risk of failure becomes very large. The fault system which had seemed promising, especially in the north of the Bafatá syncline, proved barren. These faults appear, in fact, to constitute an impermeable barrier. The water has a moderate or low mineral content.

The Mesozoic-Cenozoic strata slope north-west and west and increase in thickness in those directions, acquiring an increasingly marked marine facies. The main aquifers are found in the Maestrichtian sands, the sandy-marly limestones of Eocene-Paleocene age, and the fine Oligocene sands which contain some strata with permeable horizons - limestones, argillaceous sand, laterite - in Miocene and recent formations.

Three main aquifers have been identified, from bottom to top: Maestrichtian aquifer (mainly sands); intermediate Oligocene-Paleocene aquifer (sand in the Oligocene formation and limestone in the Eocene-Paleocene); and "surface" aquifer(s) (argillaceous sand and laterite), usually unconfined.

The Maestrichtian formations have been explored and exploited most intensively and at the shallowest depth close to the elongated zone where they outcrop (central and southern regions of the country) and at the sites of a number of deeper wells installed to obtain large yields at Bissau and Farim. The maximum tapping depths of 195-260 m have been achieved on the island of Bissau.

The map annexed to this paper indicates the location of the boreholes and the specific yields (l/s/m of downdraw) and the available yields, which are fairly high. The productivity declines in the direction of the Silurian formations owing to their lesser depth and the transition from sand to compact sandstone.

The attached diagrammatic map shows the isobaths of the top of the Maestrichtian formation and gives an idea of the depths to be drilled if this formation is to be exploited. This top layer deepens regularly and continuously towards the north-west and west. Towards the north-west it cannot be reached at less than 500 m below zero. In contrast, to the east of the 550 m contour line running north-north-west and passing through Quinhamel, Binar, Binta and Dungal, it can be reached at less than 200 m.

The Maestrichtian aquifer is for the most part confined, except at rare points where it outcrops. This artesian effect produces water spouts (of less than a metre) at only two points - Farim and Mansoa - with static water levels of +6m and +7m.

Water levels above ground level are found at the majority of the other installations in the centre and west of the country. However, the water level is below ground level in the southern part of the coastal zone, probably owing to the lower vertical permeability of the formations and their consequent smaller recharge. The highest levels are found at installations located roughly on the Farim-Mansaba-Mansoa line, and this might indicate the proximity of a more favourable recharge zone in the fairly thick shallow-lying Paleocene limestone resting on Maestrichtian sands.

The aquifer's main direction of flow would seem to be towards the north-west and west; this is confirmed by the chemical composition of the water.

The Maestrichtian water has a low mineral content (less than 500 ppm of dry residue) in the centre of the country. In the south (Bolama) the water contains 1,500 ppm, the maximum allowable under WHO drinking-water standards. The low salinity of the Maestrichtian aquifer even close to the coast indicates that the influence of sea water is practically non-existent, except at some points such as Bedanda (4,000 ppm of chlorides). However, the concentrations are low at Catió and Cufar, which have dry residues of 235 and 370 ppm respectively.

Lastly, the Maestrichtian water of mainly bicarbonate and calcium sulphate facies can be classified as very soft to soft, except in some coastal areas. The sulphate content is low, except at a few points (Bolama and Sul regions).

Some test pumping has been carried out in the Maestrichtian aquifer, with observation of the downdraw and recovery, in single wells or in coupled wells, with one serving as piezometer.

The levels are affected by the tide close to the estuaries or ocean. The water must therefore be exploited with care.

On the basis of lithological information and the permeability data obtained, a diagrammatic map has been prepared in which the country is divided into three areas fairly suitable for drilling operations:

- Extreme south with K below 10^{-6} m/s in argillaceous sandstone with low specific yields (0.2 to 0.3 l/s/m), and high mineral content owing to the slow flow rate of the ground water;

- Centre south with K of about 10^{-4} m/s. A more favourable elongated zone has been identified in this region: Buba-Empada-Catió. This formation consists of fine to coarse sands. This zone is favoured by the aquifer's direction of flow;

- To the north of this area with K of about 10^{-5} m/s. The aquifer is less favourable owing to the presence of layers of clay and fine sand.

The Oligocene-Eocene-Paleocene aquifer is found in white sand-sandstones, usually of very fine granulometry, at the base of the Oligocene formation and in the Eocene-Paleocene limestone-sandstone formations, which are not separated by any sizeable impermeable layers. Accordingly, the whole unit behaves as a single aquifer; this is confirmed by the static water levels and the chemistry of the water.

In the north-east of the country the present installations penetrate no further than the shallowest Oligocene strata, which are sufficiently thick and productive. Elsewhere in the country they generally reach and tap the limestone series. The aquifer is 316 m thick at S. Domingos and 274 m at Cagongue, with its top at a depth of 177 m. It ought therefore to be possible to exploit the Oligocene-Paleocene aquifer west of a line running through Cumere-Mansoa-Mansaba-Caujambari and terminating at Cuntima. To the east of this area the series is either of insufficient thickness or does not occur. Unlike the Maestrichtian aquifer, the intermediate aquifer has a low yield. Only the sand-sandstone Oligocene formations, when they are sufficiently thick, can furnish high specific and available yields; this is easily explained by the irregularity of the facies of the limestone series both vertically and laterally in comparison with the high permeability of the Oligocene sands. Lastly, the yields of the boreholes in the Oligocene-Eocene-Paleocene aquifer are sufficient for the country's needs in most of the regions. The aquifer is mostly continuous, of vast extent and confined.

The aquifer's piezometric level is deeper than that of the Maestrichtian aquifer, confirming that there is no communication between the two; the great differences in permeability at various points in the limestone are matched by very marked differences between the aquifers' water levels and yields.

The highest mineral content (1,000 ppm) and the highest chloride content are found in the north-west coastal zone. These levels increase from south-east to north-west but they do not reach the maximums set by WHO. The sulphate content is higher than in the Maestrichtian aquifer, as is the hardness of the water.

In conclusion, although the intermediate aquifer is extensive and continuous, its stocks are much smaller than those of the Maestrichtian aquifer and it has less favourable hydrological characteristics and water of much poorer quality.

The upper shallow aquifer has fairly small stocks and is subject to very marked annual fluctuations which are directly linked to rainfall and evapo-transpiration. Owing to its accessibility, it is intensely exploited throughout the country by innumerable wells of various depths - usually shallow, but deeper in the north-west and west where the series is thickest. This upper aquifer is the equivalent of the "continental terminal" aquifer of West Africa: it behaves in a similar manner.

Recent sand-clay formations, usually of lateritic littoral facies and with small banks of Miocene limestone, are found over most of the country's central and western regions. The north of the country has permeable sand-clay series attributed to the continental Miocene-Pliocene era. Their yield is usually quite low and varies from place to place and season to season. The water is subject to pollution and is often heavily mineralized with a high iron content; it is very aggressive.

IV. AQUIFER OF BISSAU ISLAND

Bissau island, where the capital is located, is the part of the country for which the largest amount of ground-water data is available - a consequence of the large and rapidly growing requirements of the densely populated metropolitan zone. The deep aquifers must be tapped, for the shallow-lying surface water is of only average or poor quality. An organized study must therefore be made of the depths to be reached and of the possible yields. In the past a large percentage of the boreholes proved barren or gave inadequate yields.

The island is fairly flat. The coasts are low, as is most of the land, which has many swampy areas. There is an abundance of recent alluviums consisting mainly of fine elements (silt and mud) called "bolanhas" where rice is grown in the traditional manner.

The land below 10 m in altitude ("terras baixas") is flat with alluvial layers situated between two and five metres above sea level. Only the highest areas, which are never above 40 m in altitude (with an average of 20 to 25 m), are inhabitated. The climate is wet with two distinct seasons.

Geology

The geology is known from deep-level oil and water drilling. The whole of the sedimentary series slopes to the north-west and west, more steeply as the geological age increases. The stratigraphic chart is as follows:

Quaternary: silt, mud, laterite formations (low land);

Miocene: (upland cover) mainly clays and fine argillaceous sands, with various amounts of laterite at the top of the series; marly and sandy limestone banks at the base; total thickness: 3 to 50 m;

Oligocene: clay in the upper part, fine sands and argillaceous or marly sandstone at the base; thickness: 40 to 50 m;

Eocene-Paleocene: sandy or compact limestones, with intercallations of sand and clay, and dolomitic and sandy intercalations in the lower part; thickness: 70 to 100 m;

Maestrichtian: less marked marine facies. The whole series consists of formations of fine or coarse sand with clay-marl intercalations in the upper part, and intercallations of schist in the lower part; thickness: 250 to 300 m;

Middle-Lower Cretaceous: thick schist and sand series with sandstone intercalations. Indications of metamorphism at the base; thickness: 400 m;

Paleozoic: the base of the sedimentary series includes sericitic metamorphic schists with sandstone-quartzitic intercalations and some doloritic banks of Silenian age.

Hydrogeology

The aquifer is located under the laterite zone; it is drained by the most permeable strata of the "terras baixas" at an altitude of about 10 m. Its lower limit is the marls and clays of the Miocene base and the clays of the top of the Oligocene formation. The aquifer is subject to major fluctuations due to rainfall and evapotranspiration. It can be reached at depths of 10 to 20 m in the "uplands".

These formations are very heterogeneous. It is not therefore possible to apply rigid criteria with respect to the drilling of boreholes - except in the vicinity of springs. The yield of the waterpoints exploited by hand pumps (portable) is usually below 5 m^3/h.

The Oligocene aquifer found in the sand-sandstones is mostly confined. It forms a single unit with the aquifer in the Eocene-Paleocene marly limestones.

This aquifer is very irregular in its yield, owing mainly to its lateral and vertical variations, the permeability of the formations, the thinness of the strata and the insufficient recharge.

Drilling operations are difficult owing to the very fine granulometry of the sands and sandstones and the aggressiveness of the water. The Oligocene aquifers cause blocking of the filters and rapid deterioration of the pumping equipment at most of the tapping sites on Bissau.

The intermediate aquifer on the island has a static water level of four to six metres below sea level. The available yields rarely exceed 15 m^3/h with large drawdowns and therefore very low specific yields. The top of the Oligocene formation can be reached at depths of between 30 and 60 m, and the most permeable layers lie at depths of 60 to 100 m. The Oligocene series lies above a limestone-sandstone and marl-limestone series (Eocene-Paleocene) in which permeable layers have been identified; they are irregularly distributed and more favourable and abundant in the middle and upper parts of the series, which are less marly.

As a rule the sandstones are very compact, but the limestones have a microfissure permeability; there are few fractures of any size. The marl-limestone series includes two distinct aquifer systems separated by very marly strata. The upper system corresponds to the Oligocene stratum, the lower is fed from the Maestrichtian stratum.

The intermediate aquifer in the Bissau region has a mineral content of 700-1,200 ppm. The water is neutral or slightly acid, ferrous or very hard (40° to 70°; it is of sulphate and calcium bicarbonate facies.

Both the increased salinity of the water and the low specific yields of most of the installations are due to the poor permeability of the formations.

The Bissau installations take almost all their water from the intermediate aquifer (Oligocene and part of the Eocene-Paleocene) and they discharged 10 to 40 m^3/h, with specific yields of around 0.1 to 0.6 l/s/m (an average of about 0.3 l/s/m).

Some deeper boreholes reach the base of the Paleocene formation and supply better quality water, part of which comes from the underlying Maestrichtian stratum. This water has a higher sodium content and is more neutral, with a hardness of only 2° to 5°.

The limestone aquifer is already overexploited by many uneconomical low-yield boreholes and it does not furnish sufficient water for Bissau. The underlying Maestrichtian sands must therefore be more intensively exploited. The first Maestrichtian boreholes confirm the hydrogeological importance of this series, which is well known in Senegal.

The Maestrichtian aquifer is totally confined, with its top located in the intermediate argillaceous-marly strata of the limestone series - thin and discontinuous layers which do not separate the aquifer. On Bissau the Maestrichtian aquifer is formed by sands of very varied granulometry and compactness. In the upper layers the clay intercalations are frequent and fine. The top of the series lies between 130 and 180 m below sea level. Boreholes which penetrate 50 m into the sands can discharge 50 to 150 m^3/h with a small drawdown and specific yields of five to seven l/s/m for a six-inch diameter hole. Larger yields could no doubt be obtained from deeper boreholes with larger diameters.

In contrast to the intermediate aquifer, the hydrostatic levels of the Maestrichtian aquifer are found at positive altitudes (+2m to +3m), which seems to indicate that the two aquifers are not in contact. The water is alkaline with a very low hardness (below 5° in the Bissau region) and a low mineral content of about 300 mg/1, which is very uniform for all the wells. The main recharge zone appears to lie south-east of Bissau where the series is close to ground level.

All the data on the Maestrichtian aquifer seems to indicate that its stocks of water are considerable. However, the exploitation must be controlled and efficiently managed from both the technical and the economic standpoint.

The water levels are subject to periodic variations due to the movement of the tides. However, the chloride content of the water does not vary, indicating that the Maestrichtian formation is totally separated from the overlying series. This movement weakens and becomes more irregular towards the interior of the island.

Lastly, the behaviour and characteristics of the Bissau aquifer are similar to those found throughout the sedimentary basin in Guinea-Bissau.

V. SUMMARY AND MAPS

Practical hydrogeological regional divisions

(Potential of aquifers)

Hydro-geolo-gical zone	Hydrogeological region (Corresponding administrative region)	Description of exploitable aquifer Stage	Lithology	Aquifer area	Drilling (m)	Dynamic water level (m)	Available unit flow (rarely) (l/s)	Total mineral content (g/l)	Degree of knowledge (number of test holes)	Possible uses
	I-1-western	N_2-Q	sand	± cont.	25-35	15-20	1-2(5)	0.1	fairly good (17)	village water supply
Zone I	Cacheu, except east of Bigene and bula sector	N_1	limestone	cont.	50-100 (E→W)	25-35	1-2(5)	<0.5 south >0.5 north	good(16)	village water supply (if N_2-Q absent)
		P_3	sand	cont.	100-150 (E→W)	15-25	5-10 (20)	<1 S and N >1 centre	insuff.(11)	fairly big consumption esp. irrigation.
		P_1-2	limestone	cont.	150	about 20	up to 20	1	poor (1)	occasional (if P_3 insuff.)
West	I-2-central Oio, except extreme east; east Cacheu; Biombo and Bissau	N_2-Q	sand	discont.	25-35	15-20	0.5-2 (5)	0.1-0.2	insuff.(10)	village water supply
		N_1	limestone	SW	30-50	25-30	0.5-2 (5)	0.2-1	fairly good (16)	village water supply (if N_2-Q absent)
		P_3	sand	cont.	50-100 (E→W)	15-25	1.5(15)	0.1-1 (E→W)	very good (60)	village water supply and big consumption
		P_1-2	limestone	cont.	50-150 (E→W)	20-35	1.5(10)	0.1-1 (E-W)	very good (65)	village water supply (if upper aquifer absent) and fairly big consumption
		K_2m	limestone sand sandstone	cont.	100-200 (E→W)	20-35	15-25(40)	0.2-1 (E→W)	good, esp. south(36)	towns, indust., agri. very big consumption)
	I-3-eastern (East Oio, west Bafatá)	N_2-Q	sand	discont.	25-35	15-20	0.5-1	0.1	poor(6)	village water supply
		K_2m	sand	cont.	50-100 (E→W)	10.25	10.20( ) (E→W)	0.2	W good (17) E poor	big consump., esp. irrigation
	I-4-southern (Quinara, Tombali, extreme east of Bolama	N_2-Q	sand	± cont.	20-30	15-20	0.5-1	<0.1	fairly good (60)	village water supply
		K_2m	sand sandstone	cont.	50-100 (E-W)	15-30	2.5(15)	<1 north >1 south	insuff.(12)	fairly big consumption
	1-5 Bijagos (Bolama)	N_2-Q	sand	discont.	20-35	15-20	1	<0.1	poor(3)	village water supply
Zone II	II-I-north (Gabu, except Boé sector; Bafatá - far NE part)	c.t.	sand alterite	±cont.	25-35	15-20	0.5-1(3)	<0.2	fairly good (80)	village or live-stock water supply
		Pz. Pc	sandstone schist	discont.	40-50	15-20	0.2-0.5 (4)	0.1-0.5	insuff.(40)	village or live-stock water supply
East	II-2-south (South Gabu, south-east Bafatá, east Quinara, north east Tombali	c.t.	sand alterite	discont.	25-35	15-20	0.2-0.5	<0.5	poor(4)	village or live-stock water supply

VI. REFERENCES

SMN — O clima de Portugal - fasc. XIV - Norma climatológica de Ultramar correspondentes a 1931 - 60 - ed. do Serviço Meteorológico Nacional, Lisbon, 1965.

Teixeira (A. J. Da Silva) — Os solos da Guiné Portuguesa - Estudos, ensaios e documentos No 100 - ed. da J.I.U., Lisbon, 1962.

Teixeira (J. E.) — Geologia da Guiné Portuguesa - Curso de Geologia do Ultamar - ed. da Junta de Investigação do Ultramar - vol. I - p. 55-103, Lisbon, 1968.

LIBYAN ARAB JAMAHIRIYA
MORPHOLOGY

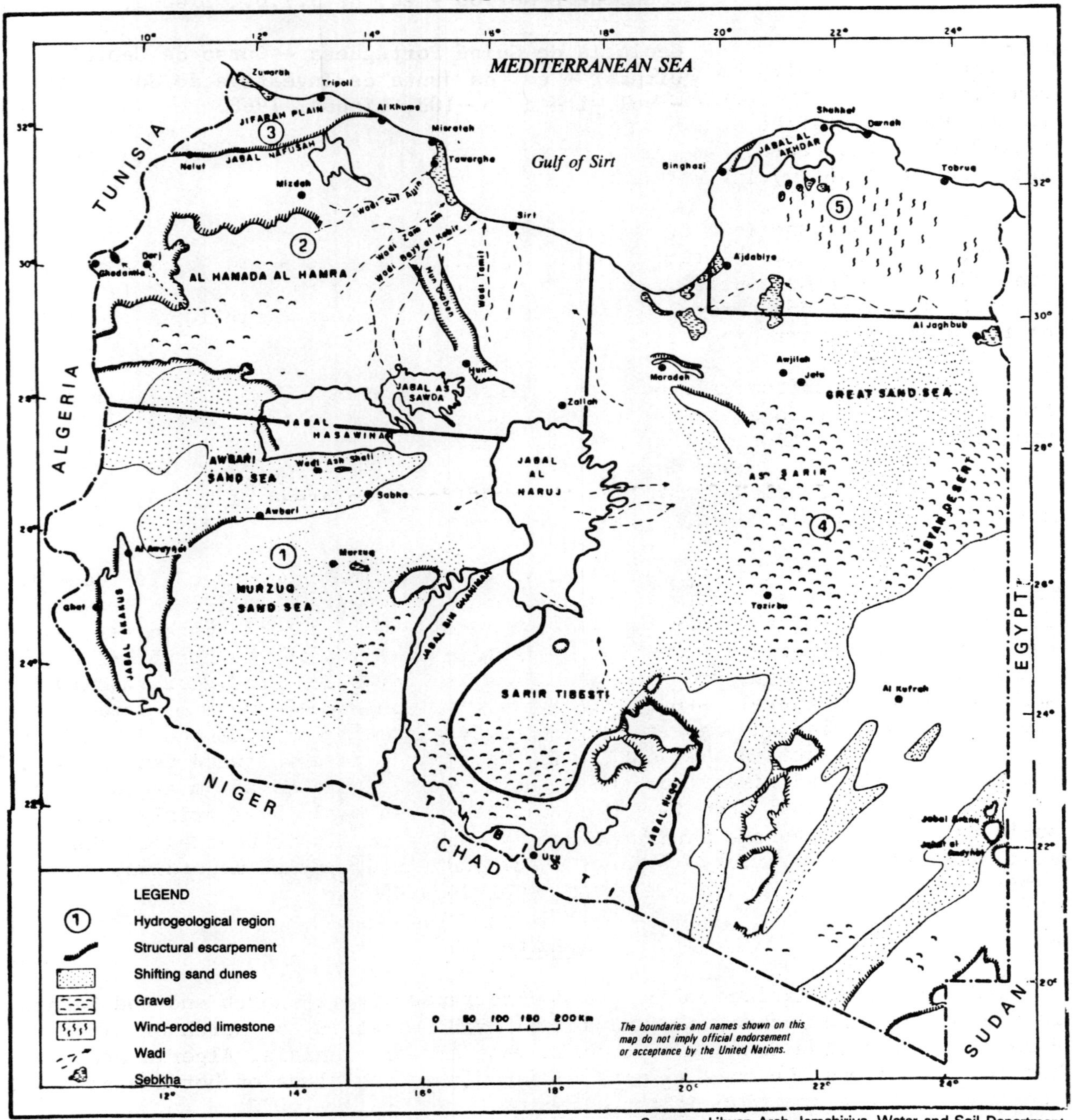

3436.2x

Source: Libyan Arab Jamahiriya, Water and Soil Department

LIBYAN ARAB JAMAHIRIYA

Area: 1,775,600 km^2

Population: 3.36 million (United Nations estimate, 1983)

I. BACKGROUND

Topography

Libya is a country of vast plains. It has 1,900 km of Mediterranean coastline. The main mountains are the Jabal Nafusah and the Jabal Akhdar in the north along the coast, the Jabal Sauda (803 m) and the Jabal Haruj (1,200 m) in the center, and the Toammo (1,022 m), Tibesti (2,300 m at Kamet) and Archenu (1,460 m) in the south.

The Jabal Nafusah (600 to 800 m) dominates the triangular coastal plain of Gefara, which has much better climatic conditions and soil and water resources than the rest of the country. It is therefore densely populated and has the capital, Tripoli, and several major urban centres. This plain supplies most of the country's farm production.

A similar situation, although on a lesser scale, is found in the north of the country, where Jabal Akhdar (500 to 800 m) dominates the coastal plain of Benghazi where the town of that name is located.

South of the Jabal Nafusah lies the vast desert plateau of Hamadah al-Hamra (500 to 600 m).

The Murzuq basin ("Edeyin"), which is covered with sand dunes, is found in the south-west of the country. Its northern part is known as Edeyin Awbari. In the south-east of the country lies the immense Kufrah basin which reaches the Egyptian frontier and rises to an altitude of 100 m in the north and 800 m in the south. In the north are found expanses of gravel and sand dunes and in the south expanses of gravel, stones and rocks with many small areas of sand dunes. There are other flat and desert areas: the Syrt desert north of the Jabal Haruj and the Sarir Tibesti to the south. The first area is a westward extension of the Kufrah-Sarir and rises from the coast to an altitude of 250 m at the foot of the Haruj. The second area is linked to the Kufrah basin by a fairly narrow strip of land between the Haruj and Nuquay mountains; it is bounded in the south and west respectively by the Tibesti and Bin Ghunaymah mountains. The Sarir Tibesti desert lies at an altitude of about 500 m.

Climate

The climate is influenced by the Mediterranean in the north and the Sahara in the south. The lack of continuous and high relief in the north of Libya - in contrast to the countries of north-west Africa (Tunisia, Algeria and Morocco) - renders the coastal area vulnerable to the flows of hot dry air from the south. There are five climatic zones:

Subtropical Mediterranean zone, which is small in extent and bounded by the Jabal Akhdar.

LIBYAN ARAB JAMAHIRIYA
AVERAGE ANNUAL PRECIPITATION

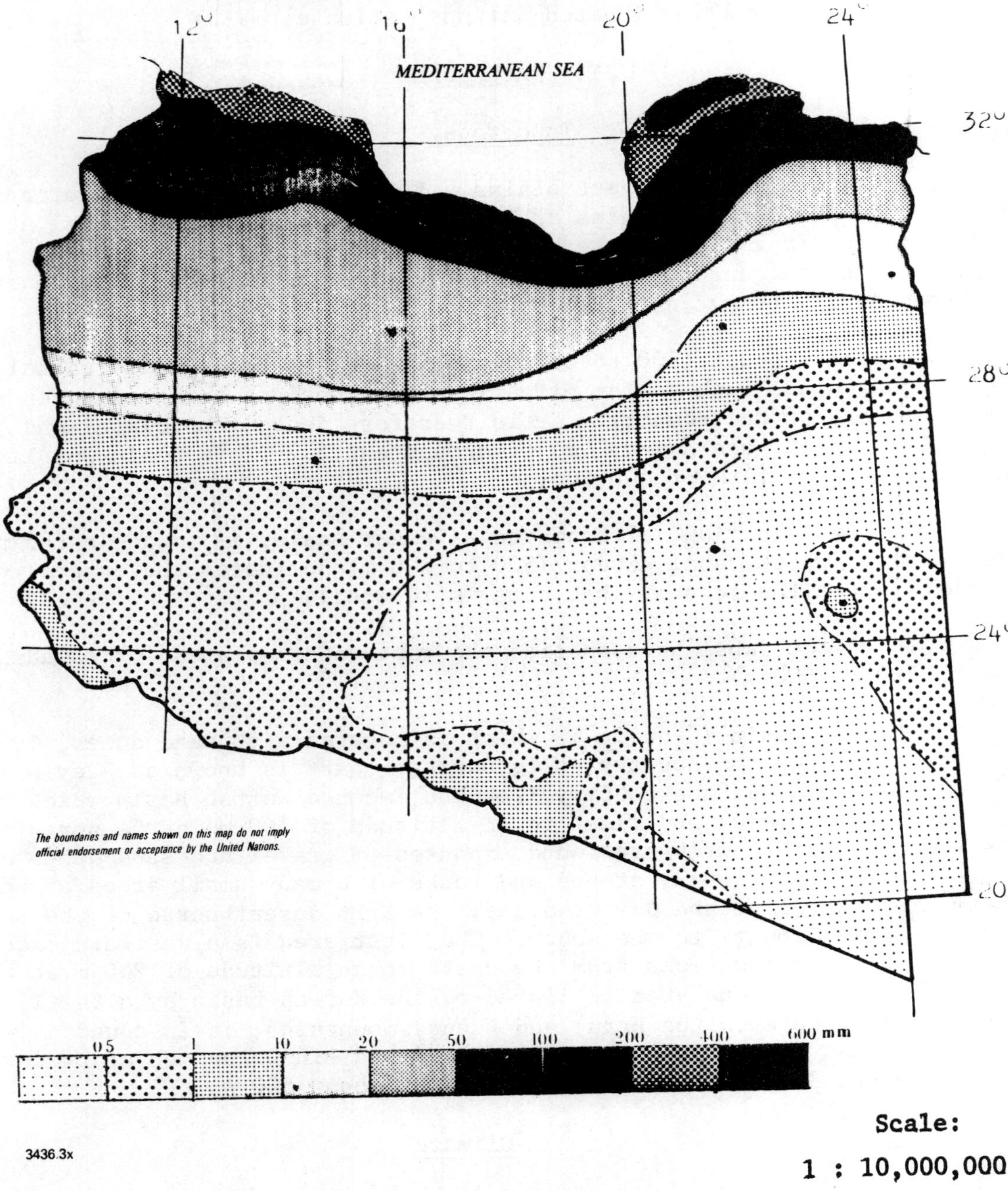

Scale:

1 : 10,000,000

LIBYAN ARAB JAMAHIRIYA
CLIMATE ZONES

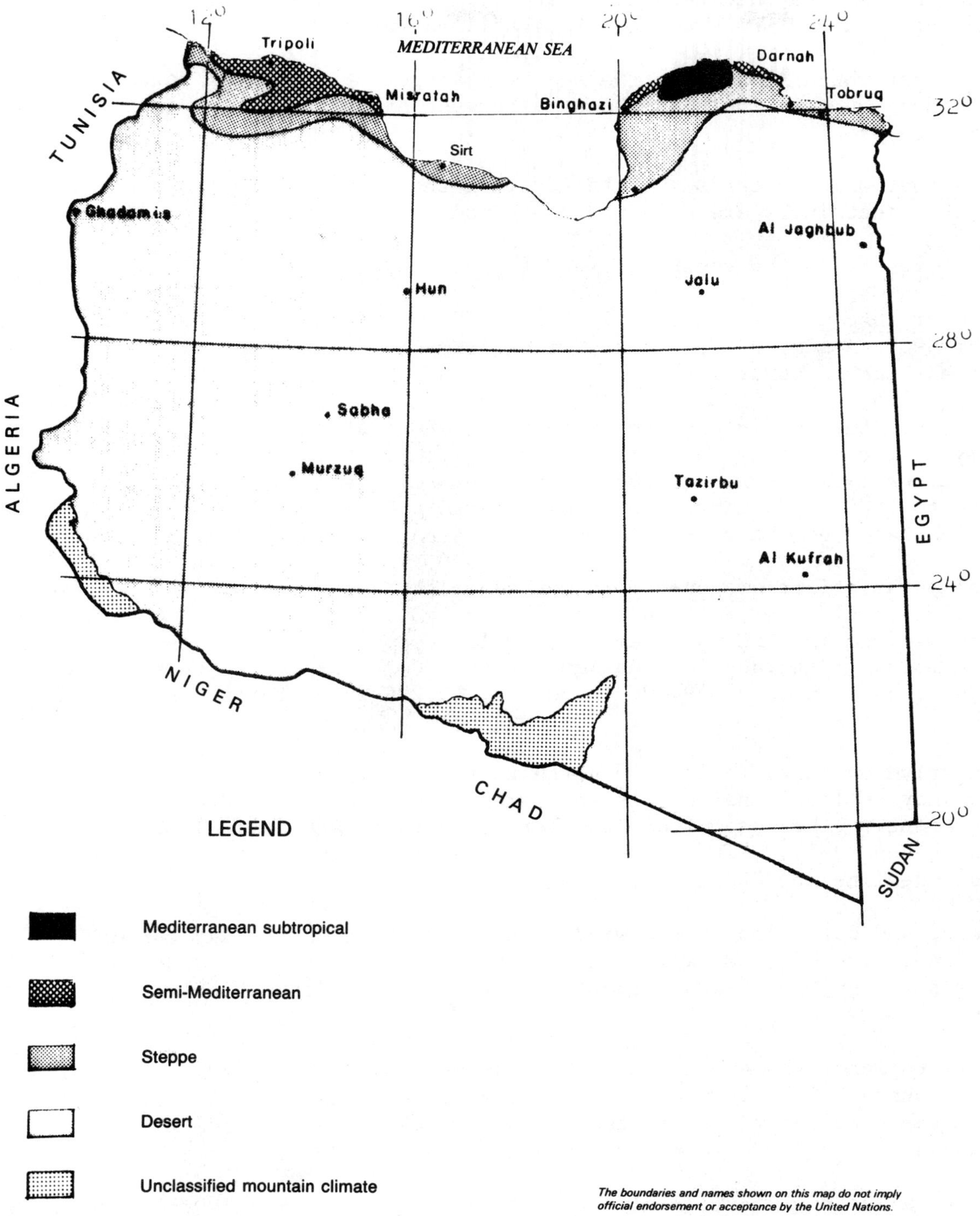

The boundaries and names shown on this map do not imply official endorsement or acceptance by the United Nations.

3436.4x

Scale
1 : 10,000,000

Semi-Mediterranean zone, with a narrow coastal strip in the west which widens in the east to include the whole of the Gefara plain and the eastern part of the Jabal Nafusah and terminates at the Town Of Misratah; it then reappears along a narrow strip between Benghazi and Darnah.

Steppe zone, which lies along the southern slopes of the Jabal Nafusah and Jabal Akhdar and also includes the area between Ajdabiya and the eastern frontier with Egypt.

Desert zone, which extends over the rest of the country with the exception of the mountainous areas in the south and south-west.

Mountain zone, in the south-west (Ghat) and south (Tibesti).

The temperature at the coast is 10°C to 14°C on average in winter, with a minimum of 0°C; in summer the hot dry winds from the Sahara can raise the temperature as high as 40°C.

Further from the sea is the steppe zone the temperature ranges are wider. At Al Aziziyah, for example, the world's highest temperature was recorded: 58°C in September 1922. However, the average for July is 24°C to 28°C, as against 10°C to 14°C in January. Temperatures of -3°C have been recorded in winter. It can snow on the upper slopes of the Jabal Nafusah and Jabal Akhdar.

The daily variations are even more marked in the desert zone. The averages range from 26°C to 41°C for July and from 8°C to 16°C for January. At Ghadamis the absolute minimum is -2°C in November, -5°C in December, -6°C in January and -8°C in February; the absolute maximum is 45.7°C in April, 46.8°C in May, 50.6°C in June, 49°C in July, 49.8°C in August, 45.2°C in September and 45.8°C in October.

The relative humidity falls from north to south. The highest levels occur in winter. They decline considerably in summer. The humidity ranges from 70 to 80 per cent on the coast from December to February and falls to only 50 per cent in June-July. The figures for Sabha range from 40 to 50 per cent in winter and decline to 20 per cent in summer.

The force and direction of the winds often vary from day to day throughout the seasons. The winds blow from the north and east from May to October and from the north and south-west from November to April. They blow at around 5 to 15 knots.

The evaporation rates are high, with a maximum in June-July and a minimum in December-January. The average annual values measured in the Piche evaporometer increase steadily from north to south, with 1,700 to 2,500 mm on the coast and 4,000 to 5,000 mm towards the interior (6,100 mm at Kufrah).

The rain falls in winter, mainly from October to March, and the amount declines towards the south. The maximum annual amounts are found along the Jabal Akhdar (400 to 600 mm); they decline to 200-400 mm towards Misurata in the west and Ajdabiya-Darnah in the east. The 100 mm isoyhet is parallel (150 km) to the coast which it follows for almost the whole length of the Gulf of Syrt. The rainfall is vestigial at best south of the 30th parallel N.

The opposing effects of the Mediterranean and the Sahara cause great irregularity in the rainfall pattern. Brief and violent storms are the rule, with as much as 100 to 140 mm of rain falling in a single day.

Hydrographic system

Libya has little surface water and this source provides only a small part of supplies.

The country has no permanent watercourse. The coastal wadis flow for short periods during the winter months: these flows use beds gouged in the Gefara and Benghazi plains to reach the sea. Some wadis flow from the same mountains (Jabal Nafusah, Jabal Akhdar) but towards the south.

Some wadis are larger in terms of the size of their basins and length of their courses; they rise at the eastern end of the Mamada plateau and reach the coast between Misratah and Sirt. They include Wadis Sawajjin, Bay al Kabir and Zamzam.

The Jabal Nafusah has 30 hydrographic basins with a total area of 7,500 km^2 and an annual flow of 77 million cubic metres. These basins receive 1,700 million cubic metres of rainfall a year. The wadis have a maximum flow of between 10 and 250 m^3/s for short periods (a few hours). The flow coefficients are in the order of 2 to 20 per cent.

Towards the east there are three wadis with basins larger than 1,000 km^2. Several others have basins smaller than 100 km^2. The sediment discharge can be as high as 132 $t/km^2/year$ with a mean annual flow coefficient of three per cent. Several earth dams have been built to store the flood waters.

There is a large number of small springs, but many of them have dried up in recent years on the Jabal Nafusah and Jabal Akhdar. The country's two main springs are the Ain Tawurgha east of Misurata and the Ain Zagyanah east of Banghazi. The Ain Zagyanah is a karstic spring which yields 90 million cubic metres a year; the Ain Tawurgha yields about 60 million cubic metres a year. Studies are being carried out to determine the origin of the water of these two springs and their vulnerability to sea-water intrusion.

II. GEOLOGY

Libya has geological formations ranging in age from Precambrian to Quaternary.

Precambrian. The Precambrian metamorphic basement rock includes quartzites, phyllites, marbles, schists, gneiss and intrusive rocks which outcrop in the south and centre of the country (Tibesti) and in the south-east. There are also small outcrops along the Algerian-Libyan frontier, and in the Bergaf area (Jabal Hasawinah) where the Murzuq (south) and Ghadamis (north) basins meet. Everywhere else in the country the basement rock lies at depths of between 500 and 5,000.

Cambrian-Ordovician. This formation is represented by sandstones of mainly continental origin resting in discontinuity on the basement rock with thicknesses of 500 to 1,500 m. These sandstones are average to coarse grained; they are conglomeratic at the base, usually quartzitic and with intersecting stratification. They outcrop in the south-west and centre of the country in the vicinity of the contact between the Murzuq and Sarir-Tibesti basins and they attain their maximum thickness in the middle of the Ghadamis-Hamadah basin. They are not found north of the Jabal Nafusah. In the east of the country, north of the 28th parallel N, these formations are either absent or found at very great depths.

Silurian. The Silurian formations occupy a vast area of the Jabal Tadrart on the western edge of the Murzuq basin and a narrow strip of land east of the Jabal Bin Ghunaymah on the eastern edge of the Murzuq basin. They also outcrop on the eastern slopes of the Tibesti and north of Kufrah. The Lower Silurian formations are 200 to 300 m thick. They include green schist-sandstones and finely bedded sandstones known as "Tanezzuft schists"; the Upper Silurian formations consist of solid sandstones with intersecting stratification. The series disappears south of the Jabal Nafusah and north of Kufrah.

Devonian. Devonian sandstones and schists overlie the Silurian formation in the western basins of the Murzuq and the Ghadamis. The Upper Devonian formation contains more schist but in the Kufrah basin the Devonian formations are more sandy and less thick. The Devonian formation is not found in the Sarir and Sirt basins.

Carboniferous. This formation is very thick and overlies the Devonian in continuity in the Murzuq and Ghadamis basins. It includes schists and fine sandstones interbedded with sandstone and limestone and it is 1,200 m thick in the middle of the Ghadamis basin. It is more sandy in the Kufrah basin and is not found in the other basins.

In the Murzuq and Ghadamis basins there is a sequence of continental sediments of Paleozoic-Mesozoic age covering in discontinuity the Paleozoic marine sediments. This is the "Post-Tassili" group of fine-grained sandstone; it is of variable thickness up to 600 m. The coarse continental sandstone known in the region as "Nubian sandstone" overlies these formations in discontinuity in the Murzuq and Kufrah basins: it outcrops extensively in the south of the country. It is more than 1,000 m thick in the middle of the two Mesozoic-Cenozoic basins. In the north-west, in the Ghadamis and Gefafa basins, are found sedimentary formations deposited in shallow water ranging in age from Middle Triassic to Paleocene. The Triassic formations consist of Kurrush sandstone, Aziziyah limestone and Abu Shayeah sandstone. These formations increase in size towards the north and can provide important aquifers. They are overlain by Jurassic rocks (Abu Ghaylan gypsum and Tokbal limestone) and Lower Cretaceous rocks (Kiklah sandstone). These formations form the Gefara group. The Nafusah group includes the Upper Cretaceous formations of Sidi al Sid Nalut, Tigrinna and Mizdah, which consist of limestones, dolomites, marls and a little gypsum. The Paleocene deposits outcrop in the Ghadamis basin; they include marls, limestones, sedimentary schists, with gypsum in places, chalk, dolomites and clays. In the Sirt and Sarir basins thick layers (3,000 m) of Tertiary rock (Paleocene-Miocene) overlie the Cretaceous formation. The Lower Middle Eocene formation alone can be up to 1,700 m thick in the Sirt basin and it includes dolomites, anhydrite, argillaceous limestone and, in places, chalky limestone and sandstone.

Basalts. Volcanic rocks are found extensively south of Ghargan and Jabals Sawda and Haruj (over an area of 30,000 km^2) and Jabal Nugayy (Tibesti). They are mostly olivine basalts and phonolites of Oligocene and Early Quarternary age.

Quaternary. The Quaternary deposits include immense expanses of sand dunes in the Murzuq, Kufrah and Sarir basins, as well as loess and alluviums in other regions, and coastal mountains consisting of eolian calcarinites and sebkha formations.

III. GROUND WATER

Administrative organization and research

The Department of Water and Soil (DWS) is the sole government body responsible for research, development and management of ground-water resources. It reports to the Secretariat of Agricultural Reclamation and Land Development (SARLD) and has four technical sections:

i) Water resources: supervision and technical support of drilling operations, hydrogeology, hydrology and geophysics;

ii) Dams: study of sites, construction, management and maintenance;

iii) Soils: soil studies, including tests and cartography;

iv) Irrigation and drainage: this section was set up only recently.

Water supplies for urban areas are the responsibility of the Secretariat of Utilities. The Secretariat of State for Electricity is responsible for the desalination plants which supply several coastal towns.

Before 1972 the Ministry of Agriculture was responsible for water-resources research and exploitation for both domestic and agricultural supplies. The General Water Authority (GWA) was set up in 1972 in response to the population growth and increased demand for water; it has become the sole body responsible for water for the whole country, with regional offices at Benghazi, Sabha and elsewhere. The country has been divided into five water regions, each serviced by a team of geologists and hydrogeologists. A drilling section and a geophysical section have also been established. The drilling section is responsible for the preparation of the technical specifications for purchases of drilling equipment and the supervision of the works. The geophysical section is responsible for geophysical prospecting and the diagraphic testing of wells. The hydrogeological section carries out studies and prospecting operations either with its own resources or by subcontracting to study companies or international bodies. It has a specialist group responsible for the collection and interpretation of data, particularly with respect to fluctuations in static water levels and water quality, as well as a hydrogeochemical laboratory.

In co-operation with the United Nations Development Programme, a large group of water-resources and hydraulics experts was invited to initiate and run study and works programmes for their Libyan counterparts, most of whom are young graduates without field experience. These experts also act as technical advisers.

Several institutional changes have since been made. For example, the GWA acquired the status of ministry in 1977 and became the Ministry of Dams and Water Resources; the Department of Water and Soil was established later. Dams and soil have thus been brought under the authority of a service which previously was concerned only with water.

Many hydrogeological studies were made before 1977 by the Agricultural Development Council (ADC), a body established to develop the land and launch agricultural projects, for example the study on the western part of the country carried out by the French Study Group in Libya (GEFLI). Petroleum research has also provided abundant data on ground water.

With a view to exercising effective control over the drawoff of ground water, the Department has prepared and is obtaining approval for strict legislation on water use and issue of drilling licenses. This legislation is also designed to limit the risks of ground-water pollution and it prohibits any increase in the extraction of water in certain sectors, along the coast in particular, in order to prevent contamination by sea water.

Modern techniques are employed in the study and evaluation of ground-water resources. Advanced technologies must be used, for water boreholes in Libya penetrate to 1,000 m and even deeper. Boreholes have had to be drilled to 2,000 m and deeper in order to study certain aquifers. Complete diagraphic studies are made of the majority of the deep boreholes, together with test pumping, chemical analyses and paleontological dating, in order to determine the properties of the aquifers and make the necessary correlations with the wells in the vicinity. Isotopic analyses are also made of the recharge, flow and age of the ground water and of losses through evaporation. Good results have been achieved recently by studies on the intrusion of sea water in the coastal zones and on evaporation from sebkhas. Mathematical models have also been used to estimate the water resources and simulate the future drawoff and its quantitative and qualitative impact.

The DWS has organized several kinds of training programme in close co-operation with FAO; training in universities abroad (United States, United Kingdom and elsewhere); training on the site or in study offices in Libya itself; and short courses in Libya or abroad. A water resources institute has also been set up in Tripoli to train senior technicians.

Hydraulic regions

Jones (1971) divided the country into 19 hydrogeological regions, mainly on the basis of geological criteria; each region has several aquifer systems which have various and complex interrelationships. Pallas (1978) introduced a more detailed classification based on hydrogeological factors and for practical reasons he reduced the number of regions to five:

- Murzuq basin;

- Jabal Nafusah, Sawfajjin basin, western Sirt and Hamdah;

- Gefara plain;

- Sarir and Kufrah basins;

- Jabal Akhdar system.

Murzuq basin - area: 300,000 km^2

1. Lower Paleozoic aquifer: Cambrian-Ordovician, Silurian and Devonian (the Lower Silurian formation known at "Tanezzuft schist" is not water-bearing). This aquifer extends towards Algeria, Niger and Chad. The Cambrian-Ordovician sediments outcrop extensively in the Jabal Fezzan. The aquifer is recharging in the whole of the centre of the basin. It consists mainly of sandstone of varied texture and composition interbedded with schists, particularly in the upper part.

The water-bearing strata in this system consist of the following formations:

- Hazawinah, Hauaz, Melezz and Memouniat (Cambrian-Ordovician, thickness 200-1,400 m);

- Tadrart, Ouan Kasa, Uwaynat, Wannin (Devonian, thickness 40-600 m).

The piezometry declines from south-west (700 m) to north-east (250 m).

Owing to the lack of rainfall (7 to 21 mm/year), the recharge is practically nil; this has been demonstrated by isotopic analysis, which has shown that no tritium is present. Carbon-14 dating indicates an age of 6,000 to 14,000 years.

The natural depletion of the aquifer is probably in the order of 0.1 mm/year; this occurs through evaporation in the sebkhas and the exploitation of wells along Wadi Ash Shati where the slope of the aquifer is 1.3 per cent as against 0.2 to 0.3 per cent in the rest of the basin. The transmissivity values are around 10^{-5} for the artesian aquifer.

The quality of the water in the Paleozoic aquifer is generally good to very good. The dry residue is usually below 1 g/1 and sometimes below 150 mg/1. However, the water quality deteriorates towards the north-west.

The Cambrian-Ordovician and Devonian aquifers have similar properties; this seems to indicate that they are interconnected (the similarity even extends to the isotopic composition). The water is mostly of the sodium chloride type, very corrosive and fairly warm (35°C to 40°C.

2. Carboniferous aquifer: this consists of a thick impermeable stratum of argillaceous schist between the previous aquifer and the aquifer lying above it, which is described below.

3. Upper "Nubian" aquifer: this aquifer is situated in the basin's centre and is covered with Quaternary eolian deposits in the form of sand dunes and a "sand sea".

It consists of coarse and thick sandy conglomerates intersected with clay strata. The thickness declines from 1,000 m in the middle to 200-300 m at the edges.

The aquifer is apparently not recharging, owing to the lack of rainfall and the heavy evaporation. Carbon-14 dating indicates that the water is about 21,000 years old. The ground water flows towards the north-east. The transmissivity values ranges from 1.3×10^{-2} to 2.8×10^{-3} m^2/s. The storage coefficient is in the order of 2×10^{-4}. The porosity of the Nubian sandstone is 20 to 40 per cent. The water is more saline in the shallow wells owing to evaporation and the recycling of irrigation water. The facies is of the sodium chloride type with a dry residue of 160 to 180 mg/1 for the deep wells and 1 to 4 g/1 for the shallow wells.

Jabal Nafusah, basins of Sawfajjin, western Sirt and Hamadah

This is a group of hydraulically interconnected sub-basins located in a zone which is very arid, except for a strip in the north where the rainfall is 200 to 300 mm/year. The following are the main aquifers:

1. The Cambrian-Ordovician aquifer is an extension of the lower aquifer of the Murzuq basin; it is located in the central part of Hamadah at great depths and cannot therefore be exploited; it is exploited in the south, where it is not recharging. It plays an important role in the recharge of the Lower Cretaceous sandstone aquifer and the Upper Cretaceous limestone aquifer. The transmissivity ranges from 10^{-2} to 10^{-3} m^2/s. The storage coefficient is 10^{-1} in the unconfined zone and 10^{-3} to 10^{-5} in the recharge zone. The water quality is exceptionally good in the south but it deteriorates towards the north in step with the decline in the hydraulic gradient.

2. The Lower Cretaceous aquifer or Kiklah aquifer is by far the most important reservoir in the north-west of the country. It consists of fine to coarse sandstones interbedded with silt, schist and clay, and it is recharged from the belt of Lower Cretaceous outcrops which runs along the western edge of the Jabal Nafusah and by the Cambrian-Ordovician aquifer. It is also thought that part of the recharge may come from the Algerian Atlas.

The aquifer is exploited by boreholes at depths of 700 to 1,000 m which are either artesian of sub-artesian springs. The water quality of the Kiklah aquifer is generally good and its dry residue does not usually exceed 1 to 2 g/1.

3. The Upper Cretaceous aquifer or Mizdah aquifer is well developed south of the Jabal Nafusah, central and eastern Hamadah and the Misratah area. It consists of limestone interbedded with marl and is 100 to 300 m thick. In the Al Jufrah area it is exploited for the irrigation of two farming areas by means of artesian boreholes. In this area the Mizdah aquifer is in contact with the Paleozoic aquifer. The ground water flows towards the north-east. The transmissivity is 3.6 to 10^{-3} m^2/s and the storage coefficient is around 10^{-3}: the aquifer's capacity declines towards the west. It is recharged by direct infiltration of rainwater and surface runoff on the southern slopes of Jabal Nafusah, the northern slopes of Jabal Gargaf, Jabal Sawda and Jabal Haraj, and by underground contact with the Paleozoic and Lower Cretaceous aquifers. The water quality of the Mizdah aquifer varies greatly throughout the

area. At Jufrah the dry residue is fairly low at 1.3 g/1, but it increases towards the west, reaching 4.4 g/1 in the Hamadah. The water has a low pH value; it is corrosive and warm. The upper Cenomanian Nalut aquifer in the north-east also deserves mention. However, this aquifer has poor properties and it is very saline in the north and north-west.

4. The Cenozoic aquifers (Eocene, Oligocene, Miocene) include thick deposits of limestone and dolomites separated by thick layers of clay, marl and argillaceous schist in the western Sirt basin in the north, and by layers of argillaceous marl and schist in the western Sirt basin north and east of the Hun graben. The dry residue is usually greater than 5 g/1, except in a few places where perched aquifers are found.

In the coastal zone between Misratah and Khums there is a Miocene-Pliocene Quaternary aquifer of somewhat better quality which is used for small-scale irrigation and drinking water. It is recharged by rainfall and by lateral contact with the Nalut aquifer. Excessive pumping has caused a deterioration in the water quality through sea-water intrusion.

This multi-layered aquifer discharges through the Tawargha and Ka'am springs, which have yields of 63 and 11 million cubic metres a year respectively, through evaporation from the big coastal sebkhas of Tawargha and Mishah, and through a large number of boreholes.

The Gefara plain - area: 15,000 km^2

This triangular plain occupies the north-west of the country between the Jabal Nafusah in the south and the Mediterranean in the north. It is of great economic importance, for it contains 40 per cent of the country's population and many of the urban areas, including the capital Tripoli. Much of the plain is irrigated and it supplies more than half of the country's farm production. The main aquifers are:

1. The Quaternary aquifer which constitutes the "phreatic" aquifer and also includes formations of Upper Miocene, Pliocene and Quaternary age in central and northern Gefara; these formations are 30 to 150 m thick; they are hydraulically interconnected and behave as a single unit. The aquifer's saturated layer is 10 to 90 m thick. Close to the coast the aquifer's transmissivity is in the order of 2×10^{-3} to 10^{-2} m^2/s and in the interior 2×10^{-2} to 10^{-1} m^2/s; the storage coefficient is 6×10^{-2} to 10^{-1}. The wells penetrating the Quaternary aquifer yield 20 to 60 m^3/h.

The static water level is declining steadily at 5 m/year in the areas of intensive irrigation. Sea water is therefore intruding along a coastal strip two to five kilometres deep between Sabratah and Tajura, 20 kilometres east of Tripoli.

In the central part the water has a dry residue of 1-2 g/1; this concentration increases towards the west owing to the influence of the gypsum and limestone intercalations at the base of the aquifer which yield heavily sulphated water. Concentrations of nitrates in excess of 45 mg/1 are common, and vertifical drainage from the underlying Middle Miocene aquifer and the presence of sebkhas have the effect of increasing the salinity to over 5 g/1 between Sabratah and the Tunisian frontier.

The Quaternary aquifer in the western Gefara plain is fairly small, with a saturated layer 10 to 30 m thick. It is also confined to a narrow strip running towards the west and it has good-quality water with a dry residue below 1 g/1.

The aquifer is directly recharged by rainwater and surface runoff from the wadis. The lateral contact in the south and the vertical drainage from the Middle and Lower Miocene formations are also thought to contribute to the recharge.

2. The Miocene aquifer includes the Middle and Lower Miocene formations of central and northern Gefara.

- Middle Miocene: the aquifer lies between two clay strata. Its top is at a depth of 10 to 120 m and it is 125 to 200 m thick. The average transmissivity is 5×10^{-2} m^2/s and the storage coefficient is 1×10^{-3}. The aquifer is recharged by lateral contact to the south. The water quality is poor, with dry residues of 3 to 4 g/1.

- Lower Miocene: the aquifer is formed by sandy or dolomitic limestone in western and central Gefara. It lies at a depth of 250 to 390 m to 485 m west of Sabratah, where it is 80 m thick and has a transmissivity of 5×10^{-4} m^2/s. The dry residue is 2 to 4 g/1. Towards the west the salinity exceeds 6 g/1 and the aquifer is not exploited despite its artesian properties. The aquifer is recharged laterally from the Triassic limestones to the south.

In eastern and central Gefara, where it seems to be in hydraulic contact with the underlying Upper Triassic aquifer of Abu Shayba, the aquifer consists of sandy limestone and conglomeratic sandstone. It lies at a depth of 150 to 200 m. The transmissivity is around 5×10^{-3} m^2/s and the dry residue ranges from 2.5 to 4.5 g/1, with a high sulphate content. Further east the aquifer declines considerably in thickness and the water quality improves.

3. The Triassic aquifers, of which there are two:

- the sandstone aquifer of Abu Shayba of Upper Triassic age has a maximum thickness of 350 m and underlies the eastern and central Gefara plain. It has a transmissivity of 5×10^{-3} m^2/s and a storage coefficient of 10^{-2}. The water quality is generally good, with a dry residue of less than 2 g/1, a high iron content and increased concentrations of salt and sulphates close to the coast. The aquifer becomes unconfined in the south close to the piedmont of Jabal Nafusah.

- The Aziziyah aquifer (Middle and Upper Triassic) is formed of dolomitic limestone and is exploited in the southern part of central Gefara where it is unconfined and has a transmissivity of 2×10^{-2} to 5×10^{-2} m^2/s and a storage coefficient of 5×10^{-2}. The secondary porosity is lower towards the south and the clay intercalations larger; the transmissivity is therefore lower here. The average yield of the wells is 70-110 m^3/h for the unconfined part and 30-50 m^3/h for the artesian part. The dry residue ranges from 2 g/1 in the south to 4 g/1 in the north. North of the Aziziyah fault the aquifer slopes down towards the north through a series of ladder faults and reaches a depth of 900 m near Tripoli. In eastern Gefara, it lies at a depth of 200 to 500 m and has a transmissivity of 1 to 5×10^{-3} m^2/s and a dry residue of

1.7 to 2.5 g/l. Other Lower and Middle Triassic sandstone aquifers are found in the Gefara plain. However, they are of no economic value owing to their great depth and heavy salinity.

Kufrah and Sarir basins

These basins cover a large part of eastern Libya and extend towards Chad and Egypt. They are limited in the north by the 30th parallel N and in the east by the 18th meridian E. This region is divided into two sub-basins running along the 25th parallel N:

- Hydrogeology of the Kufrah basin. The basin's hydrogeology was first studied in detail in a small zone around Kufrah. Lithological data collected during scattered oil-drilling operations are available for the rest of the basin.

A Cambrian-Lower Cretaceous aquifer with a maximum thickness of 3,000 m at its centre occupies the whole of the basin. It consists of continental sandstones interbedded with clay and silt. At present only the continental Mesozoic aquifer of Nubian sandstone (Triassic-Lower Cretaceous) is exploited, supplying water to two irrigated areas and to the inhabitants of Kufrah. The aquifer is probably saturated for a depth of 1,000 m, but the wells rarely exceed 300 to 400 m. The static water level in the Nubian sandstone is only a few metres below ground level.

The ground water flows north and north-east and the static water level falls by 400 m for an unknown reason. The boreholes have fairly high discharges (135 to 300 m^3/h) and high specific yields (10 to 50 $m^3/i/m$). The transmissivity ranges from 300 to 350 m^2/day and the storage coefficient from 1.1×10^{-4} to 1.5×10^{-2}. The regional drawdown due to pumping is closely monitored by means of a network of observation wells located close to or within the pumping zones. A maximum drop of 25 m was observed in the central part of these zones up to April 1985; the drop declines rapidly away from the centre and is only 20 to 25 cm at a distance of 30 km. It is due mainly to the drawoff of ground water: 90 to 120 million cubic metres a year. Pumping began in 1968 but the drawoff did not reach the above amounts until 1975. The phreatic aquifer, which is up to 60 m in depth, is sometimes considered as a separate unit owing to its hydrochemical properties. It has long been exploited by local farmers. In fact, it behaves in the same manner as the underlying aquifers and its water level has declined by nine metres (up to April 1985). The water is of excellent quality. For 137 deep wells the mineral content does not exceed 180 to 300 mg/l. It is usually higher in the phreatic aquifer, with a maximum of 8 g/l depending on the location and depth of the well. Wells located close to sebkhas or lagoons are obviously more likely to be saline. The water of the deep aquifer has a low pH value and a high CO_2 content; it is therefore corrosive; the free CO_2 content is 34 to 57 mg/l near Kufrah, thus threatening the wells and their superstructures and increasing the construction and maintenance costs.

- Hydrogeology of the Sarir (or Sirt) basin. This basin is situated north and west of the Kufrah basin. No research has yet been carried out in the Sarir Tibesti basin. The following remarks apply only to the northern part (Sarir Calancio) which has been studied since the end of the 1960s with a view to establishing two tapping fields for irrigation purposes. At present (1985) additional tapping fields are being developed to supply water to the coastal zone. This basin is younger than the Kufrah basin (Paleocene-Miocene), but the aquifers are limited to the post-Eocene formations; two of them contain fresh water:

i) The Post-Middle Miocene aquifer (PMM), which can be up to 200 m thick, consists of medium to coarse grained sands merging into calcareous sandstone with fine intercalations of clay.

ii) The Lower and Middle Miocene aquifer (LMM) in the Maradah formation, which can be 150-880 m thick. In the north it consists of marine limestones and clays with evaporites, and in the centre of alternate formations of clay, limestone, sand and coastal sandstone.

Fluvial sands and thin strata of sandstone and clay predominate in the south and south-east.

The Oligocene aquifer is 240-730 m thick and consists of coarse sands and sandstone interbedded with clay (non-marine facies) in the south and south-west.

Further north, calcareous sandstone, limestone, dolomitic rocks and clays with some evaporites (marine facies) make up the bulk of the series. This aquifer is normally in hydraulic continuity with the LMM in the exploitation zone in the south and south-west.

The Post-Eocene reservoir is therefore a multi-layered system. The vertical gradient between the PMM and the LMM is less marked in the south but it increases towards the north owing to the larger amounts of schists and carbonates. In both the aquifers the ground water flows towards the north or north-east. The piezometric maps indicate a drop of 200 m from south to north over a distance of 500 km. The hydraulic gradient is therefore in the order of 4×10^{-4}. The discharge of the Post-Eocene reservoir is effected through sebkhas and pumping. The recharge comes from underground migration from the Kufrah basin (Nubian sandstone and Cambrian-Ordovician aquifer). The hydrogeological properties have been determined from the results of several test pumpings. There are large variations in the transmissivity coefficients although most of the wells are located within a relatively narrow strip of territory running north to south. In the south, wells dug to a depth of 300 m and lined between 150 and 300 m draw water from the LMM; they indicate current transmissivity values of about 1,300 m^3/ day, but higher and lower values are also found. The storage coefficient (artesian aquifer) is 5×10^{-4}. In the north, the wells drilled in the LMM give transmissivity values of 750 to 1,000 m^2/day with a maximum of 10,000 m^2/ day for wells located in a band one to five kilometres wide and running south to north for a distance of 50 km, which in some places is considered to be a buried fossil bed.

All the boreholes in this part of the Sarir basin penetrate only part of the way into the aquifer. The test holes designed to reach relatively greater depths for the big Sarir water development project penetrate the Oligocene aquifer, crossing the saturation zone over a depth of 400 to 500 m. These holes indicate transmissivity values of 1,100 to 6,300 m^2/day. Table 1 lists the hydrogeological properties of the fresh-water aquifers in the Sarir and Tazirbu regions.

The PMM aquifer in the centre and north of the basin is unconfined in the south but becomes artesian in the north as the silt and clay intercalations grow thicker. Artesian conditions are found in the extreme north, with water levels of two to five metres above ground level at Maradah (25°N and 20°E). Elsewhere the depth to water in the Post-Eocene formations ranges from a few metres to more than 80 metres depending on the altitude of the ground level. Pumping in the southern part of the centre of the basin causes gravity drainage from the PMM to the LMM aquifer, but except in the buried fossil bed this phenomenon is less marked towards the north owing to the changes of facies. Vertical drainage may occur from one aquifer to the other, in particular in the central area where the LMM/Oligocene aquifer has a head of water 15 m higher than in the shallow LMM and PMM aquifers; the well hydrograms indicate hydraulic continuity.

The water quality in the Post-Eocene aquifer varies greatly both vertically and horizontally, depending on the lithological and topographical changes and the mixing of water and on the drainage between the main aquifers. As a rule the salinity increases from south to north in the direction of the underground flows. North of the 28th parallel N there is a sharp change in the water quality, with a high concentration of gypsum and anhydrite in the formations which increases the dry residue to 5 g/1. The topographical depressions and sebkhas in the north produce high concentrations of salts, particularly in the shallow aquifers. South of the 28th parallel N the water is usually sweet with a salinity value of less than 2 g/1 in the productive zone, i.e. the zone lying between 150 and 300 m deep. But "pockets" of very saline water exist in the more southerly areas, and dry residues of 5 to 8 g/1 have been found in the phreatic aquifer (at 3C to 50 m below ground level).

The Jabal Akhdar system

The Jabal Akhdar is located in the north-east of the country, north of the Sarir basin. It is an east-west anticline and its formations date from the Upper Cretaceous and Tertiary eras. The main aquifers are found in the fractured limestone rocks of the Eocene and Miocene eras which are heterogeneous, anisotropic, karstified and hydraulically interconnected at the regional level. The water level falls by more than 400 m over a fairly short distance north and south of the Jabal's axis. The crest of the range is a watershed for both surface and ground water. The flow is north and south in the central part and becomes radial with an easier gradient in the east and west. The aquifers are recharged exclusively by direct infiltration of rainwater and surface runoff.

The Miocene aquifer is unconfined. It consists of limestones and dolomites covered by Oolithic limestones and calcarenites. East of Benghazi the transmissivity ranges from 10^{-2} to 10^{-4} m^2/s with higher values in the karstified areas.

Table 1

Hydrological properties of fresh-water aquifers in the Sarir and Tazirbu region
Exploration and exploitation data (El Ramby 1980, Jones 1985)

Region or project	Aquifer and lithology	Hydrological properties: Sandstone depth (m)	Depth to water (m)	Transmissivity (T) (m^2/d)	Storage (S)	Hydrochemical properties: Dry residue (ppm)	Chlorides (ppm)	Sulphates (ppm)	Production: No. of wells	Discharge (m^3/h)	Since project start-up or anticipated annual drawoff
SARIR											
Sarir projects 1. Sarir-South	Post-Eocene (Oligocene, Middle Miocene)	200-300	35-74	1,240	$5x10^{-4}$	513-2,496	92-525	29-34	168	273.6	950 hm^3 since 1975
2. Sarir-North	sand, clay, sandy limestone, dolomites; fluvial, deltaic, estuarial and shallow marine lagoonal formations	200-300	26-51	250-9,000	$5x10^{-4}$	720-5,204	188-2,989	207-647	83	273.6	188 hm^3 since 1980
3. Sarir water project		400-500	64-83	1,090÷6,300	$5x10^{-4}$ $3x10^{-2}$	1,350-1,570	319-496	254-379	150	277	360 hm^3/year
4. Jazlu/ Awjilah	Post-Middle Miocene, sand, clay, argillaceous sand	80-110	33-82	1,000-2000	$0.1x10^{-3}$ $20x10^{-3}$	1,378-1,806	325-375	440-650	33	87.8	8.3 hm^3/year in 1985 25 hm^3/year thereafter
5. Maradah (farm project)	Lower Miocene, limestone, sandstone, argillaceous schists	150	artesian (2-5m ags)	550-2,080	$6x10^{-3}$ $4.4x10^{-1}$	2,840-3,330	930-1,020	770-880	12	90-144	3,153 m^3/year
TAZIRBU											
1. Tazirbu project	Tertiary, sandstone interbedded with clay and limestone (in upper part) followed by interbedded clay and sand of Nubian sandstone series	500	7-22	180-450	$3x10^{-4}$	210-510	21-255	8-141	120	345	360 hm^3/year envisaged

The dry residue is in the order of 1.3 to 2.5 g/1 and it is tending to increase as a result of the more intensive pumping in the Benghazi plain which has led to sea-water intrusion along the coast.

Elsewhere the Eocene aquifer consists of chalky and marly limestone. It is unconfined in the area of the Jabal and partly artesian or semi-artesian elsewhere. Generally speaking, the aquifer has low transmissivity values (10^{-3} to 10^{-5} m^2/s), except in the heavily fractured zones.

The water of the Eocene aquifer is of good quality with a dry residue of between 0.6 and 1.2 g/1. Higher concentrations are found in the Benghazi plain. South of the Jabal Akhdar the water quality deteriorates quickly and the dry residue increases to 5-10 g/1. Salt contents in excess of 10 g/1 are not uncommon in the vicinity of the coastal and interior sebkhas which act as natural outlets for the ground water.

IV. EXPLOITATION OF GROUND WATER

Although the intensive research into new water resources for the ambitious development plans is continuing, the authorities are launching new projects for exploitation of ground water. New tapping areas are being brought into service close to the main towns and land recently brought under cultivation. The water is either drawn off close to the utilization points or channelled to the areas where it is needed. Vast projects are being implemented to bring ground water from the interior of the country to the coastal areas. This is the result of several years of study on the part of the Department of Water and Soil working in co-operation with several bilateral and international co-operation bodies.

Drilling activities

A water-drilling section was set up in 1972 in the General Water Authority to prepare water-drilling programmes and supervise the works. Its staff includes experienced drilling engineers and drilling supervisors. The main unit is based at Tripoli with local offices at Benghazi and Sabha. An FAO team of drilling experts consisting of an engineer and three supervisors has been giving technical support to the section for about 10 years.

A three-year degree course for technicians covering several subjects in the area of water-resources exploitation, including drilling techniques, was started towards the end of the 1970s. The graduates are taken on by the DWS, drilling companies, municipalities, regional farm services and other State bodies responsible for water exploitation.

In addition to their training in the field and in study offices, six of the drilling supervisors have received intensive training in Bulgaria. Other technicians are assigned to the tasks of monitoring the aquifers, supervising test pumping, collecting water samples for chemical analysis, and to other activities concerned with dams and surface water. Several national and foreign drilling companies have set up in Libya; most of them have deep drilling equipment. Table 6 gives an idea of the scope of the drilling activities in the basins of Gefara, Jabal Nafusah, Sawfajjin and Sirt under the direct control of the Tripoli office. Drilling operations in other areas are supervised by the Benghazi and Sabha offices. The table gives details of the drilling works carried out from 1980 to 1984 by three companies; the other 20 companies are grouped together. Several thousand wells have been drilled by small private enterprises for domestic and irrigation purposes. No data is available for the majority of them.

However, the central and branch offices of the drilling section of the DWS do have data on the drilling operations which they have supervised. Subcontractors must furnish final reports containing all the necessary information.

The rotary rigs used in Libya have a lifting capacity of 20 to 180 tons, so that they are able to drill to depths of up to 2,000 m. The deepest hole so far is 2,200 m. All drilling methods are used: mud, clean water, air-lift with normal or inverse circulation, and foam.

Some public bodies and small firms still use small cable-tool rigs, mainly in the karstic formations in the east, in shallow wells close to Tripoli and in Wadi Ash Shati.

Deep drilling techniques similar to those used in oil wells are employed in the drilling and furnishing of the wells. Wells over 800 m deep are equipped with telescopic tubing. Artesian wells are fitted with a specially designed cap to control the wellhead discharge and combat corrosion. As a rule diagraphy is used to determine the calibre, the spontaneous potential, the resistivity, and the neutron-ray and gamma-ray logs. On completion the borehole is fitted out and undergoes test pumping to determine the hydrogeological properties of the aquifer. The final test lasts 72 hours and includes pumping to increasing depths and measurement of the recovery. Water samples are taken for analysis and a final report, including all the hydrogeological, lithological and drilling data, is drafted and submitted to the inventories section.

A network of observation wells has been established in all areas of ground-water exploitation in order to monitor the fluctuations in the static water level either annually or by means of automatic recorders. The data are collected periodically and used in the preparation of hydrograms and piezometric maps.

Use

Almost all Libya's water resources come from ground-water aquifers. Agriculture consumes most water (80 per cent); next come the urban and rural populations, followed by industry. Several attempts have been made over recent decades to quantify the consumption of ground water by all users by means of direct and indirect techniques employed either separately or in combination: field surveys to establish well inventories, water meters, calculation of the water drawn off on the basis of the energy consumed (for pumping), extent and nature of irrigated areas, field surveys or interpretation of aerial photographs and satellite images, etc.

The per capita domestic consumption of water is around 250 to 300 litres a day in the towns and 100 to 150 litres a day in rural areas. The present drawoff is in the order of 278 million cubic metres.

The per hectar consumption of irrigation water is 5,000 to 15,000 cubic metres a year and it can be as high as 20,000 cubic metres a year in the southern regions. In the Murzuq basin, for example, the new agricultural projects launched in 1974-1976 entailed water consumption which reached 230 million cubic metres in 1983-1984 (see table 2 below). The figures given below do not include the water drawn off for small private farms and for the requirements of towns and industries.

Table 2

Drawoff of ground water for irrigation projects in the Murzuq basin

(hm^3)

(P.Illy, 1984)

Zone	Initial estimate	1979/80	1983/84	Forecast
Murzuq	115.55	76.40	83.50	222.30
Awbari	132.42	36.10	63.17	135.40
Sabha	17.40	14.20	19.20	20.00
Birak	145.81	41.20	68.30	151.20
Total	451.18	167.90	234.17	528.90

Almost everywhere the biggest consumption is by private farmers and it is continuing to grow at a very strong rate, as can be seen from the following table.

Table 3

Drawoff of ground water in the Sabha zone

(hm^3)

(P. Illy, 1985)

Use	1972	1977	1985
Agriculture (private)	25	55	115
Sabha farm project			
Livestock project	-	3	8
Urban and industrial supply	5	12	27
Total	30	70	150

According to Idrotécnico, in 1978 the use of water (hm^3) in the Wadi Ash Shati and Jufrah zones was as follows:

	Wadi Ash Shati	Jufrah	Total
Domestic	2.5	1.8	4.3
Private agriculture	90.0	2.9	92.9
Farm projects	40.5	26.1	66.6
Total	133.0	30.8	163.8

In a recent study which it made for the DWS on the various possibilities of supplying water to the Gefara plain FAO cited the data used in estimates of the amounts of water consumed in the plain during several periods from 1948 to 1978.

Table 4

Use of water in the Gefara plain from 1948 to 1978

(hm^3/year)

Year	Pumping for agriculture	Pumping for urban consumption	Total pumping
1948	12	12	24
1953	16	23	39
1958	80	25	105
1973	343	75	418
1978	463	95	558

Table 5 gives an estimate of the drawoff of ground water based on earlier studies, on an extrapolation of population growth, on farm and industry forecasts, and on plans for new tapping areas to supply water to the coastal zone from the Kufrah-Sarir and Murzuq basins.

Table 5

Ground-water drawoff in Libya */

Hydraulic region	1985 drawoff (hm^3/year)		Forecast drawoff in 2000 (hm^3/year)		Comments
Murzuq	465	45	1,150	95	Including water delivery project
Jabal Nafusah, Hamadah, Saw-fajjin, W. Sirt	210	60	270	130	
Gefara	600	120	700	220	
Kufrah & Sarir	370	40	1,280	75	Including water injected for oil extraction
Jabal Akhdar	100	105	190	210	Including water delivery project
Total	1,745	370	3,590	730	
	2,115		4,320		

*/ Modified, according to WSD, FAO, Idrotécnico, Pallas, Illy and others, with extrapolations based on the projects planned, the water delivery plans, and the population growth.

Table 6

Drilling operations in North Libya (western & central) 1980-1984 */

Enterprise	No. of rigs	1980 wells	1980 metrage	1981 wells	1981 metrage	1982 wells	1982 metrage	1983 wells	1983 metrage	1984 wells	1984 metrage
Local firms (Libya)	20	4	1,439	13	2,672	10	2,325	4	1,827	5	2,640
Bulgarocomin (Bulgaria)	14	2	943	14	2,672	18	8,916	16	11,045	21	15,987
Stojexport (Czechoslo.)	5	9	4,994	20	10,222	20	9,805	13	5,771	13	5,305
Others		15	4,290	105	17,616	38	14,378	28	6,533	7	1,854
Total **/		30	11,668	152	33,182	86	15,444	61	25,176	46	25,786
Average well depth		389		218.3		412		412.7		560.6	

*/ Only sites supervised by DWS

**/ These figures represent about 50 per cent of the operations in the whole country.

V. PROBLEMS

It is clear from the foregoing that the demand for ground water will double between now and 2000. The existing problems will therefore grow more acute, especially in the coastal areas and the areas in which the fresh-water aquifers are small and rest on salt water. The Gefara plain now has a huge deficit in its water balance, which was already in excess of 350 million cubic metres in 1980 and has caused a considerable decline in the water table and sea-water intrusion in the coastal aquifers. The penetration is already in the order of 3 to 5 km inland from the coast.

At Bin Ghashir, 30 km south of Tripoli, the old wells tapping the first aquifer have dried up and have been abandoned. It has been necessary to drill to depths of 250 to 300 m to exploit the second aquifer.

Elsewhere, when the upper aquifer is in hydraulic contact with an underlying saline aquifer, as is the case on the coast from Misurata to Ajdabiya, the intensive pumping from shallow wells dug along the wadis to tap the lenses of ground water recharged by rainfall have caused the level of the salt water to rise. In many oases close to sebkhas the excessive pumping has caused the intrusion of saline continental water into fresh-water aquifers.

The implementation of new agricultural projects, especially when it involves the drawoff of irrigation water from deep fresh-water aquifers, has also caused problems: as the deep aquifers provide the recharge for the phreatic aquifers, these latter aquifers have been directly affected by the decline in the water level. This is the case in particular at Ghadamis and Jufrah and in many other locations. At Jufrah, a place known for the quality of its palm trees, farmers are compelled to drill to considerable depths to find the water to save their already endangered trees, which used to grow naturally by drawing the water they needed from close to the surface.

Several natural springs in the Jabal Nafusah andJabal Akhdar have recently dried up. Until very recently they played an important role in supplying water for small settlements and livestock.

Subsidence has already taken place as a result of the extraction of ground water in the Sarir area. The over-exploitation of ground water has imposed heavy economic burdens on both private and public sectors. The drilling of very deep wells which give water of average or poor quality and at high temperature entails the installation of corrosion-resistant materials and powerful submersible pumps, and the use of large amounts of energy and advanced technologies, particularly where drilling is concerned. It also entails high operating and maintenance costs, not to mention the cost of cooling and/or treating the water when necessary. Moreover, the ground-water resources have proved insufficient to meet the requirements of the densely populated coastal areas; this has meant recourse to the desalination of sea water, treatment of waste water, construction of dams to store the waters of the wadis before they are lost to the sea, and the implementation of the largest project ever conceived for the delivery of ground water, in an amount of one billion cubic metres a year, through hundreds of kilometres of large-diameter channels from the south to the north.

At present a series of dams stores surface runoff water; the total capacity is 378 million cubic metres which will rise to 686 million cubic metres when the construction programme is completed. These dams are designed to permit the recharge of the aquifers, but so far it has not been possible to determine their degree of success in this. Jones (1984) reported that a small project consisting of 12 wells situated behind the dams on Wadi Tilal West and Wadi Haniyah in the Sirt area could give good reresults in only ten years, maintaining a lens of fresh water above the Oligocene-Miocene saline aquifer. This experimental project, although on only a small scale, is of considerable importance for this region which has little fresh water.

A barrier formed of artificially recharged wells is planned as a means of halting the intrusion of sea water. The idea is to use treated waste water to recharge the wells, for the potential production of the treatment plants is about 60 million cubic metres a year and should exceed 140 million by about 2000. At present waste water is used directly for irrigation purposes close to Tripoli and at Az Zawiyah and in other places after three phases of treatment: primary and secondary phases with layers of biological filters and a tertiary phase with rapid gravity filtration through layers of sand, followed by chlorination.

VI. CONCLUSION

Ground water is Libya's main water resource. There are aquifers for all geological periods from the Cambrian to the Quaternary. In the north the resources are partially renewable by direct infiltration of rainwater or by surface runoff. However, it has not yet been possible to determine whether the aquifers are recharging in the south. As a rule the ground water flows from south to north and the aquifers are hydraulically interconnected at the regional level. The water quality varies greatly from place to place and is a major obstacle to the exploitation of ground water.

The excessive exploitation of ground water in the coastal area has led to a rapid decline in the water level and a very large deficit in the water balance, not to mention the intrusion of sea water and the deterioration in the water quality. Several solutions to this situation are envisaged: in the short term, strict control of drilling operations, limitation of irrigated areas, prohibition of crops which consume large amounts of water, prohibition of irrigation in summer except in the early morning and late evening in order to reduce losses through evaporation, high tariffs to reduce water consumption, and promulgation of a new law on the use of ground water and quality control; in the long term, channelling of ground water from deep in the interior, construction of desalination plants to meet the needs of the coastal towns, use of waste water after treatment for agricultural purposes or to recharge the aquifers, and introduction of irrigation methods which use less water and measures to encourage the population of the large centres in the north to resettle in the small towns in the south.

The cost of ground water varies greatly, depending on location, depth of the aquifer, discharge, water quality, uses and construction methods. As a rule the wells are equipped with submersible electric pumps capable of lifting the water several tens of metres. The unit cost per cubic metre is in the order to 20 to 50 United States cents. The present state of knowledge, although satisfactory for the aquifers as a whole, must be improved, especially with respect to sea-water intrusion, training of technical staff, recharge of the aquifers, problems of corrosion of the installations and their design, collection and processing of data, pollution of the aquifers, karstic aquifers and springs, evaporation from sebkhas, hydrogeological cartography, control of the drawoff of ground water, level of the water and variations in its composition and quality.

Some of these topics are already under study; for example, evaporation in the Tawargha sebkha, sea-water intrusion along the Gefara coast, hydrogeological cartography and karstic springs at Ayn Zayyanah, where an attempt is being made to tap the water before it mixes with sea water in the coastal area.

VII. REFERENCES

Bellini, E and Massa, D. A stratigraphic contribution to the Palaeozoic of the southern basins of Libya. In Salem, M.J. and Busrewil, M.T. (eds). The geology of Libya. Academic Press, London, 1980, pp. 3-56.

El Ramley, I. Water resources study of Zone V - Al Kufrah and Sirt basins. Tripoli, 1983 (Ed. Jones, M.T.).

Energoprojekat Regional groundwater study: Wadi Sawf al Jin-Wadi Zamzam-Al Juffrah. Tripoli, 1977.

FAO Report on the groundwater resources of the Gefara Plain. Rome 1982.

FAO Water supply alternatives in the Gefara Plain. Rome. 1985.

GEFLI Soil and water resources survey for hydroagricultural development: Western zone - Central zone. Tripoli, 1972.

Guerre, A. Hydrogeological study of the coastal karstic spring of Ayn Zayanah, Eastern Libya. In Salem, M.J., and Busrewil, M.T. (eds). The geology of Libya. Academic press, London, 1980, pp. 685-701.

Idrotêcnico Hydrogeological study of Wadi Ash Shati, Al Jufrah and Jabal Fezzan area. Tripoli, 1982.

Illy, P. Groundwater development control of Murzuq basin. Tripoli, 1984.

Illy, P. Investigations on the hydrogeological conditions in the Sabha area. Tripoli, 1985.

Jones, J.R. Groundwater provinces of Libyan Arab Republic. In C. Gray (editor), Symposium on the geology of Libya. Tripoli, 1971.

Jones, M.T. Groundwater resources along the coastal belt of the central zone from Wadi Jarif to Bin Jawwad. Tripoli, 1984.

Meteorological Department Climatic data files.

Pallas, P. Water resources of the Socialist People's Libyan Arab Jamahiriya. In Salem, Busrewil, M.T. (eds). The geology of Libya. Academic press, London, 1981. pp. 539-594.

Salem, O., Visser, J.H., Dray, M. and Gonfiantini, R. — Groundwater flow patterns in the western Libyan Arab Jamahiriya evaluated from isotopic data. IAEA, Vienna, 1980.

Salem, O. and Talha, O. — Hydrogeological map of north-western Libya. Arabic, Tripoli, 1984.

Secretariat of Planning Surveys Department — Atlas of the Socialist People's Libyan Arab Jamahiriya. Tripoli, 1978.

Wright, E.P., Benfield, A.C., Edmunds, W.H. Kitching, R. — Hydrogeology of the Kufrah and Sirte basins. Eastern Libya. Q.J. Eng. Geol. London, Vol. 15, 1982.

LIBERIA

Area: 111,369 km^2

Population: 2.06 million (United Nations estimate, 1983)

I. BACKGROUND

Natural regions

In Liberia the altitude increases fairly regularly from the coast towards the interior. The country can be divided into four zones parallel to the coast: the coast zone itself; the hills; the plateau eroded by the water courses; and the high plateaus. As a rule the boundaries between these zones are not clearly defined and they merge gradually into each other.

The coastal zone is 10 to 15 km wide, mostly flat and covered with sand. The inland limit of the tides, usually marked by the first rapids on the coastal rivers, corresponds to the limits of these flat areas; further inland the coastal zone consists of low land flooded by the watercourses and the low hills, which slope gently up towards the hill zone itself. The 100 contour line marks the approximate limit of the coastal zone.

The hill zone consists of broad low areas alternating with discontinuous lines of hills running north-east. The low areas lie between 100 and 200 m and the hills rise to 200 and 300 m.

The plateau eroded by the river system has two levels. The lower and more extensive level lies at 300 m of altitude and the higher level at 600 m. The ridges of the high plateaus dominate this level and become more numerous in the north-east of the country.

In the north the mountainous zones contain narrow deep gorges and it is difficult to recognize the parts of the plateau lying at their base. As a result, the mountains of the lower plateau are unconsciously considered part of the high plateaus.

Climate

Liberia is situated at the heart of the zone of wet tropical forest. The average annual rainfall along the coast is five metres, decreasing to two metres towards the interior. The isohyets run roughly parallel to the coast, with deflections caused by the mountains. The only notable seasonal variation is the "wet season" from December to April. Some places, Monrovia for example, have less wet periods lasting one to two weeks, usually in August, May and November, which can be included in either the wet or the dry season, depending on the year.

There is little variation in temperature between an absolute minimum of 16°C and an absolute maximum of 30°C, depending on the month. The maximum and minimum values occur in the interior, particularly on the high plateaus. Little information is available about evaporation measured in the pan at the three meteorological stations: Kongo, Fendel and Suakoko, for long series of observations have not been made. It is not easy to estimate the daily amounts of evaporation in the pan in the wet season in Liberia, and in many cases the values represent only orders of magnitude lower than the real figures. The Penman method is therefore used to estimate the evaporation from a free surface on the basis of meteorological data from two stations: Robertsfield and Firestone-Harbel. The average annual value at the two stations is 1,633 mm but the standard gap is 19 mm/year and 32 mm/year respectively. At these two stations 45 per cent of this total amount evaporates during the dry season, 38 per cent during the wet season and 17 per cent during the transition periods (May and November).

Surface water

The whole of the country is well drained by rivers flowing to the Atlantic. The river basins open to the sea can be divided into two categories:

- The large basins (about 20);

- The small coastal basins which cover only three per cent of the country.

The first six of the 20 large basins drain two-thirds (65.5 per cent) of the country. They are, from east to west:

Number	Name of river	Area (km^2)	
		Total	In Liberia
1	Mano	7,750	5,720
2	Lofa	10,382	8,994
3	Saint Paul	20,440	11,293
4	Saint John	17,089	14,509
5	Cestos	12,634	10,070
6	Cavalla	30,225	12,398

Since its establishment in 1972 Liberia's Hydrological Service has given special attention to the quantitative evaluation of the country's surface water resources. In 1978 it published the first hydrological year book for the period 1972-1976, covering all the country's stations. Some figures on the relative flow characteristics of the six main basins are given below:

River	Location of station	Area of basin (km^2)	Depth of runoff (mm/year)	Number of years measured
Mano	Kongo	5,540	1,178	9
Lofa	New Hope	10,200	895	4
Saint Paul	Walker's Bridge	9,760	850	19
Saint John	Baila	3,857	1,085	15
Cestos	Sawolo	4,582	429	18
Cavalla	Nyaake	12,608	1,271	12

If the distribution of rainfall (annual totals) over the country is taken into account, the depth of runoff for the whole of the basin should be higher than indicated in the table above.

On the basis of these values, the total potential of the six main rivers at their mouths has been evaluated at 98 km^3 per year.

The rivers' flow varies during the year. As a rule the highest levels occur in September, and the lowest in March. Since the daily amount of rainfall can exceed 150 mm at some points, it can be estimated that for the small basins the values should be much higher than those given in the last column of the table above.

In order to evaluate the flow coefficient it is necessary to know the amount of rainfall in the basin. However, large areas of the main basins are located outside the country in regions for which data are not yet available. Accordingly, it is not possible to calculate the flow coefficients for the various basins. The sediment discharge of Libera's rivers and streams is not known as it has not been measured. Some settling tanks have been built, but the water of the rivers which flow in the vicinity of mining operations (Saint John, Lofa and Mano) contain high concentrations of solid matter. Liberia's lakes are found along the coastal zone which extends for 560 km from Grand Cape Mount to Cape Palmas. During the wet season they are in contact with the ocean. During the dry season they are separated by sandbars of varying size. In view of these seasonal changes and the geomorphological conditions, some of these "lakes" are called lagoons. Their behaviour is heavily influenced by the size of the drainage basins which lie above them. In the case of the lakes which receive their water from the biggest drainage basins, contact is made with the ocean when the first flood waters arrive at the beginning of the wet season; however, the formation of sandbars ceases at the end of the dry season. All the country's lakes are suitable for fish-farming and tourism and sports activities.

Their fresh or slightly saline surface water is often used in small quantities for domestic purposes in rural areas. The economic potential of Lake Piso, the country's largest lake, has not yet been exploited.

Geology

The whole of the country is within the Precambrian crystalline Guinean shield, which consists of igneous and metamorphic rock. The metamorphic gneiss, granitogneiss and schist, which can be distinguished from each other by geophysical and petrographical techniques, are penetrated by igneous blocks of Jurassic age; the basement rock is overlain by lateritic deposits which can be up to 30 m thick.

Sedimentary formations are found in the coastal zone in two systems of sands and conglomerates. Fluvial and deltaic deposits are found only along the coast. Most of the diabasic dykes run in the region's usual north-west direction. The streams and rivers often follow the lines of the fault and fracture zones which are found throughout the country.

II. GROUND-WATER RESOURCES

Several ministries and other government bodies are directly or indirectly concerned with water resources in Liberia. A co-ordination body was set up in 1981: the National Water Resources and Sanitation Board. The following are the government and other services involved:

1. The Liberia Water and Sewer Corporation (LWSC), which is responsible for urban water supplies. Nine towns have distribution networks;

2. The Ministry of Agriculture, responsible for irrigation;

3. The Ministry of Health and Social Affairs, responsible for sanitation;

4. The Liberia Electricity Corporation (LEC), responsible for hydroelectric power;

5. The Ministry of Rural Development, responsible for the sinking of wells and boreholes in rural areas;

6. The bodies responsible for rural development projects, three of which were in operation in the country in 1984, supplying water to rural areas, sometimes in co-operation with the Ministry of Rural Development;

7. The Hydrological Service attached to the Ministry of Land, Mining and Energy, which has overall responsibility for water resources.

The only government body with responsibility for ground-water research and study is the Liberia Hydrological Service. The LWSC and LEC are concerned only with the sale of water to the consumer.

In the past ground water has been studied on only a modest scale but for very specific purposes. A research campaign including geophysical prospecting and test drilling in the coastal sandstone indicated that the stocks of fresh water in these aquifers are extremely small. The LWSC commissioned a study company to carry out drilling with a view to supplying six towns. The methods used included the interpretation of aerial photographs, geophysical prospecting and six-inch diameter test drilling to a depth of 100 m. The boreholes were considered positive if they gave discharges of 5 m^3/h with an acceptable draw-down. They were then widened to a diameter of eight inches.

Most of the country is covered by metamorphic rock and the areas of sedimentary formation are very small; there are therefore no large aquifers. As the laterite cover has poor permeability, the ground water is found mostly in the faults and fractures and in the altered part of the basement rock. Some test-drilling results are given below.

Depth of aquifer and geological age	Location	Depth	Specific yield ($m^3/h/m$ of drawdown)
Metamorphic basement rock	Town of Buchanan	Six wells 80 to 100 m	0.25 to 2.60
Metamorphic basement rock	Town of Zwedru	Two wells 63 and 44 m	7.60 and 24.5

Some wells drilled in the faults and fractures have given fairly high yields; this may be due to the high recharge from the abundant rainfall.

The water quality in the basement fractures is generally good. Most of the wells have a neutral pH, except when they draw their water from the laterite cover (pH below 6). The specific conductivity values do not exceed 500 millimhos/cm, and the chloride values are below 100 mg/l. The combined iron and manganese content is below 0.3 mg/l. The average dry residue is 250 mg/l, consisting mainly of acid carbonate and silicate.

Very little information is available about ground water in Liberia and it covers only a few points in the country. Research is continuing but on a small scale and for specific purposes. The resources are not available for large operations requiring geophysical work and major drilling.

III. DEVELOPMENT OF GROUND WATER

The Ministry of Rural Development and the Liberia Water and Sewer Corporation are responsible for the development of water resources. As part of the rural development programme of the Ministry of Rural Development, wells are being dug and drilled and springs are being equipped in rural areas with the help of nine light cable-tool rigs and 40 drillers training in the country. There are also a few digging teams working in light ground.

The LWSC manages the water distribution networks and it is also responsible for the study of water resources for new networks. The studies are entrusted to study companies and the construction work to subcontractors. The LWSC does not have the equipment for ground-water exploration.

The country has just one drilling company which was set up only very recently; it has two big percussion rigs and employs five drillers. Up till now these rigs have been used only for study of the ground water at shallow depths with a view to the installation of large-diameter hand-dug wells and drainage galleries.

Ground water is exploited to supply urban distribution networks and to establish water points in the villages.

Five towns are supplied from ground-water aquifers. Three of them have 150-metre tubewells; the other two are supplied from large-diameter dug wells.

In rural areas many wells have been dug in the laterite surface formations; dug wells are found in the small towns and villages, no matter what their size, but only when professional drillers are established in the area. The maintenance of the hand pumps poses considerable problems in the rural areas, with the result that many of the tubewells have been abandoned. Two mining companies use ground water taken from boreholes which they drilled to meet their own needs.

Agriculture uses very little ground water simply because the rainfall is so heavy. Ground water is used to a limited extent in the "dry" season for irrigation in nurseries (oil palms, coffee and cocoa bushes, etc.).

As agriculture develops in the future, ground water might play a bigger role in irrigation in the dry season and also in drinking water supplies. The Government has launched major well digging and drilling programmes in rural areas. It is also possible that ground water will be exploited in increasingly large amounts to supply the towns.

It cannot be said that Liberia has serious ground-water problems. In view of the abundant rainfall, the heavy infiltration and the considerable recharge potential of the aquifers, there is so far no danger of over-exploitation. Sea-water intrusion might pose a problem in the coastal zone, but this has not been the case so far.

Artificial recharge is not envisaged for the moment. It might create problems in view of the high solid content of the surface water.

IV. CONCLUSION

Liberia has abundant water resources which generally meet its current requirements. However, this situation might change if agriculture developed and farmers wanted to grow crops throughout the year and if more hydroelectric plants were built.

At present the main problem is that of water quality. Most of the surface water is vulnerable to pollution, especially in the dry season. Ground water can be found throughout the country at a reasonable depth; it is used to supply all the towns of average size (over 300 buildings) and many big villages (100 to 300 buildings). Hand-dug wells are found only in the areas where there are qualified well-diggers. Most of them belong to private individuals in the towns, but they are common property in the villages.

The wells dug and drilled under the rural water-supply programme and equipped with hand pumps supply free water to villagers and feed standpipes in six big towns which have distribution networks. Only metered water supplied to private outlets is expensive.

The United Nations Development Programme, with the Department of Technical Co-operation for Development as executing agency and in association with UNICEF, has financed ground-water study and drilling projects in rural areas in Liberia.

V. REFERENCES

Few documents are available on ground water in Liberia. The data must be extracted from reports on projects to supply medium-sized towns such as Vahun, Careysburg, Saklepie, Bahn, Kahnple, Tappita, Graie, Robertsport, Foya, Kolahun and Zorzor. Where rural areas are concerned, useful reference can be made to the United Nations project reports listed at the end of this volume.

Mention may also be made of the following fairly recent reports which are kept in the archives of the Liberia Hydrological Service:

- Reconnaissance Survey for Utilization of Warner Creek for Monrovia Industrial Park;

- Shallow Aquifers in Monrovia Based on Inventory of Water Wells;

- Surface and Groundwater Resources of Cape Mount Peninsula;

- The Coastal Lagoons between Monrovia and Marshall;

- Hydrogeological Characteristics of the Unconsolidated Sediments in the Monrovia Area;

- Warner Creek Watershed, an Evaluation;

- The Captured Springs and Hand-Dug Wells in Lofa and Nimba Counties.

In all, there are some 50 to 60 reports and other documents on ground and surface water resources. However, there are no maps or reports on regional hydrogeological studies.

MALI

Area: 1,240,000 km^2

Population: 7.53 million (United Nations estimate, 1983)

I. BACKGROUND

Mali is a landlocked country of vast plains. In the south of the country laterized sandstone plateaus dominate vast expanses of hills and plains. The largest plateaus are the Mandigue in the west with its escarpment dominating the valley of the Upper Niger and falling away gradually northwards; in the south and south-east the Dogon plateau which extends eastwards in a line of residual buttes, including Hombori which is the highest point in Mali (1,155 m). These plateaus are bounded in the north by large, partly sandy plains; in the south they give way to hills covered with laterite cuirasses.

North of the Niger loop and the interior delta vast sandy plains (ergs) stretch as far as the borders with Algeria and Mauritania. The Adrar des Iforas in the north-east of Mali is a southwards extension of the crystalline massifs of the Hoggar.

Climate

Most of the country is subject to climatic conditions of the Sudano-Sahelian type with high average annual temperatures (26°C to 30°C), rainfall concentrated in May-June and September-October, and a long dry season of six to nine months.

From south to north there are four climatic zones with decreasing rainfall and relative humidity and increasing evapotranspiration.

The South-Sudanese zone bordering on Guinea and Côte d'Ivoire receives more than 1,300 mm of rain during a six-month wet season. The annual thermal range is low here.

The North-Sudanese zone lies between the 1,300 mm and 700 mm isohyets. Bamako, the country's capital, is situated at the heart of this zone. The rainy season lasts four to six months. The average temperatures are higher in summer than in winter.

The Sahelian zone lies between the 700 mm and 200 mm isohyets; it has a long dry season and short rainy season, with rain falling on only about 30 days a year. The annual thermal range is 12°C.

The South-Saharan desert zone has intermittent rainfall of up to 200 mm and very wide thermal ranges.

The interior delta of the Niger, which covers an area of about 30,000 km^2 in the Sahelian zone constitutes a small inland sea during the flood period from October to December and it then has a local micro-climate.

Like all the other Sahelian countries, since the beginning of the last decade Mali has been suffering the effects of many years of low rainfall, the harshest consequences of which have been felt by the rural and pastoral populations.

Surface water

Mali's hydrographic system extends over the south and centre of the country. The desert area in the north has only a few traces of ancient drains and there is no runoff.

The country is crossed by two major African rivers: the Senegal flowing westward in its upper basin; and the Niger flowing eastwards in its middle course. Half of the length of the Senegal lies in Mali. The river rises in Guinea on the slopes of the Fouta Djallon. Its upper course, called the Bafing, receives the two other upstream branches on its right bank - the Bakoyé and the Baoulé. Downstream of Kayes the river is swollen by the Faleme, which forms the frontier with the Republic of Senegal. The river's mean flow at Galougo is 669 m^3/s; in the high-water period (September) the flow is 2,800 m^3/s, falling to 15 m^3/s in the low-water period (May). The dam under construction at Manantali is located on a rocky shelf of the Bafing, upstream of its confluence with the Baoulé.

The Niger also rises on the northern slopes of the Fouta Djallon. Two-fifths (1,700 km) of its total length (4,200 km) lie in Mali. Its main upstream tributary is the Sankarani, on which the Sélingué dam has been built.

As it leaves the Mandigue plateau upstream of Koulikori its valley gradually widens until from Ke-Macina it becomes a vast alluvial plain which is flooded in the high-water period; this is the "interior delta of the Niger" which ends near Timbuktoo. At the beginning of the delta, near Mopti, the Niger receives the Bani; this river is 900 km long and its upper course, called the Bagoé, is swollen by the Degoé, Bafing and Banifing.

From Timbuktoo to the Labbezanga rapids on the Niger frontier the river describes a huge loop in a sandy desert area cut by the rocky shelf of Tossaye.

Owing to its great length, the morphological configuration of its valley and the climatic zones which it crosses, the Niger's hydrographic system varies in its different sections in Mali. At Koulikoro the crest usually occurs in September, whereas at the beginning of the interior delta it occurs at the end of October (Mopti) and at the outlet of the delta in December (Kalara). Further downstream, at Gao, the crest occurs in January.

The mean flow of the Niger at Koulikoro is 1,550 m^3/s.

Within the interior delta and at its edges are found systems of lakes fed by the flood waters, the most important of which is Lake Faguibine, west of Timbuktoo. Many of these lakes have become seasonal or have dried up following the recent drought and the consequent drop in the levels of the rivers.

The sandstone plateaus and the hilly areas are intersected by a dense hydrographic system of temporary streams created by the winter rainfall. In the partly sandy plains of the Sahelian zone the flow is intermittent and often disrupted by accumulations of sand, as in the Vallée du Serpent which used to drain the Nioro-Nara plain.

Geology

Mali's main geological systems are the Birrimian crystalline basement rock, the mainly sandstone and pelitic sedimentary formations of Lower Cambrian and Paleozoic age, the Permian dolorites, the Mosozoic intercalated continental deposits, the marine sediments of the Upper Cretaceous and Eocone eras, the continental terminal and the surface formations of varied origin and age.

Crystalline basement rock

This rock is well represented in the south of Mali. It also outcrops in the Kayes region and in the middle of the Adrar des Iforas. It consists of ancient folded and metamorphized sedimentary and irruptive rocks: micaschists, gneiss, greywackes, cipolins, quartzite, gabbros and andesites, as well as granites of varied mineralogical origin and composition.

Lower Cambrian and Paleozoic sedimentary formations

These formations have built up in the vast basins which extend over more than two-thirds of Mali.

In the Taoudenit basin which covers the whole of the central and northern parts of the country, the Lower Cambrian formation outcrops in a large circle at its southern edge. It consists of a mainly sandstone series, sometimes more than several thousand metres thick, alternating with sandstone bars of variable granulometry and mineralogy and with pelitic sequences which form the Sudano-Sahelian plateaus. Above the Lower Cambrian formation lies a thick Cambrian series of fine facies consisting of argilites and siltstones with some intercalations of sandstone, limestone and jasper in its upper part. Most of the Cambrian formation is over 1,000 m thick in the Nara trench.

In the central and northern parts of the Taoudénit basin the Paleozoic series includes Cambrian-Silurian sandstones and calcareous schists, Devonian calcareous and bituminous schists, and Carboniferous sandstones and limestones.

The Lower Cambrian formations in the Gourma basin are folded and partly metamorphized. The facies is mostly schistic with sandstone-quartzitic and limestone intercalations.

Permian formation

This formation is represented by doloritic volcanism, which is particularly well developed in central Mali where it intrudes into the Lower Cambrian and Cambrian formations in the form of dykes, sills, batholites and laccolites.

Intercalated continental formation

This formation covers an area of more than 125,000 km^2; as a rule it is difficult to distinguish it from the overlying continental terminal formation.

It consists of poorly consolidated continental deposits with sandstone-sand and gravel facies and sequences of multicoloured clay. It is more than

400 m thick in the Nara trench and may exceed 1,000 m at the eastern edge of the Iullemeden basin.

Upper Cretaceous and Lower Eocene formation

These sediments, of marine and locally lagunal origin in their lower part, are found in the eastern part of the country bordering the Adrar des Iforas and further south in the Sudanese straight and the Iullemeden basin. The facies are mostly limestone and marl-limestone with some fine sandstones and schists in the upper part.

"Continental terminal"

This complex Neogene series covers an area of 270,000 km^2 and is several tens of metres thick except in the south-east of the country where it is more than a thousand metres thick and may include Eocene strata. It consists of sandstone and non-consolidated sands, clays and argillaceous sands with laterized and silicified layers. In the Iullemeden basin, the basic detritic series is overlain by a clay-sand complex topped by argillaceous sandstone with large amounts of concretion and ferruginous oolites. Lignite and pyrite are common throughout the series.

Surface formations

- Laterites and ferruginous cuirasses are particularly common in the sandstone plateaus and the areas of Birrimian basement rock. They are of various ages and the oldest, which jut up from the smooth high surfaces, probably date from the Eocene era;

- Quaternary alluviums, mostly of fine facies, are common in the Niger basin;

- Fixed or mobile sand dunes cover vast areas north of the 14th parallel N;

- Lacustral tuffs and sebkha deposits are found in places in the Saharan and North-Saharan areas. The Taoudénit salt deposits are the best known and have long been exploited.

II. GROUND-WATER RESEARCH

Government bodies

The study and development of ground water are the responsibility of the National Water and Energy Office (DNHE) which is part of the Ministry of State for Capital Development. The DNHE carries out basic hydrogeological research, controls the well and borehole development projects, and collects and stores all ground-water data. A wells service was set up in 1974 in the same Ministry; it is responsible for large-diameter installation.

The specialized offices of the Ministry of Rural Development and the Ministry of Agriculture are responsible for several ground-water development projects.

Research background

The first hydrogeological studies were made by the Water Service of the Department of Public Works of the AOF, including inventories of traditional water points and sinking of reconnaissance wells and boreholes. These operations and studies were summarized in a "hydrogeological map of Sudan" (J. Roure, 1957). After Mali's independence in 1960, the research was concentrated on certain regions then given priority: Gourma, northern Taoudénit basin, and Gondo plain. These studies gave most attention to the geological aspects.

It was not until 1967 with the start-up of the UNDP/United Nations project MLI/67/507 that a systematic approach was taken to the various aspects of ground-water research and exploitation. The study methodology and the criteria for installation of wells and boreholes and the drilling techniques were gradually established. The hydrogeological conditions and the possibility of developing ground water in various lithological units were also determined. During the implementation of project MLI/74/001, which followed on from project MLI/67/507, the development techniques were tested successfully although on a small scale. The transition from the phase of reconnaissance to the phase of intensive development as part of an emergency programme to counter the effects of the drought entailed the gradual refinement of the methods used, in particular the geophysical ones, with a view to minimizing the borehole failure rate and thus reducing the unit cost of the productive installations. This phase began with the implementation of UNDP/UNICEF project MLI/76/004 which provided training in the field for a large number of DNHE's national technical staff; this project was followed by several village and pastoral water supply projects, including UNDP project MLI/82/005.

Study methodology and installation of boreholes

South of the 14th parallel N, where almost all Mali's population and economic activity are found, the main exploitable water resources are contained in small local fissure aquifers. The traditional hydrogeological methods of drilling research in fissured formations are usually used in conjunction with geophysical prospecting. The experience acquired over 15 years, during which several thousand boreholes were drilled, shows that, except in very favourable local hydrogeological conditions where the water-bearing zones are in continuity over a large area, the best results are obtained when the drilling work is carried out in close conjunction with geophysical studies. The results of each drilling are immediately compared with the hydrogeological and geophysical assumptions on which the drilling was based; these assumptions are systematically reinterpreted and taken into account in subsequent drilling operations. This procedure has improved the drilling success rate in a series of UNDP projects by more than 25 per cent to an average of 70 to 75 per cent in 1983 in a sector with average hydrogeological conditions.

The traditional hydrogeological methods include studies of the terrain (geology and geomorphology), inventory of traditional water points and study of the structural conformation by means of photogeological interpretation.

A geophysical study programme is established on the basis of the hydrogeological picture. Depending on local conditions, use is made of electrical, magnetic or seismic methods, usually in conjunction with each other.

- Electrical methods: profiling (aerial geophysical prospecting) and isolated soundings are used systematically, for they usually offer the best means of solving the problems and are easy to use. They identify discontinuities in the substratum associated with less resistant anomalies and establish a provisional lithological section for the drilling operation.

- The magnetic method using a portable proton magnetometer is essential for drilling studies in sectors where doloritic intrusions are common. These volcanic intrusions are usually marked by easily identifiable anomalies in the magnetic profiles. These profiles can be obtained and qualitatively interpreted by the hydrogeologists.

- Seismic refraction is used only for very discontinuous aquifers with small permeable zones where the boreholes must be very accurately located.

Drilling techniques

The lateritic surface formations and the underlying fissured water-bearing formations are usually of satisfactory consistency for small-diameter drilling. The pneumatic down-the-hole hammer technique has proved by far the most effective in these conditions. In most cases the holes are drilled with a diameter of 6.5" to depths of about 15 m below the top of the squifer; the hole is then fully lined with 5" or 6" metal or PVC tubing. The hole is then drilled down to its final depth with a diameter of 4.5" and no tubing is installed in this rock section of good consistency.

In basement rock areas the thick surface and alteration formations are penetrated by rotary mud rigs. The fissured substratum is drilled by pneumatic down-the-hole hammer.

Some boreholes in fissured formations, especially those designed to feed water-supply networks, are fully lined; filters and a layer of stabilizing gravel are installed in the ring space and the wellhead area is concreted.

The sandstone intercalated continental and continental terminal formations are drilled with mud rigs and the holes are fully lined and equipped with filters.

The discharge of the productive wells is measured during short air-lift production tests (one to three hours) carried out by the drilling team. In the case of wells discharging more than 5 m^3/h and therefore suitable for the installation of more sophisticated means of drawoff than hand pumps, 12 to 48 hours of test pumping is carried out by the traditional air-lift method or with groups of submersible motorized pumps.

III. GROUND-WATER RESOURCES

Ground-water resources are found in two types of aquifer: discontinuous aquifers with mainly fissure permeability and continuous aquifers with intergranular permeability. Although most of the ground-water resources are stored in continuous aquifers, their geographical location renders them much less important from the economic and social standpoint than the discontinuous aquifers, which are virtually the only ones capable of meeting the village requirements and a large part of the pastoral requirements.

Discontinuous aquifers

Crystalline and metamorphic formation of the Precambrian basement rock

The ground water exploitable by borehole is found in the fissured areas of the bedrock substratum and in the coarse-grained sands. The permeability is irregularly distributed and generally low; it depends on the degree of fracturation and especially on the proportion of clay in the coarse-grained sands and on the thickness of these formations. The specific yields of the installations are low, ranging from a few m^3/h up to 20 m^3/h. The success rate of the drilling is high, above 80 per cent, when the surface formations are thick and saturated for a considerable depth. It is much lower when the surface thickness is small or the formations saturated for only a small depth. The average depth of the producing wells is about 60 m. The intermittent recharge of the water-bearing zones causes large seasonal fluctuations in the water level (between two and six metres). The water is aggressive with pH values between 5.5 and 7.7; the salt content is usually small with a dry residue below 0.5 g/1.

Lower Cambrian formations

The hydrogeological properties of these formations depend on whether they consist mainly of sandstone series with table configuration lying under a lateritic cover (sandstone plateaus) or of mainly schistic series interbedded with limestone and sandstone in a fold structure (Gourma basin).

The sandstone plateaus contain the country's main discontinuous aquifer and more than two-thirds of the wells at present in production are located here. This aquifer is divided into sections by doloritic intrusions and shelves of solid and compact rock; it constitutes a multi-layered system with stratified permeable areas associated with a number of more heavily fissured banks or banks with intergranular permeability which are linked by networks of subvertical fractures. Another hydrogeological characteristic of the sandstone plateaus is that they are in hydraulic contact with the small surface aquifers found in the lateritic formations and the intergranular altered fringe of the substratum and the underlying fissured stratum. These surface layers act as a reservoir and the much more permeable fissured zones as a drain, so that the water can be drawn off by small-diameter installations.

The Lower Cambrian sandstones are usually water-bearing for only a few tens of metres below ground level. The main stocks of water are found at a depth of 20 to 60 m and the static water level at a depth of 10 to 25 m. The piezometric surface is interrupted by the doloritic intrusions and the blocks of non-fissured bedrock; as it is very irregular, the water usually moves vertically by infiltration or evaporation.

The seasonal fluctuations in the level are large, sometimes in excess of 5 m, with an average of 2 to 3 m. Annual fluctuations are observed and there has been a general decline of several metres in the level following the period of drought during the last decade.

The permeability is very variable and depends on the degree of fracturation. The transmissivity coefficient determined during test pumping is between 2×10^{-5} m^2/s and 3.1×10^{-2} m^2/s, with averages of between 10-4 m^2/s and 10-3 m^2/s. The heterogeneity of the water bearing zones is demonstrated by the impermeable-screen phenomena revealed by several tests.

The average yield of the producing wells is about 5 m^3/h, with a success rate of 50 to 90 per cent, depending on the area and the drilling method. The water is mostly aggressive with a low salt content and an average dry residue of 0.3 g/1. Isotopic and piezometric studies have demonstrated that this aquifer has a good recharge and large exploitable resources, even though they are irregularly distributed.

In the Gourma basin the aquifers are small and closely connected with certain limestone and sandstone strata. The depth of the static water levels is more than 50 m. The permeability is low and the yield of the producing wells rarely exceeds 1-2 m^3/h. The aquifers appear to have little recharge, if any at all. To date the failure rate has been about 75 per cent. The development potential is therefore small, even for pastoral water supplies, and requires large investments.

Cambrian schists

This mainly pelitic formation is generally impermeable and contains small aquifers linked to secondary fissure permeability. The aquifers have much smaller lateral and vertical extension than those of the Lower Cambrian sandstones.

The drilling success rate is not above 50 per cent even when combined with systematic use of geophysical methods. The exploitable stocks in the boreholes are generally found at depths below 40 m, sometimes 60 m, and the yields are small. The static water level is found at a depth of between 5 and 25 m, almost always in the fissured substratum and not in the altered fringe or the surface formations. The stocks are therefore small but the exploitable resources are not negligible, for the seasonal infiltration causes piezometric fluctuations of several metres.

The transmissivity determined during test pumping is between 1×10^{-3} m^2/s and 5×10^{-5} m^2/s, with yields of between 4 and 20 m^3/h. Impermeable screen phenomena are very common, confirming the very discontinuous nature of this aquifer.

The salinity of the water is usually below 1 g/1. However, brackish water is found in places with a salt content of up to 17 g/1. The water always has a low pH value.

The average depth of the producing wells is about 55 m. The wells are lined with tubing as far as the top of the first large deposit of water.

Continuous aquifers

Intercalated continental. This formation represents Mali's main aquifer potential by reason of the size of its stocks and its great permeability. As it is located in the North-Sahelian and Saharan zones, its hydrogeology is known only in the western part and at the southern edge of the Adrar des Iforas, where it is exploited by a number of pastoral boreholes and wells.

In the Nara trench, where the ground-water horizons are deep (between 50 and 90 m), high transmissivity values (above 10^{-3} m^2/s) are associated with the sand-gravel strata, with yields above 20 m^3/h per well. At the edges of the trench clay and silt layers predominate over sand layers and the transmissivity and the yield of the producing wells decline. In this unit the water table is practically horizontal and the seasonal fluctuations nil or very small. The water is of good chemical quality.

The producing wells can be up to 150 m deep and must be fully lined and equipped with filter blocks.

In the central part of the Tadoudénit basin, on the Khenachich plateau, the intercalated continental formation contains only small aquifers, generally of fresh water.

At the edge of the Adrar des Iforas the permeable layers of the intercalated continental formation intersect at various depths, sometimes in excess of 100 m, with stable water levels between 35 and 60 m below ground level. Discharges of about 10 m^3/h have been obtained from boreholes 150 to 200 m deep. The water usually has a low mineral content.

Cretaceous-Lower Eocene

This formation contains permeable layers in the limestone and sandstone intercalations which are exploited at the edge of the Adrar and in the Tilemsi valley, with yields of a few cubic metres an hour and high drawdowns. The salinity of the water varies greatly, from fresh to brackish.

Continental terminal

In terms of the size of its stocks, this is the second largest continuous aquifer in Mali. In the interior delta it is in hydraulic contact with the surface water and the Quaternary alluviums. Yields of several tens of cubic metres can be obtained from lined boreholes equipped with filter blocks when the sand strata are thick and the water lies at depths of less than 20 m below ground level close to the Niger; but the yields decrease further away from the river owing to the combined effect of head loss and evaporation. The salinity of the water also increases from the recharge area in the delta zone towards the Azaouad basin. This aquifer is exploited by several traditional wells with depths of 20 to 60 m depending on the distance from the river and the depth of the horizon.

In the Sudanese straight the water levels are usually not lower than 40 m, but they can be more than 100 m deep further west and south in the Gao trench where the continental terminal becomes very thick. Geophysical prospecting indicates that the water becomes brackish towards depths of 200 m.

Surface formations

The alluvial deposits in the central delta zone constitute a single hydraulic unit with the continental terminal. However, the possibility of exploiting them by means of small-diameter wells is clearly less owing to their finer facies.

The dune formations can contain small local aquifers, the size of which is difficult to determine and depends on the nature of the substratum. These formations are important only because of their capacity rapidly to absorb the short heavy showers and thus provide part of the recharge of the underlying aquifers.

The lateritic formations contain local aquifers traditionally exploited by the villagers and herdsmen of the Sahelian and Sudanese areas by means of tens of thousands of wells and sumpholes. These aquifers are in hydraulic contact with the underlying fissure aquifers.

Current studies

As part of its priority activity of establishing new water points, in 1982 the National Water and Energy Office initiated in collaboration with UNDP and the United Nations (project MLI/82/005) several studies with the medium-term goal of evaluating the country's ground-water sources and determining their location and mode of exploitation in the various regions.

Computerized records of the water points throughout the country are being prepared. They will serve as the basis for the inventory of the country's water resources, the identification of new development projects and hydrogeological summaries.

Piezometric observation networks are gradually being established in the main aquifers; they cover the various climatic zones and monitor the natural fluctuations in the water levels. In the 1983 rainy season, even though the rainfall was less than usual (60 to 70 per cent of the average), the levels showed a general recovery in the Sahelian and Sudano-Sahelian areas, with average ranges of about one metres in the aquifers of the fissured substratum (Lower Cambrian sandstone and Cambrian schist) and of almost five metres in the aquifer of the lateritic cover.

Detailed studies have been undertaken at several sites of intensive pumping with a view to estimating the local ground-water resources and the recharge in the area of discontinuous aquifers. The first results indicate that at Tioribougou the direct infiltration represented about 20 per cent of the rainfall in the 1983 season, which was estimated at 530 mm, with an effective storage coefficient in the Lower Cambrian table sandstones of about five per cent; the renewable resources can therefore be estimated, as an initial approximation, at about three cubic metres per day per hectare for the local aquifer. These detailed studies were to be continued and multiplied during the period 1983-1986 and extended to other hydrogeological and climatic regions, so that an informed approach can be taken in the establishment of the balance of ground-water resources.

IV. EXPLOITATION OF GROUND WATER

Government bodies

The Ground-Water Division of the National Water and Energy Office (DNHE) is responsible for the technical control of all drilling projects for ground-water development and it manages all the village water-supply projects. In 1984 it employed about 70 qualified persons, half of them engineers (hydrogeologists, geophysicists, drillers), and the other half geological technicians, specialist foremen and administrators. There are a further 50 contractual staff: labourers, drivers, drilling hands. The DNHE has seven regional bases: Bougouni, San, Gao, Bamako, Kita, Timbuktoo and Mopti-Sévaré.

The body responsible for drilling operations installs large-diameter wells of modern design for both village and livestock purposes. In the past it encountered a number of difficulties and the results were sometimes disappointing owing to the lack of hydrogeological reconnaissance work.

The adoption of stricter management methods and closer collaboration with the DNHE with respect to basic hydrogeological studies for the installations should enable it to play its proper role in the development of ground water. At present it has about 300 permanent and contractual staff; it has a head office at Bamako and six regional offices; it operates in 14 sectors.

Position of ground-water development in 1983

Up to 1976 most of the boreholes were drilled during reconnaissance missions to evaluate the possibility of developing the main aquifers or during implementation of small local water-supply projects.

Since 1976 a number of large regional village water-supply projects have been undertaken; their number and size have grown with time.

The work carried out under the main projects up to July 1983 is listed in table 1.

The inclusion of the boreholes installed under small local projects and those installed before 1970 produces a total of 4,300 boreholes installed in Mali, including 2,500 classified as producing, with a minimum yield of 0.8 to 1 m^3/h. Of this total, about 1,500 are at present equipped with pumps: 1,300 with hand or foot pumps; 50 with solar-energy pumps; and 150 with motorized pumps.

About 450 large-diameter wells were installed between 1974 and 1983 by the body responsible for the wells operation.

As the Rural Water Service installed a similar number of wells between 1958 and 1973, the country now has 900 "modern" village and pastoral wells (with concrete lining and standpipe).

Water-point projects in the period 1983-1986

Work is to be continued under all the big village water-supply programmes launched in the last decade. Several new projects have also been undertaken since 1983-1984. The plan was to install some 4,300 boreholes, almost 90 per cent of them for village water supplies. Given an average success rate of about 65 per cent, the number of new producing holes can be estimated at about 2,800, which means that the number of producing boreholes would have doubled by 1986. The amount of work planned under the various projects is indicated in table 2. The number of wells was also to increase by almost 600 during the same period.

In 1986 the drilling equipment available to the DNHE under the various projects should have totalled 16 to 18 units manned by qualified national technicians. The total implementation capacity was to be close to 1,400 holes a year.

Overall estimate

The number of new water points to be established in the medium term is very large and it illustrates the amount of work still to be completed and the amount of financial resources to be found, even for a small daily allocation per inhabitant (20 to 30 litres).

Village water supply. Mali has almost 10,000 rural settlements of less that 2,000 inhabitants. With one hand pump for 200 inhabitants and at least two pumps for each settlement, the number of water points needed to meet the needs of the rural population totals more than 30,000, of which less than 2,000 have so far been installed.

Table 1

Ground-water development studies in Mali up to July 1983

Project	Financing	Purpose	Drilling unit (1983)	Drilling operations: No.	Drilling operations: Average success rate (%)	Drilling operations: 1982/83 campaign	Drilling operations: 1982/83 success rate (%)
UNDP projects (1967-83	UNDP UNDP/UNICEF	Hydrogeological reconnaissance - village water supply	1 FORACO SM-70 1 AC Aquadrill 441 1 AC Aquadrill 661 1 Failing 1500	1,655	50	196	70
MALI AQUA VIVA (1976-83)	NGO/EC Central Fund	Village water supply	2 FORACO SIS-66 1 FORACO	1,300	80	157	85
HELVETAS (1977-83)	Swiss co-op	Village water supply	1 FORACO SIS-66	420	80	75	80
Bandiagara (1978-83)	CARITAS	Village water supply	1 STENUIK HST-35	234	70	57	70
JICA (1978-83	Japanese co-op	Village water supply	1 TOP-200 1 TOP-300	36	90	15	80
ODEM (1977-83)	World Bank	Pastoral water supply	1 FORACO SIS-66	74	60	9	90
Mali South 1 (CMDT) (1983)	ABEDA	Village water supply	1 POLYDRILL-500 (enterprise)	18	80	18	80
Gao-Anderabou-kane road (1977-78	UNSO/ Netherlands co-op	Village and pastoral water supply	1 Failing 1500	15	-	-	-
West Sahel (1977-82)	Saudi Fund	Pastoral water supply	-	326	45	-	-
Kayes-Yélimané (1974-75)	FAC	Village water supply	-	26	45	-	-
Bafoulabé-Kéniéba (1974)	World Bank	Village water supply	-	61	40	-	-
			Total	4,165	65	527	

Table 2

Ground-water development projects in operation in 1983-1986

Project	Period	Financing	Purpose	Works planned
1. Continuation of earlier projects				
MLI/82/005	1983-85	UNDP UNDP/EDF? V	Village and pastoral water supply	750 boreholes 6 wells
Mali Aqua Viva	1983-85	FAC EC Central Fund	Village water supply	320 boreholes 25 wells
Helevetas	1983-86	Swiss Cooperation	Village water supply	300 boreholes
Bandiagara	1983-86	CARITAS	Village water supply	200 boreholes/ wells
PRODESO	1983-84	Saudi Fund	Pastoral water	34 wells
JICA	1983-84	Japanese co-operation	Village water supply	20 boreholes
Kaarta	1981-84	USAID	Village water supply	12 wells
Mali South 1 - CMDT	1983	ABEDA	Village water supply	42 boreholes
Gao-Menaka	1983-86	ADF	Village and pastoral water supply	275 wells
2. New projects				
UNICEF (Mopti - Timbuktoo)	1983-85	UNICEF/UNDP	Health/educ. infrastructure	144 boreholes
Route du sel (Timbuktoo-Taoudénit)	1983-84	UNDP	Creation of oasis	35 boreholes
Mali South 2 - CMDT	1984-86	EC Central Fund/ World Bank	Village water supply	450 boreholes
CEAO (Kayes-Koulikoro)	1983-85	EC Central Fund	Village water supply	480 boreholes 61 wells
Liptako-Gourma	1984-85	ADB/Kuwaiti Fund	Integrated dev.	51 boreholes 95 wells
Ground-water Koulikoro-Segou	1984-85	Saudi Fund	Village water supply	400 boreholes
Kita	1984-86	World Bank	Village water supply	580 boreholes
Ground water (Gondo)	1984-86	EDF V	Village water supply	100 boreholes 80 wells
Ground water (Kati-Bamako-Kangaba)	1984-86	Italian co-operation	Village water supply	350 boreholes
ODIK (Kaarta	1984-86	Canadian co-operation	Integrated dev.	100 boreholes
			Total: 4286 boreholes: 588 wells	

Table 3

Water-point requirement for village water supplies

Administrative region	Number of settlements with fewer than 2,000 inhabitants	Water points		
		Needed	Existing	Balance
Kayes	1,356	4,260	152	4,108
Koulikoro	1,792	5,403	551	4,852
Sikasso	1,654	5,315	508	4,807
Ségou	2,109	5,883	553	5,330
Mopti	1,931	5,992	150	5.842
Timbuktoo	667	2,225	8	2,217
Gao	340	1,352	27	1,325
Total	9.849	30,430	1,949	28,481

Urban water supplies. According to the 1976 census, there were 300 settlements with more than 2,000 inhabitants. Now, some 10 years later, there are probably more than 350, as a result of population movements caused by the persistent drought. The number of boreholes needed for water supplies in these settlements depends on several factors, in particular the local hydrogeological conditions and the effective yields of the holes. The total will be about 1,500 to 2,000 boreholes.

Pastoral water supplies. The requirement for pastoral water points situated mainly in the Sahelian zone should be evaluated on the basis of the national livestock development programme. The requirement is for several thousand water points scattered over areas of difficult access for which little hydrogeological information is available. Pastoral water points also require special installations to facilitate the speedy watering of the herds: large-diameter wells, motorized pumps, storage facilities, etc.

Problems of exploiting ground water

In the short and medium-terms the exploitation of ground water should not give rise to any serious problems with respect to water quality and conservation resources.

The salinity of the main exploited aquifers is low and they are recharged periodically during the rainy season by direct infiltration or migration of surface water. Furthermore, the dispersed mode of exploitation for both village and pastoral water supplies does not have a serious impact on stocks. In the areas of lateritic surface formations, the surface aquifers traditionally exploited by village wells will have their levels reduced by the exploitation of the underlying fissure aquifers. However, this effect could be limited if pools were quickly established above the villages to facilitate the artificial local recharge of the surface aquifers.

In certain regions such as Gourma, where the aquifer is deep and has no apparent recharge, the levels may decline as a result of the exhaustion of local stocks, which appear very limited owing to the small size of the permeable zones.

V. CONCLUSIONS

In the long term the development of ground water should be capable of meeting the water requirements, in quantity and quality, of the rural population and of the transhumant livestock which constitutes one of the country's main resources. It will help to stabilize village dwellers by furnishing them with a permanent water supply regardless of climatic uncertainties and in some cases offering them additional income from the small irrigation projects associated with the modern wells.

The ground-water potential is of course limited. The deposits are contained in aquifers with poor permeability and storage properties but which are sufficient for small-scale widely dispersed exploitation and for village and livestock needs. In addition, the winter rains and the migration of surface water guarantees a sufficient volume of renewable resources throughout the part of the country south of the 15th parallel N where almost all population lives.

In recent years the ground-water projects have been concerned mainly with the establishment of boreholes and other installations which could be completed quickly. But the authorities have not been neglecting the digging of large-diameter wells, for they do not have the pump maintenance problems common in pastoral areas. In 1983 the average cost of a cubic metre of water at the wellhead was between 80 and 120 Mali francs, with an installation cost of about FM 5 million for boreholes and FM 10 to 14 million for modern wells.

The national ground-water development policy takes into consideration both the hydrogelogical conditions and the way of life of the people. Now that a local industry has been established for the manufacture of hand pumps, the conditions seem to be met for the country to take charge in the medium term of the whole of its "water chain".

VI. REFERENCES

Archambault, J. — Les eaux souterraines de la boucle du Niger (Soudan français). BURGEAP, Dakar, 1953.

Archambault, J. — Les eaux souterraines de l'Afrique occidentale. AOF hydrological service, Dakar, 1960.

Benamour, A. — Hydrogéologie de la région Liptako-Gourma. CIEN, Ouagadougou, 1972.

BRGM — Carte de planification des ressources en eau de l'Afrique de l'ouest. CIEH, 1976.

Compagnie générale de géophysique — Etude par prospection électrique et sismique de la région de Gounam, Nara, Nema, 1956.

Compagnie générale de géophysique — Etude hydrogéologique par prospection électrique et sismique dans la plaine du Gondo (Soudan). 1955.

Companie générale de géophysique — Prospection aéromagnétique - Permis de KE-MACINA.

Dars, R. — Les formations sédimentaires et les dolérites du Soudan Occidental. University of Paris thesis, 1960

Defosses, M. — Contributions à l'étude géologique et hydrogéologique de la boucle du Niger. University of Strasbourg thesis, 1958.

Helvetas. — Projet de forages hydrauliques - Zone Mali Sud, 1981.

Palausi, G. — Contribution a l'étude géologique et hydrogéologique des formations primaires du Soudan méridional et en Haute-Volta, 1959.

Reichelt, R. — Géologie et hydrogéologie du Gourma. BRGM, 1963.

Rore, J. — Carte hydrogéologique du Soudan Occidental à l'échelle du 1 : 1 000 000, 1957.

Saad, K.F. — Etude hydrogéologique de l'Est du Mali. UNESCO, 1969.

Saad, K.F. — Etude hydrogéologique du Sud du Mali. UNESCO, 1970.

Saad, K.F. — Etude hydrogéologique du nord de la boucle du Niger, UNESCO, 1970.

Project MLI/67/507 — Strengthening of government services responsible for discovering and developing ground water. Project conclusions and recommendations. UNDP, 1975.

Project/MLI/74/001 Ground-water research and development. Final project report. UNDP, 1976.

Project MLI/76/004 Ground-water research and development. Project con- conslusions and recommendations. UNDP 1982.

Wahler & Associates Etude des eaux souterraines au Mali et compilation des données, 1980.

MAURITANIA

Area: 1,240,000 hm^2

Population: 1.8 million (United Nations estimate, 1983)

I. BACKGROUND

Mauritania has an Atlantic coastline of 500 km and forms a bridge between two African subregions: the Maghreb in the north and West Africa. The country straddles these two subregions in terms of its climate, geography and population, which includes two Moorish groups - of Arab type - Berbers and blacks. It has served as a transit route for herds, mainly of camels and goats, caravans and warriors between the "Saharan ports" - the oasis markets at the edge of the Sahara south of the Maghreb - and the Senegal valley in the south, which forms the southern frontier between Mauritania and Senegal. Ground water is a vital resource, for the country has no permanent surface water apart from the river. The Saharan peoples have lived traditionally from their animals and the water supplies which they bear when the grazing land is adequate and the water points not too far apart. However, in periods of drought, such as the one which has afflicted Mauritania for some years, the traditional water points and grazing land are not sufficient; this has brought about a profound change in the lifestyle of the nomads. Furthermore, increasing urbanization, education of children, improvement of health and social conditions and the development of agriculture and industry demand increasing amounts of water, and a permanent supply can be guaranteed only through the exploitation of ground-water aquifers which are protected from the effects of current climatic conditions.

Mauritania has few mountains and is a country of vast tablelands: the Reguibat peneplain in the north, the Atlantic coastal plain in the west, and the Taoudénit basin in the east. The average altitude is 200 to 400 m with a high point of 915 m (Kediet Idjil). Four-fifths of the country lies north of the 17th parallel N, which roughly corresponds to the 200 mm isohyet, and it therefore has a climate of the Saharan type with low and irregular rainfall and large interannual ranges; in some years hardly any rain falls.

In the rest of the country, south of the 17th parallel N, the rainfall is more regular and heavier, with an annual average of 600 mm at Selibabi. The rainy season is from June to October in these regions, where the population tends to be sedentary. The population density is also quite heavy in the Atlantic plain where the Saharan influence is tempered by the trade winds and the proximity of the ocean.

As a result of these morphological and climatic conditions, no permanent watercourse rises in Mauritania, except for the Gorgol, a temporary tributary of the Senegal (right bank). In periods of heavy rainfall this right bank is also fed by considerable runoff from the Mauritanian regions of Guidimaka and Aftout, which are drained by a very dense hydrographic network. Outside these regions the hydrographic network is weak and dispersed; it has no flow to the ocean and is partly blocked by sand.

II. GEOLOGY

Geogolists usually distinguish six main structural units (see general geological chart below).

- In the west, the Mesozoic and Cenozoic sedimentary coastal basin, which consists of sand-clay and limestone formations resting on the crystalline and metamorphic basement rock. The thickness increases from east to west, reaching 1,000 m at the coast;

- In the centre, the Paleozoic edge of the Taoudénit basin (Cambrian-Siluran), which includes the plateaus of Assaba, Tagant, Adrar, Khatt and Hank, topped with limestone and sandstone and bounded in the west by high cliffs;

- At the foot and to the west of these cliffs lies an "epimetamorphic" unit (Precambrian-Cambrian), which is part of the Arc des Mauritanides. It consists mainly of metamorphic schists carried down during the period of Hercynian orogeny;

- In the north, the Reguibat ridge - Pre-cambrian granitogneiss basement rock injected with green rocks;

- At the northern borders and in the north-west of the country (Zemmour-Noir-Yetti) outcrops the southern limit of the Tindouf syncline (Cambrian-Ordovician and Devonian) which consists mainly of limestone and sandstone;

- Lastly, in the east, the Mesozoic-Cenozoic Taoudénit sedimentary basin, which is one of the largest and least known basins in the world, for it is almost entirely covered with sand dunes. The formations increase in thickness towards the east, and the basin's centre is in Mali.

III. GROUND-WATER RESOURCES

Intense hydrogeological research has been carried out in Mauritania since 1950. The extent of this research can be seen from the list of works cited at the end of this article, in particular in the area of cartography.

Mauritania's ground-water resources have been more or less extensively prospected, depending on the region. They are found in very different kinds of deposits, as can be seen from the table below.

The coastal basin has several continuous aquifers lying one above the other, including the familiar Trarza aquifer which is bounded in the east by a dry segment and in the west by a saline segment, and the so-called Maestrichtian deep confined aquifer, about which little is known.

The Paleozoic sandstones and limestones of the Tindouf syncline and the Taoudénit basin also contain aquifers which are probably continuous, but little is known about them: the pelitic aquifer of Hodh, the Atar limestones, the Oujeft sandstones, and the (assumed) aquifers of the Cambrian-Ordovician sandstones of Zemmour-Noir and the Aoun sandstones. Mention must also be made of a Taoudenit aquifer which is probably continuous but little known except at the southern edge (aquifers of Dahr de Néma and the Ouartemachet trench).

The metamorphic and crystalline rocks of the Arc des Mauritanides and the Reguibat ridge contain local aquifers in the alteration, fracturation and fissuration fringes.

Lastly, several aquifers are found in the Quaternary formations; they are mostly small and scattered throughout the territory.

IV. GROUND-WATER STUDY AND EXPLOITATION

Most ground-water activities are the responsibility of the Ministry of Water and Energy and in particular of the Water Department, which has the following services:

Water resources

- Water points inventory and control office;

- Hydrogeological office: hydrogeological studies and control of drilling and test pumping;

- Geophysical office.

Water infrastructure

- Pastoral water-supply office: this office controls the nine well-construction teams based at Rosso, Ayoun, Néma, Kiffa, Aleg, Tidjikja, Kaédi, Selibabi and Atar;

- Drilling office: this office carries out drilling operations and is responsible for the installation, operation and maintenance of pumping stations.

Urban water supply

This service is responsible for water supply and sanitation in settlements of more than 2,000 to 2,500 inhabitants; this does not include the large towns where the networks are built and managed by the National Water and Electricity Corporation (SONELEC).

Equipment

This service has its main base at Ayoun and has approximately 70 vehicles, some compressors and INGERSOLL TH 100 drilling rigs.

Mauritania has several private well construction and drilling companies:

- SAFOR: this Mauritanian company has 10 mechanical well-digging rigs and units equipped with two Salzgitter RB30 rigs (combined rotary/down-the-hole hammer on a Mercedes truck and one heavier Salzgitter RB 225, also a combined rig.

- SOMAFOR: This Mauritanian drilling company has one combined INGERSOLL TH 100 rig (a second is planned).

Mention must also be made of the SODESE-ENGINEERING study company which is particularly concerned with surveys of water use in rural areas.

There is a very large programme of water drilling and well construction and repair for 1985 and subsequent years. It is financed by several bodies including:

- The West African Economic Community (CEAO);

- The European Development Fund of the European Economic Community (EDF/EEC);

- UNDP and associated bodies (the United Nations Sahelian Office in particular;

- The African Development Bank;

- The Arab Fund for Economic and Social Development;

- The World Bank;

- The Government of Saudi Arabia;

- USAID.

The programme envisaged the installation of 76 new wells and the repair of 180 existing wells.

The country's water points total about 10,000 to 12,000 (?), distributed as follows:

- More than 1,000 boreholes;

- About 3,000 modern wells;

- Several hundred waterholes;

- Five to ten thousand traditional water points (?), most of which are temporary excavations in alluviums.

The capital Nouakchott is supplied with water from a group of 18 boreholes at Idini, almost 70 km to the east, which draw water from a fossil aquifer in sands and sandstones of the continental terminal formation.

The problem of water - and especially of ground water - is in the forefront of Mauritania's priorities owing to the persistent drought which has exhausted the shallow aquifers which are the people's traditional source of water, and as a result of the sedentarization of the nomads and the development of urban and peri-urban communities. The considerable efforts made and the assistance furnished must be co-ordinated by rational planning designed to draw maximum profit from the investments and save water resources. To this end it is necessary to update and computerize the inventory of water points.

determine the needs and place them in order of priority, and improve groundwater research techniques and well-drilling technology, with a view to rapid implementation and adaptation to local natural and social conditions at minimum cost.

It is also necessary to determine the most suitable drawoff methods, in order to keep costs low and reduce the work of maintenance, repair and replacement; these operations must involve the Government's technical services and the local people, who should be largely responsible for implementation.

This is the goal of a United Nations project financed by UNDP which started up in 1986.

V. REFERENCES

General works

Maps and notes

BRGM — 1:200,000 hydrogeological maps and notes, "Ford-Gouraud-Oumm Dferat" (Plote, H.); "Faraoun" (Moussu, H., Trompette, R.); "Atar" (Moussu, H. Trompette, R.); "Chinguetti" (Moussu, H. Trompette, R.). 1966.

BRGM — 1:1,000,000 hydrogeological map of Mauretania and note, 1968.

BRGM — Planning map for ground-water resources in Sudano-Sahelian Africa (west sheet) and explanatory note, 1976.

BURGEAP — 1:1,000,000 hydrogeological sketch map of south-east Mauritania (see note 34), 1966.

Water Department — Eleven regional maps of water resources on scale 1:1,000,000 Seminar on "Water planning and use", Nouakchott, 8-13 December, 1980. Published 1980.

Paloc, H. — 1:500,000 hydrogeological map of the south-west Mauritanian basin and explanatory note. BRGM, Dakar, 1962.

Roussel, P. — 1:1,000,000 hydrogeological reconnaissance map of Mauritania and explanatory note. Water Department - RIM, 1968.

SONADER — 1:1,000,000 map of the main dam sites in Mauritania, 1979.

United Nations — Technical documents of the project "Water planning and use" (DP/UN/MAU-77-002/1). Final report: Conclusions and recommendations (1980). Communications, final report and 1:1,000,000 maps of the seminar of 8-13 December 1980 (Nouakchott).

Bourguet, L. — <u>Etudes économiques des différents types d'ouvrage - coût de l'eau - détermination des équipements les mieux adaptés.</u> BURGEAP R 414/2 855, 1980.

Razafy, C. — <u>Evaluation des besoins et ressources en eaux</u> (see note 76), 1980.

Other basic works and documents

Paloc, H. — <u>Connaissance hydrogéologique de la Mauritanie en 1962.</u> BRGM DAK 62 A-46, 1962.

UNDP — Project "Strengthening of ground-water ressources" (MAU 67-502-2). Final report of the executing agency (Conclusions and recommendations), 1975.

BRGM, BURGEAP, Water Department, HYDRA, INTRAFOR-COFOR, LNTP, MENDES, SASIF, SIF, SNIM — Drilling and sounding reports.

Roussel, P. — Guide des eaux souterraines de Mauritanie. Hydrogeological Office - RIM, 1966.

Toupet, C. and Laclavere, G. — Atlas de la République Islamique de Mauritanie. Editions Jeune Afrique, Paris, 1977.

Regional studies

North of the country

Bezeid Ould Mahamed — Le problème de l'eau à Zouérate. SNIM-SEM-MINING Department, 1981.

Bourguet, L. — Compte-rendu de tournées dans le Hank, le Ghallaman et le Zemmour-Blanc. Water Service - RIM, 1940.

Gouzes, R. — Etudes hydrogéologiques des palmeraies de l'Adrar. BRGM DAK 66-A5, 1966.

Roussel, P. — Alimentation en eau d'Atar. Hydrogeological Office -RIM, 1965.

Simonot, M. — Zone de l'Amneker. Reconnaissance hydrogéologique. UNDP MAU 67-502/2, 1974.

Trompette, R. — Les possibilités d'existence de nappes profondes dans le paléozoïque de la bordure occidentale du bassin de Taoudeni.

Weyns, W. — Zone de l'amneker, Graret El Fras, Oued Acharim. Rapport technique. UNDP MAU 67-502/2, 1974.

Weyns, W. — Zone du Zemmour-Noir. Zone du Zemmour-Blanc. Reconnaissance hydrogéologique. UNDP MAU 67-502/2, 1975.

South-east of the country

Baretto, M. — Inventaire des points d'eau et années d'observations climatologiques 1961-62 en Aouker. BRGM 63 A1, 1963.

Bourguet, L. — Synthèse hydrogéologique et aménagement hydraulique du sud-est mauritanien. Volume 1, 1:200,000 maps, BURGEAP R 415, 1966.

BURGEAP — Développement de l'élevage dans le sud-est mauritanien. L'inventaire des points d'eau effectué en 1974-75. BURGEAP R 186-E 304, 1975.

BURGEAP — Développement de l'élevage dans le sud-est mauritanien, (texte et annexes). BURGEAP R 188-E 304, 1976.

Delpy, J. — Fauille Ksar El Barka au 1/200,000. Notice hydrogéologique. BRGM, Dakar, 1960.

Delpy, J. — Notice hyrogéologique de la feuille Tidjikja au 1/500,000. BRGM, Dakar, 1961.

Gouzes, R. — Etude hydrogéologique des sources du Tagant et de l'Assaba. BRGM, DAK 62 A 34, 1962.

Lemoine, J. — Eaux souterraines et aménagement hydraulique du Hodh et L'Aouker (Mauritanie orientale). BURGEAP R 218, 1957.

Riou, V, and Bourguet, L. — Route Kiffa-Nema. Etude hydrogéologique sur le terrain. Compte-rendu, phase 2. BURGEAP R 153-E 321, 1974.

Roussel, P. — Hydrogéologie du Tagant et de l'Aouker. Hydrogeological Office - RIM, 1965.

Bourguet, L. — Implantation de puits pastoraux sur le Dahr de Néma. BURGEAP R 339, 1963.

West of the country

Bernet, G. — Alimenation en eau de Nouadhibou. Etude complémentaire de l'aquifère de Bou Lanouar en vue de porter la production de 3,000 a 6,000 m^3/jour.

BRGM — Etude géochimique des eaux souterraines de l'Afrique de l'ouest. Nappes du sud-ouest mauritanien. BRGM DAK 67 A 12, 1967.

BURGEAP — Alimentation en eau de Nouakchott. Etude préliminaire. Rapport No. 1. Etude des ressources en eaux souterraines. BURGEAP R 353/1979, 1979.

Depagne, J. — Alimentation en eau de Nouakchott. Reconnaissance hydrogéologique dans la région d'Idini. BRGM DAK 64 A 20, 1964.

Depagne, J. — Réalisation du dispositif d'exploitation et de surveillance de la nappe de Bennichab. BRGM 69 DAK 2, 1969.

Elouard, P. — Etude géologique et hydrogéologique du Guelba mauritanien et de la vallée du Senegal. BRGM, 1962.

Elouard, P. — Synthèse des travaux de recherches d'eau dans la nappe aquifère du continental terminal de Bennichab. Geological Laboratory of the Faculty of Science, University of Dakar, Rapport No. 8. 1964.

Martin, A. — Alimentation en eau de Nouadhibou. Données nouvelles sur la nappe du continental terminal, BRGM 69 DAK 9, 1969.

Martin, G. — Etude du tracé de la route Boutilimit-Kiffa. Rapport hydrogéologique concernant les études et recherches d'eau pour la construction de la route. BCEOM-BRGM 73 RME 024 AF, 1973.

Nuovo Castoro — Projet de développment de l'élevage dans le sud-ouest mauritanien. HYDROGEO, Rome, 1975.

Roussel, P. — Etude par sondage du secteur central de la nappe d'Akjoujt. Hydrogeological Office - RIM, 1968.

Roussel, P. — Caractéristiques et possibilités de la nappe d'Akjoujt. Hydrogeological Office - RIM, 1969.

SOGREAH — Aménagement de l'Aftout-Es-Sahel. Adduction d'eau de Nouakchott. Etude préliminaire, 1978.

Hydrology

Hoorelbeck J., Dubearnes, B and Sarr, D. — Atlas hydrogéologique de la Mauritanie. Hydrological Office - RIM, 1981.

BCEOM — Etude hydrologique de la Tâmoûrt En Naâj.

Brunet-Moret, Y. — Résultats de la campagne hydrologique dans le Brakna et le Tagant en 1958 et 1959, 1959.

Deveaux, R. — Hydrologie des principaux oueds de Mauritanie en 1960. Agricultural Engineering Service - RIM, 1968.

Inial, J.L. and Hamdinou, A. — Note sur l'hydrologie de quelques oueds mauritaniens en 1964. Agricultural Engineering Service - RIM, 1964.

Jaccon, G. and Camus, H. — Etude hydrologique du Ghorfa, du Niordé et des Oualos en amont de Kaédi en 1964, 1965, et 1966. Agricultural Engineering Service - RIM, 1966.

SCET International — Aménagement hydroagricole du Oualo du Gorgol, 1975.

Rochette, C. — Le bassin du fleuve Sénégal. Monographic hydrologique. ORSTOM No.1, 1974.

HYDROGEOLOGICAL BASINS

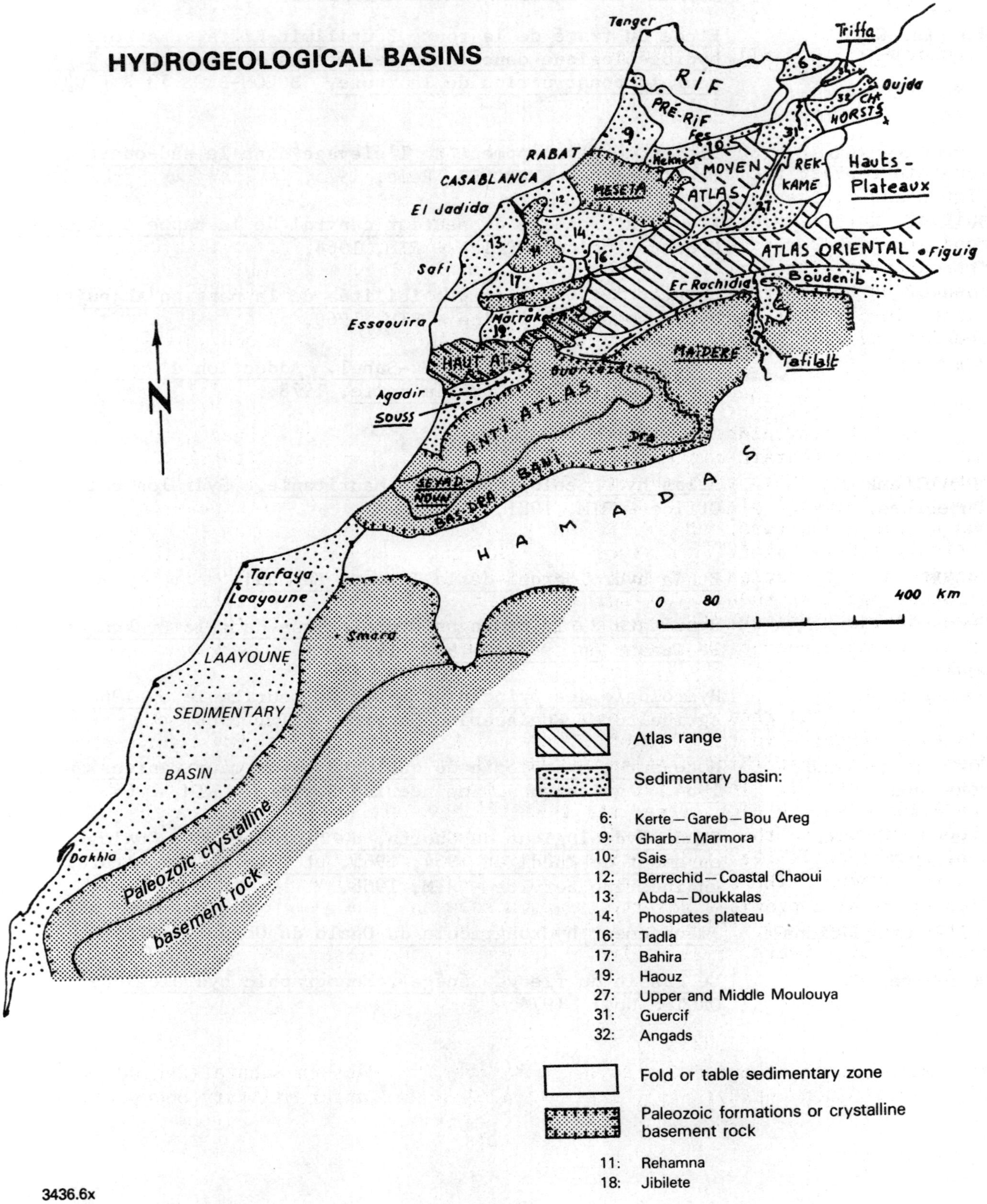

3436.6x

MOROCCO

Area: 446,000 km^2 */

Population: 22.11 million (United Nations estimate, 1983)

I. BACKGROUND

Morocco has a long Atlantic and Mediterranean coastline and two big mountain ranges: the Rif in the north with a high point of 2,456 m; and the High Atlas in the centre with a high point of 4,165 m. These conditions are suitable for a favourable distribution of sedimentary formations which provide good ground-water reservoirs. However, when the effects of low latitude, pronounced continentality and blocking of the trade winds combine, the natural conditions become very arid. The extreme south of the country has a Saharan clime; in contrast, the impermeable high-lying region of the Rif receives abundant rainfall which produces high and often torrential flood waters.

Large river basins

The Rif mountains in the extreme north of the country extend over 40,000 km^2 of uneven terrain consisting almost exclusively of marly flysch in over-thrust layers. With the exception of the limestone ridge (2,600 km^2) and a few small alluvial plains (650 km^2), the whole of the area is impermeable and has no large aquifers. While the 400 km of Mediterranean slopes receive relatively little rainfall (an average of 300 to 1,000 mm), the Atlantic slopes receive abundant annual rainfall (700 mm to more than 1,500 mm depending on altitude and longitude). The surface water is very unevenly distributed in time, drying up in the dry season and with devastating floods in some years; it is the main source of water; the main catchments are the Ouergha and the Loukkos.

The Atlantic area extends over 85,000 km^2 of mountain ranges, chains and plateaus arranged in the form of a fork which divided the country from north-east to south west (Middle Atlas and Beni Snassene) and roughly from east to west (High Atlas). The relief is high, rising to 3,364 m in the Middle Atlas and 4,165 m in the High Atlas; the land is therefore well watered on the slopes exposed to the ocean winds. The Liassic limestone formations are very well developed in the Middle Atlas and central High Atlas and contain large reserve stocks. The high altitude of the ancient impermeable massif of the High Atlas also provides reserve stocks throughout the year owing to the slow melting of the snow. These chains are the "water towers" which feed the country's big rivers, especially the permanent ones (Moulouya, Sebou, Oum er Rbia).

---

*/ Morocco has asserted its sovereignty over the Western Sahara (266,000 km^2 and 100,000 inhabitants), which it administers under military occupation.

The basins of the Atlantic slopes of the Atlas (75,000 km^2) are fed by the Atlas chains. Between the broad outcrops of the impermeable Paleozoic basement rock - for example, from north to south, the massifs of Meseta, Rehamna and Jbilet - lie vast detritic or limestone sedimentary basins containing thick confined and unconfined aquifers, such as the aquifers of Meknès-Fès, Rharb, Tadla and Haouz. These plains and plateaus are suitable for the development of large-scale modern irrigation, even though they already receive large amounts of rainfall: 500 to 600 mm/year for Rharb and Meknès-Fès in the north, and 400 to 200 mm/year for Tadla and Haouz in the south. Almost two-thirds of the irrigated land is found here.

The basins of the eastern slopes of the Atlas are all drained by Wadi Moulouya, the only permanent river in this vast area (56,000 km^2) which is poorly watered, for it is sheltered from the west winds by the Atlas chains; the depressions, such as the valleys of the middle and lower Moulouya-Chott Rharbi and Tigri, receive less than 200 mm/year, and the precipitation is hardly any higher on the high eastern plateaus, even though their altitude is above 1,000 m. Almost all the water resources in this area come from the mountains in the south and west in the form of flood waters; the minimum flows are small and used entirely for irrigation; the aquifers are small and heavily exploited, in particular through the artesian system of Ain Beni Mathar.

The basins of the Mediterranean slopes (5,000 km^2) are better watered owing to the screen effect of the Horst and Beni Snassene chains which block the way to storms from the Mediterranean; 70,000 ha are under modern irrigation in the plains of the lower Moulouya as a result of the construction of the Mohammed V Dam on the Moulouya. The other basins are of moderate size.

The basins of the South-Atlas area (460,000 km^2) have climates of the pre-Saharan and Saharan types; the aridity is tempered in some places by the altitude - the Anti-Atlas mountains which rise to 3,300 m - or by exposure to the influence of the Atlantic, as in the Sous plain. The precipitation is low, usually less than 200 mm/year, with less than 50 mm/year in the far south of the country in Tarfaya province, Saquiat El Hamra and Wadi Eddahab; the precipitation is also extremely irregular. The southern slopes of the Atlas give rise to rivers which produce sudden and violent floods (Sous-Drâa-Ziz-Guir); the spreading ground of these rivers has been cultivated since time immemorial but they are now regulated by dams. The ground-water aquifers are essential to the survival of these regions, even though they are often small in size with only intermittent recharge; the alluvial reservoirs are always the most important, for they are regularly recharged by flood waters. All these regions have serious problems with respect to the quality of the water, which is often polluted by leaching and evaporation.

Climate

Morocco forms a sort of climatic island at the edges of the temperate and subtropical zones, surrounded by the Mediterranean in the north, the Atlantic in the west and the desert Saharan areas in the south and south-east. Generally speaking, the climate is of the semi-arid type with some variations:

- The Mediterranean climate is apparent in the rainfall and hydrological patterns in the north of the country, more specifically in the Mediterranean and coastal fringe, along the axis of the Rif and in the Loukkos and Ouerrha basins;

- The influence of the ocean is clearly felt on the Mediterranean coast as far as Al Hoceima and in the central Atlantic plains, in particular in the northern part;

- A subtropical desert climate prevails in all the regions south and each of the Atlas. It is more intense further south and further inland.

The north-west part of Morocco, in particular the northern area, which enjoys the influences of both Mediterranean and Atlantic, is the best watered (500 to 1,800 mm).

The rest of the country receives less precipitation: 200 to 400 mm in the southern Atlantic area, eastern Morocco and Sous, and 40 to 200 mm in the pre-Saharan and Saharan areas.

As a rule the precipitation tends to decline from north to south under the influence of latitude, and from west to east under the continental influence. However, it increases with altitude and is heaviest on the exposed slopes in the north and west which intercept the wet winds. The main feature of the precipitation is its extreme irregularity within any year and from one year to another. The rain falls during a fairly short period: four to five months in the north, one to two months in the centre and centre-west, and on only five to ten days in the Saharan area. The irregularity of the precipitation is even more marked south of the Atlas where the amounts can vary in a ratio of 1 to 10 and even higher in some places. Periods of drought are common.

In the mountains the snowline is at 1,000-1,500 m; the precipitation ranges from 500 to 2,000 mm according to latitude, altitude and the exposure of the slopes. The snow can lie for several months (from November to July) but there are no eternal snows.

Outside the mountains and the pre-Saharan and Saharan areas, the average annual temperatures are around 18°C. (January is the coldest month); the average minimums are below 10°C, sometimes below 0°C in the mountains. In contrast, frost is rare in the Atlantic plains. July and August are the hottest months; the maximum average temperatures are above 30°C except on the Atlantic coast. The absolute maximum can reach 45°C.

The evaporation is very high and increases towards the interior. The annual figures (measured in the Piche evaporometer) are 800 mm at Tangiers, 2,200 mm at Marrakesh and 4,800 mm at Ouarzazate.

Surface water

Surface water makes up three-quarters of Morocco's water resources. Before 1960 little was known about most of the wadis, except for the catchment basins of the Oum er Rbia, Sebou and Moulouya and some of their main tributaries. The first measuring stations were established during hydroelectric development studies.

Floodwaters

In 1983 the observation network had more than 200 stations. A very big effort was made to equip most of them with limnigraphic recording systems. Some stations are incorporated in a radio-controlled flood warning network.

The hydrological system of Morocco's watercourses is of the torrential type, for the precipitation is concentrated in a very short period and is extremely heavy.

Accordingly, Morocco's wadis have rapid floods with fairly high flow coefficients. The hydrological system is therefore marked by extreme annual and interannual irregularity. Owing to the periodic droughts, particularly in the south and east of the country, installations have been built to regulate and store the flood waters, which represent 65 to 90 per cent of the total annual flow.

The flood waters usually occur from September to April. They are particularly high from December to February, during which period the daily rainfall figures can be very much higher than usual.

The pre-Saharan basins can have floods as early as September in some wet years. In the Moulouya basin the highest crests usually occur in spring when the snow melts.

Spring floods also occur in Atlantic Morocco, most of the water coming from the melting snow.

However, the commonest feature of the flood pattern is the heavy concentration of the flow resulting from the violence and rapidity of the phenomenon. For example, the peak flow is generally in excess of 1,000 m^3/s at the sluice gates of the S. Mohamed Ben Abdullah Dam, near Rabat.

During storms, particularly in the Rif and the pre-Rif and on the slopes of the South Atlas, the rainfall can be as heavy as 40 mm/h and even as high as 100 mm/h. This feature of the rainfall pattern is found most often in the upper basins of the pre-Saharan wadis where the exceptional floodwaters often cause major damage.

Examples of exceptional flood waters

Wadi	Station	Area of catchment basin (km^2)	Maximum flow observed (m^3/s)	Date	Design flood with frequency 1/10,000 (m^3/s)
Sebou	Ain Timedrina	4,387	1,050	10/10/66	2,900
Loukkos	Koudiet El Rhorfa	1,840	3,800	23/1/77	8,600
Moulouya	Melga El Ouidane	48,666	7,200	27/5/63	13,500
Bou Regreg	Md. B. Abdallah dam	9,800	5,000	2/63	12,000
Oum er Rbia	El Massira	28,500	3,200	1963	3,850
Massa) Sth. Atlas area	Youssef Ben Tachfine	3,780	2,800	15/12/57	5,500
Drâa)	Mansour Eddahbi	15,000	5,250	1949	7,000
Ziz)	Hassan Eddakhil	4,385	5,000	6/11/65	10,000

Lower-water period

The minimum flows are provided mainly by drainage from the aquifers. The alluviums absorb and replenish the water according to their porosity, thickness and variations in their slope. This is particularly the case in the Sous, Guir, Ziz, Ghéris, Drâa and Msour. Accordingly, the watercourses which drain catchment basins which have little ground water are not permenent. In contrast, those which cross areas where there are big aquifers have permanent low-water flows. This is particularly true for the Oum er Rbia and Sebou, which continue to flow even during long dry periods.

Low-water flows

Wadi (flow in m^3/s	Station	Minimum median monthly flow at 0.5 frequency (m^3/s	Minimum observed low-water flow (m^3/s)	Observation period	Module (m^3/s
Sebou	Ain Timedrina	7.2	2.70	1932-80	21
Moulouya	Melga El Ouidane	3.0	0.82	1965-79	33.8
Bou-Regreg	Sidi Mohamed Ben Abdelah	0.5	0.01	1927-80	20
Oum er Rbia	El Massira	41.8	10	1941-70	99
Drâa	Mansour Eddahbi	-	0.10	1931-55	13.4
Ziz	Hassan Eddakhil	1.5	0.0	1948-80	5.3

The flow is of course affected by drawoff and storage upstream of the measuring stations.

Sediment discharge

Erosion is of great importance in Morocco, as in the whole of North-West Africa; it has a particular impact on the basins drained by torrential wadis. For example, in the Rif area where the land is particularly vulnerable to erosion and the steep slipes are exposed to the weather, the sediment discharge can be as high as 3,500 m^3/km^2/year.

The sediment discharge occurs mainly during the flood period, with concentrations often in excess of 50 g/1. In some basins it can be as high as 100 to 150 g/1 during exceptional floods.

The following table gives the annual amounts of sediment discharge measured at the sluice gates of certain big dams.

Sediment discharge

Dam (wadi)	Lalle Takerkoust (N'Fis)	Im Fout (Oum er Rbia)	Nakhla (Nakhla)	El Kansera (Beht)	Mohamed V (Moulouya)	Hassan Addakhil (Ziz)	Sidi. Moh. Ben Abdallah (Bou Regreg)
Initial volume (10^6 m^3)							
	77.7	25	8.7	290	726	369	493
Average annual silting (10^6 m^3)							
	0.45	2.3	0.13	0.4	8.8	1.7	3

For all the dams which will be in operation in 2,000, with a total capacity of around 20 billion cubic metres, the annual average amount of sediment discharge may reach 70 to 80 million cubic metres a year, which would mean an accumulation of more than one billion cubic metres of silt in the dams over the next 20 years.

II. GROUND WATER

Government bodies responsible for water resources

The Water Department is part of the Ministry of Capital Development and it includes a water research and planning office.

At the central level this office has a management and planning section and a water resources section, which have specific responsibility throughout the Kingdom for the prospecting, study and evaluation of surface-water (hydrology service) and ground-water (hydrogeological service) resources.

The office has eight regional water offices (of which five are at present in operation) working on the study, exploitation, control and management of water resources within the natural limits of the big catchment basins (Sahara, Sous-Massa, Tensift, Oum er Rbia-Bou Regreg, Sebou, Loukkos, Moulouya, Nekor, Guir, Rheris-Ziz).

Background of hydrogeological research

Water research from 1956 to 1982 is summarized in the following table. A total of almost 9,300 boreholes and wells has been installed by the Department, representing a total depth of 456,000 m. These works have provided a potential supply of 45 m^3/s, i.e. almost 1.4 billion cubic metres a year.

After a slow-down from 1964 to 1968 the research work was given a considerable boost during the 1968-1972 five-year plan and again during the 1973-1977 economic and social development plan with the introduction of the pneumatic drilling techniques and the interpretation of aerial photographs as the basis for location of the wells, especially in the areas of hard rock in the Sahara.

During this period a large water research programme was undertaken in the Saharan provinces following their restoration to Moroccan authority in 1975, with a view to supplying water to a number of settlements. A total available yield of 12.9 m^3/s was obtained during 1973-1977, i.e. 400 million cubic metres a year, in the various regions of the Kingdom.

This effort was continued during the 1978-1981 three-year plan, especially in the Saharan provinces, with an average rate of 2.2 m^3/s.

The programme of the 1981-1985 economic and social development plan envisaged strengthening the research effort in the deep aquifers, in the arid zones and in particular in the Saharan provinces, with a view to evaluating the potential of the various aquifers. Particular emphasis was given to the mobilization of water resources to meet the long-term needs of the rural and urban population.

Water research, 1956-1982

(boreholes and wells)

Year	Number of boreholes and wells (exploi.+recon.+piez)	Depth drilled (exploi.+recon.+piez)	Available yield (m^3/s)	Cumulative available yield ($10^6 \times m^3$/year)
1956	302	5,154	1.15	36.2
1957	197	1,291	0.11	39.7
1958	131	1,300	0.37	51.3
1959	298	3,701	0.93	80.6
1960	158	4,359	0.40	93.2
1961	73	2,965	0.42	106.4
1962	198	8,900	0.35	117.4
1963	92	17,362	0.20	123.7
1964	204	10,746	2.84	213.2
1965	66	1,984	0.56	230.8
1966	531	23,184	0.74	254.2
1967	200	13,004	1.06	287.5
1968	145	10,557	2.11	354.0
1969	207	11,250	1.90	414.0
1970	196	10,878	2.58	495.2
1971	106	6,185	1.81	552.4
1972	109	4,585	1.70	605.8
1973	55	5,656	1.49	652.7
1974	258	15,787	1.92	713.2
1975	636	30,410	2.82	802.0
1976	744	38,106	3.55	914.0
1977	910	41,634	3.20	1,015
1978	619	30,970	2.82	1,104
1979	252	12,358	1.53	1,152
1980	612	33,460	2.07	1,217
1981	1,047	60,625	2.72	1,303
1982	927	56,704	3.89	1,425.8
TOTAL	9,273	463,125	45.24	14,354.3

Work carried out from 1974 to 1982

(reconnaissance and monitoring)

Year	Number of installations				Depths drilled (m)			Number of piezometric measurements
	Exploit.	Reconn.	Piezo.	Total	Exploit.	Reconn.	Piezo.	
1974	77	149	22	248	5,847	7,072	2,868	18,666
1975	140	478	18	636	11,154	18,357	900	21,732
1976	313	367	64	744	18,330	9,772	2,308	19,237
1977	513	390	7	910	22,720	19,061	80	19,746
1978	221	368	30	619	14,649	15,295	1,026	16,558
1979	215	35	2	252	9,992	789	1,577	12,742
1980	150	459	3	612	11,364	21,796	300	11,940
1981	290	729	28	1,047	14,602	42,505	3,518	13,142
1982	289	593	45	927	17,783	38,678	243	11,061
TOTAL	2,208	3,568	219	95	126,441	173,325	12,820	
(% per category	37	59	4	100	40	54	4	

Main aquifers: description, location, properties

In Morocco the main permeable formations in almost all the big stratigraphic divisions are capable of being aquifers. The reason for this lies in the very ancient and extensive tectonics which caused heavy overlapping of the various formations and created mountains where the precipitation is now very heavy.

Table 1 shows the stratigraphic situation of the main aquifers, with their dominant lithological facies and geographical location.

Table 2 gives the physical properties (depth, quality and stocks) and the hydrodynamic properties (permeability, transmissivity and storage coefficient) of the most heavily exploited water-bearing formations.

Rate of exploitation of ground water

The drawoff of ground water is in the order of one-third of the renewable resources (but almost a third of these resources come from upstream underground flow), i.e. about three billion cubic metres a year.

This drawoff represents a little less than half of the recharge of each of the Mediterranean and Atlantic areas, which alone provide more than two-thirds of the country's drawoff.

However, very little ground water is exploited in the Atlas (8 to 12 per cent), for 90 per cent of the water drains towards the downstream basins of the Mediterranean, Atlantic and South-Atlas areas.

However, the ground water is more heavily exploited in the South-Atlas area (81 to 100 per cent, although only 12 per cent of the total average recharge is drawn off. Before the recent big development programmes this very dry region could never count on permanent surface water resources and it was therefore forced to use ground water (large numbers of springs, rhettaras in the piedmont areas, and wells in the small alluvial plains).

III. EXPLOITATION OF GROUND WATER

Government bodies

The following government bodies are concerned with the exploitation of ground and surface water:

- The National Drinking Water Office (ONEP) of the Ministry of Capital Development is responsible for drinking-water supplies for the towns and large centres, at present 250 in number; it manages itself or in partnership the water tapping, treatment and distribution installations, etc. In co-operation with the Water Department the ONEP prepares the balance sheet of the resources mobilized and the existing and forecast requirements and it plans and implements the projects entrusted to it. Almost two million people have been supplied with water by the ONEP;

- The "water distribution boards" of the Ministry of the Interior are responsible for water supplies for the biggest towns;

- Nine regional agricultural development offices (ORMVA) of the Ministry of Agriculture and Agrarian Reform are responsible, *inter alia*, for the management of the water resources needed for the development of the large irrigated areas (a total of almost 600,000 ha). It manages both the dams (Massa, Tafilalt, for example) and the pumping stations. More than nine million people have been supplied with water in this way. Small and medium-sized systems at present irrigate more than 200,000 ha, but most of the resources are supplied by the natural flow of surface waters. All these bodies have competent staff, sometimes domestically trained (ONEP), and the necessary equipment.

Drilling enterprises

Morocco has many drilling enterprises. The National Office for Petroleum Research and Exploitation (ONAREP) and the Mining Research and Participation Office (BRPM) extend their own specific prospecting activities (petroleum, minerals) to ground-water operations: deep drilling by ONAREP, shallow wells and boreholes by BRPM.

There are also at least six private companies capable of performing all kinds of work with modern and efficient equipment. The works commissioned or supervised by the Water Department alone represented almost 57,000 linear metres in 1982 (wells, percussion or rotary drilling, pneumatic drilling) with depths ranging from a few tens to several hundreds of metres.

Use of ground water

Agriculture

From the quantitative standpoint ground water plays a less important role than surface water in agriculture. However, in addition to the big irrigated areas supplied with water from dammed reservoirs, there are other areas where the ground-water potential is sufficient for intensive cultivation. This is true of the market-garden zones which are developing around the big urban centres, for example along the Atlantic coast from Kenitra to Safi and in the Sous plain.

Urban and rural supplies

The resources at present used for drinking water are supplied in equal parts by ground and surface water. However, the drawoff of surface water is concentrated in a few developed basins (coastal area of Rabat-Casablanca, Beni Mellal, Berkane), mainly to meet the needs of a few towns, whereas ground water is used to supply the needs of all the secondary centres as well as some large towns located at some distance from the big rivers.

The drawoff for a few large towns which use only ground water is given below (millions of m^3/year).

Town	1981	1982
Fès		52.8
Marrakesh	29.6	26.0
Agadir	11.3	
Oujda		7.5

The increased exploitation of ground water is creating problems with respect to the exploitation itself and the global resources.

IV. PROBLEMS

Exploitation problems

The study of piezometric variations in the aquifers in Morocco over the past 25 to 30 years shows that several big aquifers are being over-exploited, i.e. that the draw off has been exceeding the recharge, sometimes for more than 10 years.

However, this development has sometimes been planned as part of an integrated surface-water development programme, the intention being that the surface water should eventually recharge the aquifer. Sous is an example of this.

Large drawdowns of up to several metres in 10 years have been observed in several aquifers; this affects the shallow traditional installations which have only a small depth of water, with the result that the wells or "rhetarras" (underground galleries similar to the kanats of Central Asia) must be deepened from time to time.

In the coastal area the development of market gardening irrigated by water pumped from the small coastal aquifers is clearly causing an increase in salt-water intrusion.

Some aquifers (especially alluvial) have had their average level reduced by several metres as the result of the commissioning of an upstream dam/reservoir (Drâa, Tafilalt). On the other hand, the plains and valleys below such dams are protected against the sometimes devastating floods and the quality of the ground water is improved by the reduction of direct evaporation from these aquifers.

Lastly and in contrast, the development of irrigation networks fed by surface water taken from the reservoirs of the big dams has sometimes produced considerable recovery in the water levels of the aquifers, in particular in Beni Moussa (10 to 20 m in four years) and Triffa. It has therefore been necessary to install drainage networks and step up the pumping of ground water.

Resource problems

The ground-water resources used in Morocco represent 55 per cent of the renewable resources; 18 per cent of these resources are lost to the sea or through evaporation, and the rest is drained by the watercourses and can be used downstream as surface water.

The prevention of these losses would provide an additional 10 to 20 per cent of the amount used at present, but this could not be done without special and costly development.

The present effort is therefore focussed on the identification of new resources as yet unknown (and therefore to be explored) or under-exploited, such as the deep aquifers. The aquifers are often recharging, with water of variable quality. However, the development of these resources would require sophisticated research techniques and deep and therefore costly installations.

The preservation of the exploited resources entails rational management, especially with respect to prevention of waste, which can be achieved only by repair of the distribution networks and improvement of the irrigation techniques. Surveillance measures are also in operation to protect the quality of the ground water, for there is growing awareness of the vulnerability of the aquifers and of the need to prevent any pollution.

Although the integrated management of ground and surface water has long been achieved at some sites - such as Charf el Akab since 1958 to supply Tangiers - it has still not become the rule. In 1970 a programme for the integrated development of the Sous plain envisaged increased over-exploitation, which had already reached a very high level (-60 hm^3/year - a threefold increase in 50 years) and acceptance of an average drawdown of about 20 m for the aquifer, which would be made good at a later stage by regulation of the surface waters. This initial programme was revised in 1980 to reduce the over-exploitation and bring forward the commissioning dates of the dams, as pumping was becoming very expensive owing to the series of increases in the cost of energy for the pumping operations. However, the introduction of integrated management is being impeded by financial and technical difficulties (in particular with respect to the identification of the areas and methods of artificial recharge), as well as by legal problems.

V. REFERENCES

All the information, published or not, on the hydrogeology and ground water of Morocco may be found in the following documents published by the National Geological Service:

- Bibliographie hydrogéologique du Maroc (période 1905-1957). Notes et Mémoires du Service géologique du Maroc, No. 142 (1958), 64 pp.

- Bibliographie hydrogéologique analytique du Maroc (période 1958-1968). Notes et Mémoires du Service géologique du Maroc, No. 220 (1969), 180 pp.

- Bibliographie des resources en eau du Maroc (période 1969-1978). Notes et Mémoires du Service géolique du Maroc. In publication.

- A general summary of the state of knowledge of water resources, published between 1971 and 1977: Ressources en eau du Maroc: Vol. 1 (1971); Vol.2 (1975): Vol 3 (1977). Publication of the National Geological Service.

Mention must also be made of the map of the aquifer systems of Morocco on scale 1:1,000,000 (northern province) with an explanatory note, published by the Water Resources Division of the Water Department of the Ministry of Capital Development.

ANNEX

Table 1

Description and stratigraphic situation of Morocco's main aquifers

Period		Facies	Extent of exploited aquifers
Quaternary		Recent alluviums: continues	Drâa-Ziz
		discontinuous	Moulouya-Trourit Oujda-Maidère
			Anti Atlas
		Sandstone/sand/clay	Gharb-Mamora-R'Mell
		Volcanic	Angads (Oujda)
		Limestone, marly limestone	Sais-Ain-B-Mathar-Tadla
		Sandy limestone	Abda Doukkalas-Chtouka-Tarfaya
		Conglomerates	Triffa-Haouz-Sous
Pliocene-Villafranchian		Lacustral limestone	Bahira-Seyad Noun
		Sandstone	Essaouira syncline
		Lancustral limestone/conglom	Tafilalt-Berrechid- coastal Chaouia
			Kerte-Garet-Bou Areg
		Marly lime/clay/shell sandstone/conglomerates	Sous
Pliocene		Detritic continental	Ouarzazate basin
		(Marl-flysch) */	Rif-Gharb
	Upper	Conglom/lacustral lime	Bahira
Miocene	Tortonian	Molasse, lime., marl	Sais-Charf El Akab
	Middle & Lower	(Marl-flysch)*/	Atlantic and Mediterranean basins
Eocene	Middle & Lower	Phosphated lime series	Atlantic basins: Bahira-Haouz
Paleocene		Sand and sandstone	Laayoune basin (Sahara)
Cretaceous			Laayoune basin (Sahara)
Upper	(Maestr.	Sand and sandstone	Rif
	(	(Marl)*/	Atlantic basins
	(Senonian	(Marl and gypsum) */	Middle Atlas
		Limestone (karst.)	Atlantic basins: Tadla. Phosphate plateau
Middle	(Turonian	Limestone and marly lime	Sous-Essaouira-Errachidia-Boudenib
	(Cenomanian	Dolomite	Berrechid)
	(Albian	Sandstone	) Laayoune basin
Lower	(Hauterivian	Dridrat limestone (karst)	Doukkalas-Sahel de Safi Azemour
Jurassic			
Upper	Batho-Bajo	(Marl)*/	Rif and pre-Rif
Middle	(Bajoc.	Zehroun sandstone	Pre-Rif
	(Aalian	Lime. dolomitic massifs	Middle Moulouya-Rekkame
			High plateaus
Middle and Lower Liassic		Dolomitic limestone often karstified	Moulouya (Angads) - Horsts chain Guercif-Causse-Oral High Atlas

Table 1 (continued)

Period	Facies	Extent of exploited aquifers	
Permian-Triassic			
Upper	(Gypsum clay) */	Berrechid-Rif-Sahara	
Lower	Sandstone/conglomerates	Rif (locally)	
Carboniferous	Schist/quartzite	Discontinuous aquifer Fissured areas	Jibilete
Devonian	Schist/lime/sand/conglom.		
Cambrian-Ordovician	(Hercynian granite schist/quartzite		Bani-Maidère + Meseta
Cambrian	Schist/grauwackes		Rehama
Low. Georgian	Dolomitic lime/sandstone		Anti-Atlas Maidère
Lower Cambrian Adoudounian	Karstified dolomitic lime	Anti-Atlas: Goulimine Ras-Draa - Seyad Noun-Maidère	

*/ Main aquicludes - impermeable strata capable of recharging underlying aquifers.

Table 2

Properties of Morocco's main aquifers

Hydrogeological properties

	Location	Geology*/	Depth of top or upper level (m)	Thickness (m)	Specific yield (1/s/m)	Permeability (10^{-4} m/s)	Hydraulic conductivity (km/day)	Average trans-missivity	Average storage coeff.	Renewable stocks ($10^6 m^3$)	Salinity g/1
Med. basins	Nekor	Q/PQ	5-40	10-200	0.2-20			1-60	2-6		1.5-3.5
	Triffa	Q	1-40	10-100				10.60			0.5-11
	Kerte	PQ	10-50	50-200		0.04-1.8		0.1-34		9	1-8
	Garet	PQ	5-40	10-80		0.2-10		0.1-140	2-4		2-8
	Bou Areg	PQ	5-40	5-60		0.001-72		8-50			2-8
	Angads	Q	15-75	10-100		0.1-10		1-8			1-3
Atlantic basins	R'Mell Larache	PQ	5-10	20-100		0.5-1.5		3-18	0.3-08	2	0.5
	Meknes-Fest: uncon.	PQ	5-70	10-50		0.1-500		0.02-100	1.5-3	650-1,300	0.2-14(2)
	con.	Liassic	20-300	200	10.30						0.2-0.8
	Gharb: unconfined	Q	1-5	10-30		2-5		2-30	0,002-0.02)		(1)-2-10
	confined	PQ	10-30	50-200		0.5-5		5-100	0.005-0.01)		1.5
	Mamora	Q/PQ	1-50	0-100		1-5		1-80	0.1-3.5)		0.2-1.2
	Dradere-Souere	PQ	3-40	30-70	1-4	0.5-5			)		
	Temara	PQ	5-50	0-10	<0.5			1	0.4)		0.5-1
	Berrechid	PQ	5-50	5-30		1.4-17		3-30	5	500-800	1-3
	Abda-Doukkalas	Q/PQ	40-80	10-20		0.1-6		0.1-1	0.8-5	150	1-3-(8)
	Tadla: B. Amir	PQ	2-20					1-50-(190)	2-4.5		2-5(8.5)
	B. Moussa	PQ	2-50	50				1-150	1-5		0.3-2-(8)
	confined	Turonian	100-800	20-50	1-20			(1)-20-300)	2 @ 4.10-5	10@50000	0.5-2
	Bahira	PQ	1-40	20-200		0.05-20		0.01-500			0.5-2(10)
	Haouz	PQ	10-70	10-500				2.40	1-8	50@100000	0.4-3.5
South of Atlas	Sous	PQ	5-100	100				0.2.200	0.8-10	40000	0.3-2
	Chtouka	PQ	15-60	20-150	1-6						0.6-1.4
	Tiznit	Q	5-30	5-30	0.6-7						0.6-2.5-(10)
	Middle Drâa	Q	5-15	10-40		1.5-25		7.50	2.20	350-500	0.4-10-(18)
	Tafilalt	Q	5-25	1-20		10		10-20	2-6	100-250	0.8-70
	Laayoune	Paleocene-Cretaceous	0-1.500	100-1,000	0.3			2		2,000-6,000	1.5-10

*/ Q - Quaternary; PQ - Pliocene-Quaternary

NIGER

Area: 1,267,000 km2

Population: 5.77 million (United Nations estimate, 1983)

I. BACKGROUND

Niger lies at the heart of West Africa, roughly equidistant from the Mediterranean (Tunis) and the Gulf of Guinea (Lagos). Most of the country is in the South-Saharan desert zone, north of the 15th parallel N. South of this line the climate is of the subtropical Sahelian type. Most of the population lives here. The country is generally flat, with altitudes between 250 and 450 m. The centre-north is occupied by the Air mountain range, which has a high point of almost 2,000 m.

In the north-east the Djado plateau (500 to 700 m) forms a sort of threshold between the vast high massifs of the Hoggar in Algeria and the Tibesti in Chad.

The only large watercourse is the Niger which runs for about 450 km through the south-west corner of the country. The south-east frontier runs through Lake Chad.

Niger is generally a very dry country: its northern half receives less than one metre of rainfall in 10 years (it would be wrong to talk of an annual average of 100 mm). In the south a narrow strip along the Nigerian border receives 400 mm of rainfall. Between these two areas the rainfall ranges from 100 to 400 mm, higher on the uplands. Accordingly, 87 per cent of the country is desert or steppe.

II. GEOLOGY

The Precambrian basement rock outcrops in the south-west as far as the Niger valley and to a lesser extent in the Zinder region; it is covered by the following sedimentary formations:

Sandstone-clay intercalated continental: Agades and Tagama formations;

Marine Cenomanian: hard limestone;

Marine Turonian: yellow sandy dolomitic limestones overlain by sandstone;

Argillaceous Upper Cretaceous;

Paleocene-Eocene: chalky limestones and schists.

This sedimentary series forms a subsidence basin of considerable thickness (several thousand metres) in the central area, north of Tahoua. East of the Air, in the Ténére-Chad subsidence basin, the sedimentary series is less well known. Tablelands of Primary formations outcrop in the north-east in the Mangeni-Djado: Ordovician sandstones and limestones, calcareous sandstones and clays of the Carboniferous and Devonian eras. The formations of the continental terminal outcrop over vast areas in the north-east of the country, as far as the Tahoua region. The north-west and east are covered with sandy formations, in particular the Ténéré-Bilma, which is one of the most inaccessible regions of Africa.

III. GROUND WATER

Ground water is a vital resource in Niger in view of the aridity of the country and the agro-pastoral and mining occupations of its people.

Hydrogeological studies and research have been carried out on a large scale, as have well-digging and drilling programmes, sometimes to great depths. Some older aquifers have been studied on mathematical models.

Niger has both continuous aquifers and discontinuous aquifers of only local importance.

Hydrogeological studies are the responsibility of the Water Resources Department of the Ministry of Water and Environment.

Continuous aquifers

Continental terminal of the Niger basin

This is a series of Oligocene-Miocene argillaceous sandstones with permeable layers of sand. The aquifer is about 100,000 km^2 in extent and is mostly unconfined but with artesian conditions in places.

Owing to the lithological variations, the aquifer's productivity is not uniform over this vast area. Some sections consist entirely of clay and are therefore barren. The aquifer is recharged by infiltration of rainfall of about 13 mm a year on average. This figure varies with latitude. The specific yields are low and decline from the outcrop areas at the edge of the basin towards its centre. The western continental terminal has two areas, one on each side of the river. Deep aquifers under pressure are found east of the river. The eastern continental terminal contains under the surface aquifer recharging aquifers in the lower and middle sands which form a multi-layered unit with the surface aquifer.

The renewable resources are estimated at a maximum of 50,000 m^3/km^2. The total recharge is about 1.2 billion cubic metres. The total amount of exploitable reserves is between 40 and 100 billion cubic metres. The storage coefficient is around eight per cent.

Intercalated continental

The intercalated continental and the Hamadian continental form the multi-layered aquifer of a large continental series of sands, limestones and argillaceous sands of Lower Cretaceous age and 500 to 700 m thick. The Hamadian continental includes sandstones and sands, which are often coarse-grained and correspond to the Middle and Upper Cretaceous lagoonal and marine formations, which consist of clays, limestones and silts. In Western Niger the intercalated continental and Hamadian continental form continuous deposits. The aquifer is unconfined in its eastern part and becomes confined towards the west beneath the Lower Cretaceous or Tertiary clay series, which can be up to 600 m thick; artesian conditions are found in the deep valleys. The average recharge of the unconfined aquifer is estimated at 3 mm/year, and the precipitation is below 400 mm; however, the stocks of water are large owing to the great extent of the outcrops of the unconfined aquifer over an area of 200,000 km^2. The series has large water-bearing sandstone strata; these are the Tegama sandstones of Lower Cretaceous age and an average thickness of about 500 m, which lie beneath a cover of 500 to 600 m which makes them difficult to tap. This unit lies on Irhazer clays above the aquifers of the Paleozoic formation and the Agades group; in the in the south and centre it is in direct contact with the Precambrian basement rock but the extent of the contact is unknown.

The renewable resources of the unconfined aquifer total 650 million cubic metres a year and the storage coefficient is 5 to 10 per cent.

The exploitable resources total 100 to 200 billion cubic metres.

The area of the confined aquifer is about 150,000 km^2. The exploitable stock is between 50,000 and 500,000 m^3/km^2.

Sandstones of the Agades or Tchirezrine group of Triassic and Jurassic age 1/

These sandstones form a large unconfined or recharging aquifer with an area of 30,000 km^2 in the unconfined part and 20,000 km^2 in the confined part, which can produce artesian springs in topographical depressions. The facies becomes mainly argillaceous towards the east, while in the west the series rests in discontinuity on Paleozoic sandstones containing saline water. The outcrops are small in extent, and the recharge is provided mainly by the intermittent and meagre runoff from the Air. The amount of recharge is very small in comparison with the stocks. No fluctuation in the level has been observed. The exploitable stocks are about 15 to 25 billion cubic metres.

Telous sandstones

These sandstones lie north of the 18th parallel N (25,000 km^2) and form a rather poor aquifer, usually recharging (no artesian conditions). The water table lies at a depth of 30 to 80 m. The exploitable stocks total about one to two billion cubic metres.

Paleozoic sandstone

These sandstones are found between the basement rock of the Air and the Telous sandstones; they are separated from each other by thick clay series. Four aquifers have been identified.

The Ordovician aquifer (5,000 km^2). Although it is not yet exploited, this Ordovician sandstone aquifer represents a considerable potential resource in view of its area and volume, its permeability, its excellent chemical properties and its large stocks of artesian water. The biggest problem is its great depth, for half of the aquifer lies at a depth of more than 800 mm. Most of the aquifer is in Algeria and it projects only slightly into Niger.

The Upper Devonian aquifer. This sandstone series is more argillaceous than the previous one. It can be exploited in the small coarse-grained sandstone layers at the edge of the Air, but it is of limited importance in comparison with the Lower Visian aquifer which lies above it.

The Carboniferous aquifers, including:

- The Farazekat sandstone aquifer (Lower Visian - 30,000 km^2), mostly confined and artesian; the producing wells have artesian discharges of 3 to 4 m^3/h (Fouka) and about 30 m^3/h (Tessellamane); the water is fresh in the northern part of the aquifer, saline in the middle part;

- The Tagora sandstone aquifer (Namurian) contained in two sandstone strata separated by clays. Little is known about this reservoir despite the many holes drilled in it.

The direct recharge of these aquifers, which are as a rule confined, is very small in comparison with the exploitable stocks.

The exploitable stocks are in the order of 500,000 to 1 million m^3/km^2, and the storage coefficient is four to eight per cent.

Formations in the east of the country ranging from Upper Cretaceous to continental terminal.

This area corresponds to the sedimentary basins of Termit-Agadem and Bilma.

Little is known of the structure of these basins; apart from the sandstone ranges of Termit-Agadem and Bilma, their surface is covered by a thick and continuous erg.

The intercalated continental formations of the Termit-Agadem basin lie under Middle and Upper Cretaceous formations, which consist of clay for a great depth, with sandstone at the top.

The existence of water points at the foot of the Termit range indicates a continuous aquifer at the level of the upper sandstone sequence.

The situation is much the same in the Bilma basin, except that in the north of the basin the Cretaceous formations rest directly on the basement rock or Primary formations.

The Bilma sandstone formation, like the Termit sandstone, probably contains a continuous aquifer.

In the absence of concrete data on the dimensions of the aquifers probably contained in the Cretaceous sandstones and on the hydraulic properties of these sandstones, it is at present difficult to determine the effective reserves.

The exploitable stocks are estimated at 500,000 to 1 million m^3/km^2.

Tafassasset sands (Tenere)

No water point has been identified here. Geophysical data suggest the possible existence, beneath the sand cover, of a tectonic trench filled with sediments which would contain confined water.

In the absence of concrete data and in view of the rainfall of these regions, it is not possible to indicate any usable resources.

Formation of the Koramas basin

The water-bearing complex of the Koramas basin (105,000 km^2) can be described as a two-layered aquifer.

The upper layer consists mainly of silt with some sandy clay and it contains an aquifer recharged by rainfall in the alluvial valleys and topographical depressions. This aquifer exists in practice and can be exploited only within and in the immediate vicinity of these valleys and depressions. This layer contains considerable renewable resources but they are difficult to tap because the exloitable stock is small. The lower layer consists of clay-sand formations of the continental terminal and contains a continuous aquifer which is not directly recharged by rainfall but does have exploitable stocks; here again, they are difficult to tap owing to the poor permeability of the terrain.

The renewable resources are about two billion cubic metres and the infiltration is estimated at 127 mm/year.

Pliocene-Quaternary

The Lake Chad area has a heterogeneous Pliocene-Quaternary formation consisting of a fluvial/lacustral series and dune formations. It contains an unconfined aquifer in its upper part. Beneath this aquifer there is a sand aquifer containing the middle pressurized aquifer of the Chad formations which gives good artesian discharges in the vicinity of the lake.

The top of this aquifer lies at a depth of 250 to 300 m below ground level. There is probably also a third aquifer, at an even greater depth and under higher pressure.

The unconfined aquifer has an area of 100,000 km^2. Its exploitable stocks are about 25 to 60 billion cubic metres and the recharge around 300 million cubic metres. The middle aquifer is about 25,000 km^2 in area. The exploitable stocks are 5 to 10 billion cubic metres.

Discontinuous aquifers

The crystalline basement rock outcrops or suboutcrops in various parts of Niger: in the south-west it forms part of the "Voltaic shield", in the north part of the Air range, and in the centre-south part of the Maradi and Zinder ranges.

In Niger the Voltaic shield terminates in the Liptako mountains which consist of schists, quartzites and green rocks with injections of granites and green rocks; the series is of Middle and Upper Precambrian age. The water resources are small and are found only in the alteration zones.

In the Air the conditions are even less favourable owing to the obstruction of the fissures by argillaceous material and the decomposition of the granites and gneiss. Water resources are found only in the alluvium areas. The presence of metamorphic quartzites and limestones gives some hope of finding ground-water resources in places.

In the Zinder mountains there are arkosic sandstones which contain water in their depressed areas where they are charged by surface runoff following storms. These formations cover some 130,000 km^2 and the annual recharge is in excess of 300 million cubic metres.

IV. EXPLOITATION OF GROUND WATER

In January 1986 the number of modern installations - i.e. large-diameter wells lined with concrete and equipped with a filter block - and of boreholes equipped with pumps totalled about 10,000, two-thirds of them wells and one-third boreholes. The present construction rate is about 1,000 to 1,200 new installations a year. At this rate, the total estimated requirement of 25,000 installations would be met by 2000. Of the 850 water points brought into service in 1985, 740 were boreholes equipped with pumps. The well-construction capacity is about 400 installations a year, but this rate is declining owing to the slowness and cost of the work. In contrast, some 3,000 boreholes are to be installed in 1986-1988.

The maintenance of the pumps poses all kinds of problems and cannot yet be satisfactorily guaranteed. In 1986, for example, 50 per cent of the pumping installations were non-operational.

Accordingly, the Government's water policy is focussed on three goals:

- Completion of about 15,000 installations to meet the requirements for 2000;

- Action with the people to ensure that they have the will and the means to maintain and repair the pumps and keep them in operation;

- Development of small-scale irrigation in the villages if the ground-water resources are available and if it is technically and economically possible.

Many friendly countries are helping to finance technical co-operation projects for the development of ground water: Belgium (boreholes/manual pumps), Denmark (boreholes/manual pumps), Italy (deep boreholes), Switzerland (hydrogeology), the European Communities (small projects: 600 wells), France (provision of experts, small-scale irrigation, hydrogeological studies), Canada (small-scale irrigation by motorized pump), Japan (supply of equipment), and Netherlands (small-scale irrigation, manufacture of hand pumps).

The Ministry of Water and Environment carries out the country's water policy, especially with respect to ground water. It works through the Department of Water Resources (DRE), the Department of Water Infrastructures (DIH), and the regional offices which are being established at Zinder, Maradi, Tahoua and Diffa.

V. REFERENCES

Biscaldi, R. Reconnaissance hydrogéologique du Liptako-Sud (Niger). Rapport de fin de campagne 1961-1962. Report BRGM DAK 62 A 39, Archives of Ministry of Public Works Mines and Water, Niamey, 1962.

Biscaldi, R. Hydrogéologie du Damagarim et du Mounio (Niger). Report BRGM NIA 63 A 3, Archives of Ministry of Public Works, Mines and Water, Niamey, 1963.

Biscaldi, R. Reconnaissances hydrogéologiques du Liptako-Sud (Niger), 2ème partie novembre-décembre 1962. Report BRGM DAK 64 A 3, Archives of Ministry of Public Works, Mines and Water, Niamey, 1964.

Biscaldi, R., Greigert, J., Joulia, R. and Pirard, F. Recherches et exploitation d'eaux souterraines dans la République du Niger. Compte rendu des sondages et forages executés de juillet 1962 à juillet 1963. Report BRGM DAK 64 A 5, Archives of Ministry of Public Works, Mines and Water, Niamey, 1964.

Bockh, E. Contribution à l'étude hydrogéologique de la zone sédentaire de la République du Niger. Report DAK 65 A 20, BRGM and Bundesanstalt für Bodenforschung (B.f.B), Hanover. Archives of Ministry of Public Works, Mines and Water, Niamey, 1965.

BRGM République du Niger, Carte de reconnaissance hydrogéologique du Niger sud-oriental. Echelle 1/1,000,000, 1964.

BRGM Atlas des eaux souterraines du Niger: Greigert, J. - Etat des connaissances, mai 1978, Vol. 1, 9 sections; Bernert, G. - Bibliographie à la date du 30 Avril 1978, catalogue des forages sélectionnés, Vol. 2, 1 section, 1979. Comité Interetats d'Etudes Hydrauliques/BRGM Notice explicative de la planification des ressources en eau souterraine de l'Afrique Soudano-Sahélienne (Section: Niger), 1976.

FAO Etude de la mise en valeur du Dallol Maouri (Niger). Report FAO/UNDP SE 281 NFR 8.

Faure, H. Reconnaissance hydrogéologique du Niger Oriental. Mission 1960. Report, BRGM, Dakar, 1960.

Gagnière, G. and Vadon, J. Etude pour le développement des régions de socle hydrogéologique de l'Ouest Nigérien. Rapport provisoire de fin de première campagne. Report BRGM 73 NIA 002, Archives of Department of Mines, Geology and Water, Niamey, 1973.

Greigert, J. Rapports des mission géologiques au Niger (1948-1950). Report, Department of Mines, AOF, Dakar, 1951.

Greigert, J. — Note sur les recherches d'eau executées dans le cercle de Taouha (Niger). Report, Department of Mines and Geology, AOF, Dakar, 1952.

Greigert, J. — Introduction à la connaissance hydrogéologique du bassin occidental du Niger. Rapport fin de campagne 1955-1956. 3ème partie. Les principales nappes du bassin occidental du Niger. Report, Department of Mines and Geology, AOF, Dakar, 1957.

Greigert, J. — Recherches et exploitation des eaux souterraines dans la République du Niger. Compte rendu des sondages et forages executés de juillet 1963 à juillet 1966. Report BRGM DAK 66 A 7, Archives of Ministry of Public Works, Mines and Water, Niamey, 1966.

Greigert, J. — Atlas hydrogéologique de l'Adar Doutchi. Report BRGM 68 ABI 004 NIA, Archives of Serv. G.R., Niamey, 1968.

Greigert, J. — Les eaux souterraines de la République du Niger. (2 volumes). Report BRGM 68 ABI 006 NIA, Archives of Ministry of Public Works, Transport, Mines and Urban Development, Niamey, 1968.

Greigert, J. and Sauvel, C. — Modernisation de la zone pastorale nigérienne. Etude hydrogéologique. Report BRGM 70 ABI 006 NIA, Archives of Department of Livestock, Niamey, 1970.

Greigert, J. — Modernisation de la zone pastorale nigérienne. Proposition d'un programme de construction de captages d'eau dans la zone de l'élevage permanent des bovins. Ministry of Rural Economy, Department of Livestock. Report BRGM 73 NIA 001, 1973.

Greigert, J. — Etude pour le développement du socle de l'Ouest nigérien. 2ème campagne: implantation de puits par sondage à la tarière mécanique. Report BRGM 75 AGE 009, 1975.

Joulia, F. — Note sur l'hydrogéologie du Damergou (Niger). Report, Department of Mines and Geology, AOF, Dakar, 1953.

Joulia, F. — Hydrogéologie des régions à l'ouest et au sud de l'Air. Report BRGM DAK 65 A 16, Archives of Ministry of Public Works, Mines and Water, Niamey, 1965.

Marget, J. — Aquifère du continental intercalaire du bassin sédimentaire du Niger (Mali, Niger, Nigéria). BRGM bulletin 2/III, No.2, pp. 177-183, 1982.

Pirard, F. — Reconnaissance hydrogéologique du Niger oriental. Rapport de fin de travaux 1960-1963. Essai d'esquisses hydrogéologiques. Report BRGM DAK 64 A 11, Archives of Ministry of Public Works, Mines and Water, Niamey, 1964.

UNDP/FAO — Commission du Bassin du Lac Tchad (Cameroun, Niger, Nigeria, Tchad), Rapports divers 1969-1973. United Nations Development Programme, Rome, 1973.

Reichelt, R. — Géologie et hydrogéologie de la bordure sud-est du Gourma. Rapport de fin de campagne 1961-1962. Report BRGM DAK 62 A 29, 1962.

Sauvel, C. — Hydrogéologie de la haute Maggia (Niger). Report BRGM NIA 66 A 3, Archives of Ministry of Public Works and Mines, Niamey, 1966.

Sauvel, C. — Rapport de fin de campagne 1965-1966. Hydrogéologie de la vallée de Keita (Niger). Report BRGM NIA 66 A 3, Archives of Ministry of Public Works and Mines, Niamey, 1966.

Sauvel, C. — Hydrogéologie des vallées sèches de l'Adar Doutchi (Niger occidental). Thesis, Montpellier University, 1968.

Tirat, M. — Hydrogéologie du continental terminal Rapport préliminaire. Report BRGM DAK 62 A 42, Archives of Ministry of Public Works, Mines and Water, Niamey, 1962.

Tirat, M. — Contribution à l'étude hydrogéologique du continental terminal. Report BRGM 64 A 1, Archives of Ministry of Public Works and Mines, Niamey, 1964.

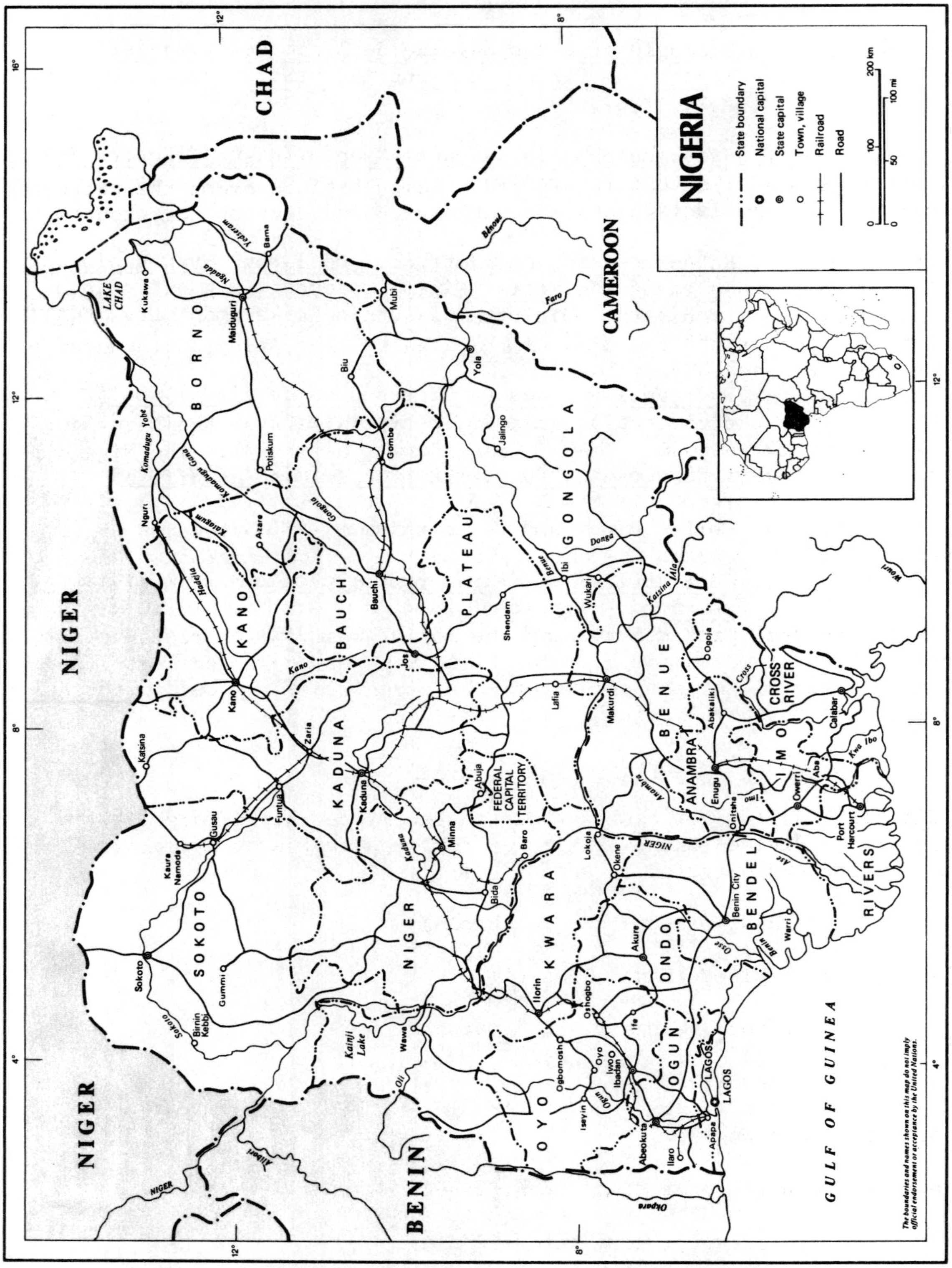

MAP NO.3115 Rev.2 UNITED NATIONS
FEBRUARY 1985

NIGERIA

Area: 923,770 km^2

Population: 89.02 million (United Nations estimate, 1983)

I. BACKGROUND

Nigeria has the biggest population in Africa, with about a fifth of the continent's people. In 1984 it occupied eleventh place among the petroleum producing countries, twelfth place for reserves and seventh for exports.

This West African country is shaped like a trapezium, with its short base formed by the 800 km of coastline on the Gulf of Guinea, half of which is occupied by the Niger delta. The long base of about 1,000 km touches Lake Chad in the north. The trapezium's highest side, i.e. at the country's eastern end, is about 800 km long.

The heart of the country is crossed by the valleys of the Niger and its main tributary, the Bénoué. These rivers form a capital Y. The Niger is one of the longest rivers of Africa. Its basin affects nine countries.

In Nigeria the land rises steadily from the coast northwards up to a maximum altitude of 400 m. The east is occupied by the Cameroon plateaus, which are 800 to 1,400 m high on average, and up to 2,400 m in places.

In the central part the mass of the Jos plateau, which rises to 1,781 m, dominates the vast surrounding plains. The southern part receives 1,000 to 1,500 mm of rainfall a year; the south-east receives more, but in the drier regions in the north the rainfall is below 500 mm.

Hydrographic system

Nigeria has 11 river basins which together are shaped roughly like a tree. The country is divided into eight zones for the purpose of hydrological observations and exploitation of water:

1. Northern Niger (Sokoto-Rima basin);

2. Central Niger (Kaduna basin);

3. Upper Bénoué (Gongola and Tomba basin);

4. Lower Bénoué;

5. Southern Niger;

6. Western littoral zone (Ogun, Osse);

7. Eastern littoral zone (Cross basin).

The heaviest flows are from June to October, with the crests in August-September. The flows are usually regular, but with occasional rapids when the water flows directly over the crystalline basement formations without any alluvial cover.

As they cross the margin of the crystalline basement rock and the sedimentary basin the water courses lose force and spread out in vast flood plains known in the north as "fadamas".

In the north the high evapotranspiration rates reduce still further the downstream flows, but this phenomenon is offset by the draining of ground water towards the surface which maintains a permanent minimum flow. Some typical data are given in the table annexed to this paper.

In the southern part of the country, i.e. in the low-lying sedimentary areas of the eastern and western littoral zones, the climate is of the south-tropical type with:

- A long wet season from March to July;
- A short dry season from mid-July to mid-August;
- A short wet season in September-October;
- A long dry season, when the harmattan blows, from November to February.

The country's hydrographic network consists mainly of the Niger-Bénoué system and its many tributaries; the Niger discharges about 220 billion cubic metres of water a year through its delta. There is also a dense network of rivers and smaller watercourses, most of them permanent, with a total combined annual flow of 200 billion cubic metres.

Geology

The country consists of Precambrian formations of the "base complex" which outcrop over half of the territory. These are zones of gneiss, mica-schists schists and amphibolites, extensively penetrated by ancient granitic batholites - hard, mostly tectonized sequences forming the Goura and Kagoro mountains of the Jos plateau - and by younger granitic batholites.

In the centre, in the basin of the Niger and the Bénoué, the ancient basement rock is covered by marls, sandstones and clays, saliferous in places, of the intercalated continental (500 m), overlain by the marine Upper Cretaceous formation, which consists of thin Cenomanian-Turonian limestones, clays and layers of carbon. The Cretaceous sediments increase in thickness towards the south in the direction of the vast coastal subsidence zone.

The Eocene period is represented in the southern sedimentary basin; it includes clay strata which can be up to 1,000 m thick. In the south-west and south-east the sediments are less thick (100 m) and include sandstone and phosphated sand layers.

The marine Eocene period has also been identified in the north (Sokoto province).

The continental terminal covers a large area in the north-east, i.e. in the Chad basin, with argillaceous sands, sandstones and diatomites to a depth of 700 m. In the south the continental terminal includes two series: the lignite group (sandstones and clays with lignites for 25 m), covered by the

"Benin" argillaceous sands of Pliocene-Pleistocene age (250 m).

Some areas of basaltic volcanism of uncertain age are found in the centre and east.

The main sedimentary series include the Sokoto basin, the Chad basin, the middle Niger, the Bènoué trench, the eastern and western low-lying land in the south, and the coastal plains, including the Niger delta.

II. GROUND-WATER RESOURCES

Ground-water bodies

The following bodies have overall responsibility for water resources:

- The Federal Department of Water Resources;
- The Water Boards and Water Corporations of each of the states of the Federal Republic of Nigeria;
- The National Electric Power Authority (NEPA) which manages the hydro-electric power stations;
- The Ministries of Agriculture of the Federal State and the federated states;
- The Federal Department of Meteorology;
- The local office of UNICEF.

Only the first three bodies are directly concerned with ground water exploration.

Background of hydrogeological research in Nigeria

The Mining Service of Southern Nigeria made the country's first geological studies from 1903 to 1913.

In 1938 geological and geophysical work was carried out by petroleum-research teams (Shell d'Arcy). At that time a number of geologists prepared maps of the sedimentary formations, notably du Preeze (1945) and Simpson (1947).

Of all the major geological and hydrogeological studies covering vast areas of the country, mention must be made of those by Anderson and Ogilbee (1973), Barber (1965) and White (1972), Dessauvagie (1975), Deswardt and Casay (1963), du Preeze and Barber (1975), Hazell (1961), Jones and Hockey (1968) Kiser and A. Kingbelin (1966), Kiser (1968), Kogbe (1976), Kowal (1972), Miller and others (1968), Raeburn and Tatam (1930), Raeburn and Jones (1964) and Rayment (1965).

The training of new or experienced staff can easily be carried out in the field by assigning them to the study and operational companies working in the country. Some young people have been trained at the National Institute for Natural Resources at Kaduna. The Institute's Department of Ground-Water

Research was reorganized in 1985; it was to be strengthened so as to be able in the future to train all categories of staff in the field of ground water.

Prospecting methods

Various methods of geophysical prospecting are used in Nigeria for ground-water research, including:

- Airborne geophysical prospecting;

- Electro-resistivity, mostly using the Schlumberger apparatus;

- Seismic refraction;

Electrical resistivity methods are the most commonly used owing to their low cost and easy operation.

The geophysical prospecting has concentrated on the crystalline basement rock, in particular since the start-up of the national water-drilling programme in 1981.

Main aquifers

Nigeria's main aquifers are found in the sedimentary formations which cover 524,000 km^2 of the country and range in age from Upper Cretaceous to Recent. The various basins are described below.

1. Sokoto basin

This basin is in the north-west of the country and consists of an area of more than 146,000 km^2 covering most of Sokoto and some parts of the Kaduna states. It is a south-east extension of the great basin of the Iullemenden which covers vast areas of Mali, Niger and Algeria and part of Mauritania. The depth of the water table in the formation ranges from 15 to 75 m for the unconfined aquifers., Elsewhere, artesian conditions occur in the sediments of several formations. Artesian aquifers are found at depths of 75 to 100 m close to the eastern edge of the Sokoto basin; the depth increases towards the west to a maximum of 390 m near the frontier between Nigeria and Niger. The depth of the water below ground level ranges from 2 to 54 m.

2. Chad basin

This basin covers an area of about 130,000 km^2 in the north-east of Nigeria. It includes the whole of Borno State and part of Bauchi and Kano; it has 53,000 km^2 of artesian aquifers.

The Chad formation consists mainly of non-consolidated fluvial and lacustral sands, together with silts and clays which were probably deposited during the Pliocene-Pleistocene era. The sequence includes intercalations of large lenses of grey plastic clay which can put the aquifers under pressure, i.e. create artesian conditions in some parts of the basin. Three main aquifers have been identified to the east (Borno State): the "upper", the "middle" and the "lower".

a) "Upper" aquifer. This aquifer is located at 30 to 100 m below ground level at Maiduguri. The depth to ground water varies in this zone; it is close to zero in the Chad area and about 75 m near Maiduguri. It is generally less along the river and fossil valleys.

b) "Middle" aquifer. This aquifer of the Chad formation has been identified over an area of almost 30,000 km^2 in the eastern part of the Chad basin. This is the area which has been most intensively studied and exploited both for urban water supplies and for irrigation.

The depth from ground level to the top of the aquifer ranges from 230 m around Maiduguri in the north-east to 350 m near Lake Chad. The thickness of about 43 m increases rapidly towards the north. A thickness of 98 m has been recorded at some points.

The head ranges from zero at Maiduguri to 18.3 m below ground level near the lake; the free flow ranges from 1.2 m^3/h for a head of 5.5 m to 12 m^3/h for a head of 15 m.

c) "Lower" aquifer. This aquifer consists of 105 m of medium and coarse sands and clays at a depth of 425 to 530 m. It is intensively exploited in the Maiduguri region.

3. Bénoué valley

The Bénoué basin runs north-east/south-east from the Cameroon frontier to the confluence of the Bénoué with the Niger.

It consists mainly of Cretaceous sediments and covers a vast area in the states of Banchi, Borno, Gongola and the Plateau. The valley, which is commonly and appropriately known as the "Benoue trench", covers an area of 80,000 km^3, including the vallye of the Gongola. In comparison with other regions of the country there has so far been little ground-water exploration and exploitation in this region. The reason for this is that the aquifers, in particular in the Gongola basin, have only average yields and the potential for ground-water exploitation is very small. Most of the aquifers are unconfined. However, artesian conditions are found in the Cretaceous formation in the Yola-Numan area.

4. Middle basin of the Niger

This basin covers an area of 37,000 km^2; the dominant feature in the east is the river Niger which crosses this part of the country from west to east. Cretaceous sediments of Nubian sandstones predominate here. There are many springs. Artesian conditions with low heads have been found near Pategi.

5. Low-lying sedimentary terrain in the western littoral area

This area is bounded in the south by the coastal alluviums and terminates in the north at the edge of the crystalline basement rock, covering an area of about 38,000 km^2; it runs through Abeokuta and to the north of Ijebu Ode (Ogun State) and to the north of Ore (Ondo State) as far as the western banks of the Niger near Agenebode (Bendel State).

The area consists of Tertiary and Cretaceous sediments with vast alluvial deposits along the main watercourses, which run north to south. The thickness of the alluvial deposits is variable but it can be as much as 15 m. These deposits are up to 8-10 km wide in the flood plains.

The free aquifer is usually found at a depth of 0 to 20 m. A free aquifer is also found in the areas where the other sediments outcrop, but artesian conditions may occur towards the south in the direction of slope of the three formations (Ilaro, Abeokuta and Ewekoro). The Ilaro formation has the best and most heavily exploited aquifer. The Ewekoro formation consists mainly of argillaceous limestones and schists and its yield is small. The Abeokuta formation contains clays but also considerable amounts of sand and silt.

6. Low-lying sedimentary terrain in the eastern littoral area

This area is the eastern equivalent of the area described above; it has been combined with that area into a single "hydrogeological province" with an area of 65,000 km^2 running through the states of the rivers Imo and Cross. It consists mainly of sandy sediments, generally with high porosity. The ground water is found at a depth of 15 m or less to 30 m.

7. Anambra basin

This basin is shaped like a mushroom and lies east of the Niger, mainly in Anambra State between the east bank of the river and the meridian of Enugu and between the Idah-Ankpa plateau in Bénoué State and roughly the latitude of Onitsha in the south. It consists mainly of Cretaceous sediments and the main aquifer is found in false-bedded sandstones; the aquifers are unconfined in the north and become steadily artesian towards the south. The depth of the static water level ranges from 60 to 150 m.

III. EXPLOITATION OF GROUND WATER

The bodies mentioned above are responsible for the exploitation of ground water. The Federal Department of Water Resources in particular has a very comprehensive training programme for its staff.

Local water-drilling companies and branches of foreign companies operating in Nigeria are more numerous than elsewhere in Africa, and some of them have a very large range of equipment. The most commonly used drilling rigs are of United States (Ingersoll-Rand) or Swedish (Atlas Copco) manufacture.

The ground-water requirements differ greatly, from those of urban and rural communities to those of the industries located mainly in urban centres such as Lagos, Warri, Port Harcourt, Maiduguri, Ilorin, Ondo, Kaduna, Bauchi and Sokoto. In the north ground water is widely used for irrigation to offset the decline in rainfall over the past ten years.

IV. PROBLEMS

The lack of control over the exploitation of ground water has serious consequences for the whole country. It has led to overexploitation of the resources in Borno State, where boreholes must now be drilled to great depths in order to reach the lower aquifer (described above under "Chad basin").

Another major problem is the increased salinity of the coastal aquifers caused by sea-water intrusion resulting from excessive and uncontrolled pumping. This phenomenon has been observed close to Lagos, at Mahin (Ondo State) and in some parts of the coastal zones of Bendel State and the River States.

Very serious consideration is now being given to the possibility of the artificial recharge of ground-water aquifers, especially in the areas affected by drought.

Like many other developing countries, Nigeria has very little hydrogeological information about vast areas of the country. Accordingly, test pumping in new boreholes cannot be used as a means of forecasting the long-term behaviour of the aquifer, even though it does indicate the safe yield for the moment. A yield which is abundant when the borehole is first tapped can decline if additional boreholes are brought into use without any control. The Chad basin in the area of Maiduguri is an example of this. Other regions of the country also have many exhausted and abandoned boreholes. This problem is connected with the problem of the operation and maintenance of the installations. Concerted efforts are being made to ensure that all activities affecting water resources, particularly ground water, are subject to uniform national legislation applicable in all the states. This will ensure, inter alia, that the exploitation of the country's water resources will be equitably distributed and rationally controlled.

V. CONCLUSION

The economic and social value of Nigeria's ground water is summed up in the following quotation from Peter Sutcliffe (1982): "Water constitutes a large part of the environment in which man lives and it has an enormous influence on the other elements of the environment. It is a heritage, a basic element of the planet on which we live, essential to life itself. Water must therefore be respected."

The economic and social importance of ground water in this part of the world is difficult to evaluate in general terms, for the cost of developing this resource varies according to lithology, depth, and type of equipment used. It should be given greater weight in the preparation of measures for the rational management of the country's water resources. The most important and urgent step is to adopt water legislation placing the exploitation of aquifers under strict control and ensuring uniform and proper management of water resources.

There are considerable possibilities for ground-water exploration and exploitation, in particular in the sedimentary zones which give high yields. Even in the places where the crystalline basement rock suboutcrops or lies at shallow depths studies are being made of the fracture zones of the bedrock or altered rock which can help to recharge the aquifers. Increased use is also being made of geophysical and remote-sensing methods in geological studies.

VI. REFERENCES

Du Preez, J.W,	The Distribution of Groundwater in Northern Nigeria - Geological Survey of Nigeria, Report No. 1188, 1961.
Du Preez, J.W. and Barber, W.	The Distribution and Chemical Quality of Groundwater in Northern Nigeria - Geological Survey of Nigeria. Bulletin No. 36, 1965.
Geological Survey of Nigeria	Note on the Development of Groundwater in Water Deficient Areas in Nigeria. Report No. 1306.
Kogbe, C.A.	Geology of the Upper Cretaceous and Tertiary Sediments of the Niger Sector of the Illemmeden Basin (West Africa). Geol. Rundschau 62, 1, pp. 197-211. Stuttgart 1973.
Kogbe, C.A.	Outline of the Geology of the Lullemmeden Basin in Northwestern Nigeria. Geology of Nigeria, pp. 311-338. Elizabethan Publishing Co., Lagos, 1976.
Terahound Consultants	A Summary of the Hydrogeology of the Sedimentary Basins of Nigeria, 1980.
Sonuga, J.O.	Keynote Address at the International Seminar on Water Resources Management Practices, Ilorin, Nigeria, 1984.
Sutcliffe, P.	Water Resources and the Environment - A Comprehensive Review. Proceedings of the 3rd APD-IAHR Congress, Bangdung, Indonesia, 1982.

NORTHERN REGION

Station	Basin	Area of catchment basin upstream of the station (km^2)	Flow (m^3/s) Maximum	Minimum	Mean	Storage Coefficient	Specific yield ($m^3/s/km^2$)
Jebba	Niger	631,960	3,740	370	1,292	0.33	0.002
Wuya	Kaduna	65,527	4,738	3	523	0.13	0,009
Baro	Niger	730,380	6,822	332	1,947	0.29	0,003
Makurdi	Bénoué	304,325	14,627	200	2,982	0.20	0,010
Umu Aja	Bénoué	333,619	18,425	46	3,425	0.19	0,010
Lokoja	Niger	1,087,000	27,140	712	5,737	0.21	0,005
Shintaku	Niger	1,088,000	24,110	970	5,512	0.23	0,005
Idah	Niger	1,103,200	26,790	840	5,613	0,21	0,005

DIAGRAMMATIC MAP OF GROUND-WATER RESSOURCES OF SENEGAL

Scale: 1:1 500 000

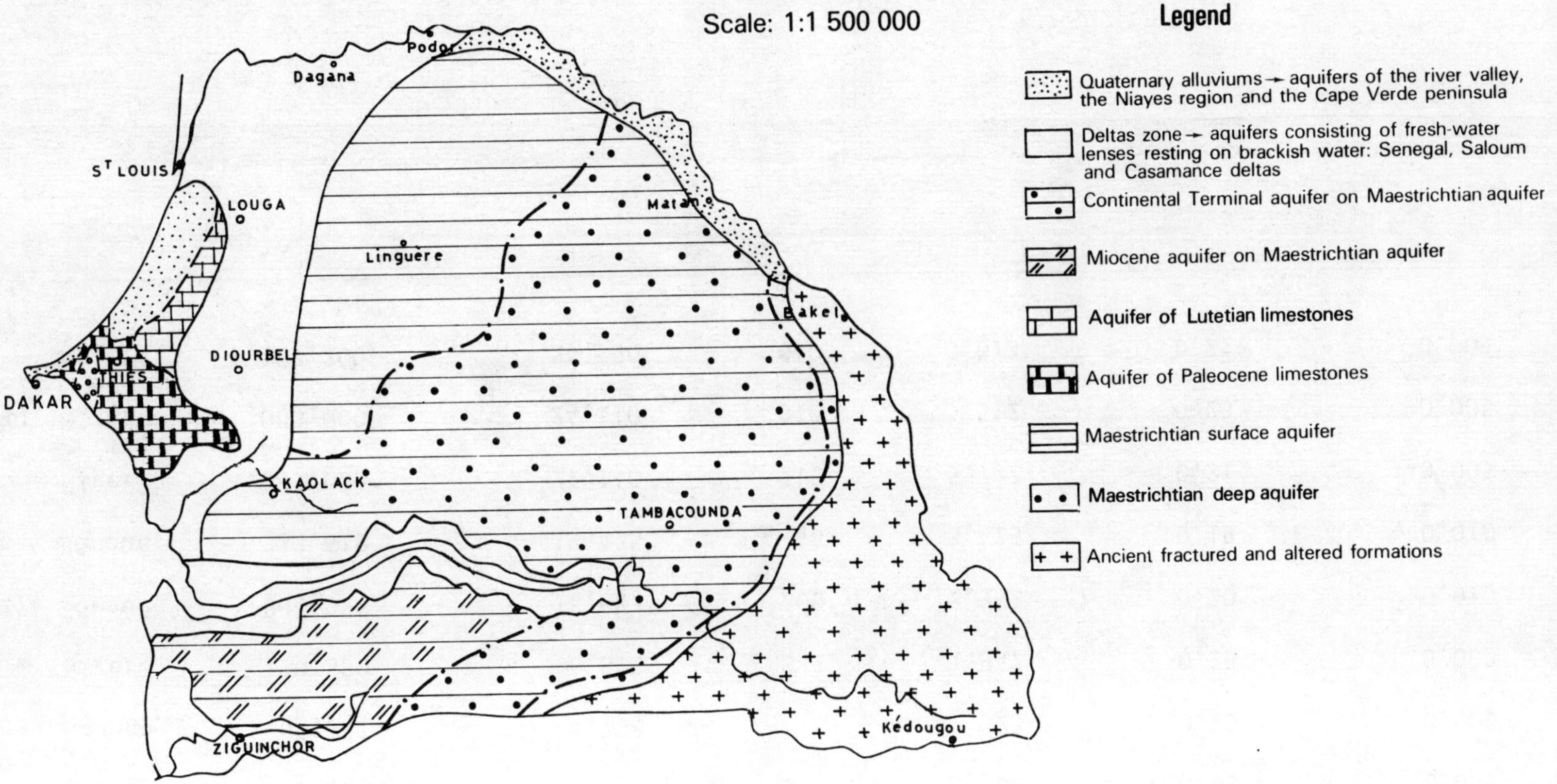

Source: Sénégal, Département de Géologie de la Faculté des Sciences de Dakar

3436.9x

SENEGAL

Area: 196,200 km^2

Population: 6.32 million (United Nations estimate, 1983)

INTRODUCTION

As Senegal is a Sahelian country with little rainfall, ground-water research and exploitation have always been one of the main concerns of the authorities. The country's few rivers are heavily influenced by the tides which, in the absence of anti-salination dams, carry salt water far into the interior.

Ground-water aquifers are therefore of very great importance. Many hydrogeological studies have therefore been carried out throughout the country, the first ones nearly 70 years ago. One of the main tasks of the Hydrogeological Service is to prepare a hydrogeological summary from the results of these studies. This work began in 1976 but had not been completed in 1985.

I. BACKGROUND

Senegal is situated in the extreme west of the African continent, in the part of Africa closest to the western hemisphere, and it has an Atlantic coastline of 600 km.

It is a very flat country with no relief above 100 m except in the south-east and the far west: the ancient mountain ranges in the east (400 to 450 m) and the volcanic formations of the Cape Verde peninsula.

Climate

Senegal's climate is determined entirely by the flows of air connected with high pressure zones:

- The Azores anticyclone which causes maritime and continental trade winds to blow in the north-west of the country;

- The Arabian anticyclone which generates the harmattan, whose effects are felt as far as the country's west coast;

- The South Atlantic anticyclone which brings the monsoon.

The limit of these masses of tropical air (trade winds, harmattan and equatorial air (monsoon) forms the intertropical front, the extreme positions of which determine the two main seasons in Senegal: a cool dry season from November to March, followed by a hot season until May and a wet hot season from June to October.

The rainfall has the following temporal and spatial characteristics:

- A gradual increase in rainfall from May-June to August, with the maximum usually in August and a very rapid decline in October;

- A sharp drop in the total annual rainfall from south to north, from 1,500 to 350 mm; for example, the town of Ziguinchor receives an average of 1,547 mm, as against 335 mm at Podor;

- Subdivision of the country into two parts: north of a line through Dakar-Kédougou the rain is delivered mainly by storms, and the rainy season always comes later to the coastal section of this area; south of this line the rainy season is longer and the rain has a more regular distribution.

While this diversity provides the southern part of the country with a more reliable water supply, it discriminates against the northern part, which has suffered drought for most of the past 15 years.

At the coast the highest average monthly temperature is only 27.5°C at Dakar, 28°C at Saint-Louis and 28°C at Ziguinchor. The temperature ranges between day and night are equally low (5.2°C in July and 7.1°C in February at Dakar). The seasonal variation is also small (average annual range of 4°C at Ziguinchor, 6.9°C at Saint-Louis and 7°C at Dakar. The temperatures increase towards the interior; for example, the average monthly temperature is 30°C at Diourbel, 31.7°C at Kolda, 32°C at Podor, 32.5°C at Tambacounda, 33°C at Kédougou and 33.6°C at Matam. The daily thermal ranges are also very wide: 7.9°C to 16.8°C at Thiès, 9.7°C at Podor, 9.7°C to 19.2°C at Matam and 9.3°C to 22.4°C at Kédougou. The widest ranges occur in the dry season when the nights are cold and the days sometimes very hot, whereas in the rainy season the atmospheric humidity causes a considerable reduction in the thermal ranges. In contrast, the seasonal thermal ranges differ little from those recorded at the coast (7°C at Kolda, 7.7°C at Tambacounda, 7.8°C at Kédougou, 8.8°C at Podor and 10.2°C at Matam).

The evaporation as measured in the Piche evaporometer is very high in the dry season and it increases towards the interior. It averages 1,000 mm at Dakar, 1,500 mm at Saint-Louis and 1,200 mm at Thiès. In contrast, it tends to decline and standardize during the rainy season owing to the high humidity.

Surface water

Watercourses

The watercourses follow a tropical pattern with an annual crest after the maximum rainfall in August-September and a fairly long low-water period. The rivers dry up completely from February-March until July. Senegal has four main rivers: the Senegal, the Gambia, the Casamance and the Siné-Saloum.

The Senegal is an international river 1,790 km long with its source in the Fouta Djallon at an altitude of 800 m; it has a catchment basin of 270,000 km^2 and its main tributary in Senegal is the Falémé which drains the eastern part of the country.

The river follows a clearly tropical pattern, with a high-water period lasting three months (August, September, October) and a long low-water period. The mean maximum flow is 4,900 m^3/s with large variations. The mean annual minimum flow is about 800 m^3/s in 1925. The 100-year flow is close to 10,000 m^3/s. The sediment discharge is high at the beginning of the flood period and can be as high as 1 g/l of solids in suspension.

The flows are poor in the low-water period (2 to 3 m^3/s) and during this period sea water penetrates up the river in the flood plain, sometimes as far as 450 km inland from its mouth.

The cumulative flow in an average year is 24 billion cubic metres and in a dry year 13 billion cubic metres.

The Gambia is 920 km long with a catchment basin of 75,000 km^2 in Senegal, Gambia, Guinea and Guinea-Bissau. Little is known about the hydrology of this river and it is the subject of studies by the Organization for the Development of the Gambiar River (OMVG). A maximum flow of 2,000 m^3/s is common in the Tambacounda region; the minimum flow is small, in the order of 250 to 100 l/s. The total annual flow is approximately 10 billion cubic metres. The tides carry salt water up the river in the low-water period.

The Casamance is 300 km long and has a relatively small catchment basin. The river is in fact an anastomotic estuary which penetrates deep inland and receives the runoff from several basins, principally the Casamance and the Soungrougou. In the dry season the salt water of the estuary penetrates quite far inland. The hydrological regimes of the Casamance and its main tributaries have been studied for many years under dam construction projects.

The valleys of the Siné-Saloum drain small basins complete devoid of fluvial recharge. They terminate in an estuary which has very high salinity even after the rainy season.

Lakes and waterholes

The largest lake is Lake Guiers which constitutes the downstream part of the valley of the Ferlo, a fossil tributary on the right bank of the Senegal. This lake is shallow (four to five metres) and is connected by the Taoué to the Senegal, which feeds it every year with its flood waters. It contains 750 million cubic metres of water at the +3m level. Since the construction of the Richard-Toll dam the lake can be maintained at its maximum depth; the water is used to irrigate the plantation of the Sugar Corporation and the SAED and to supply Dakar. For this purpose 40,000 to 50,000 m^3 is drawn off every day by the National Water Exploitation Corporation of Senegal (SNEE). However, the lake loses 70 per cent of its water through evaporation. Studies being carried out by ORSTOM and the Institute of Environmental Sciences will give a better picture of the lake's hydraulic regime.

II. GEOLOGY

Senegal lies on the Precambrian basement rock of Africa, which rises up in the east of the country and sinks steadily towards the west to a depth of almost eight kilometres at Dakar. This Birrimian formation is made up of metamorphized and granitized rock and is covered in discontinuity in the south-east by quartzitic sandstones and Lower Cambrian limestones and in the west by Cambrian-Ordovician limestones and sandstones.

Apart from these small crystalline outcrops, most of the country is occupied by sedimentary formations which make up the sedimentary basin of the Senegal. This basin is a vast monoclinal structure of Mesozoic and Cenozoic formations which slope westwards. This structure develops complications in the west, at the level of the Cape Verde peninsula, which is a zone of continental fold. This region was subjected to abrupt tectonics which produced the Ndiass and Dakar horsts and various volcanic phenomena, of which the most recent is the early Quaternary volcanism of the Mamelles, which occurred 1.5 to 0.8 million years ago.

Most of the country's stratigraphy has been through studied on the basis of data collected during a large number of oil and water drilling operations. The main subdivisions are, from bottom to top:

- The Upper Jurassic, identified by the deepest soundings in the Ndiass area; this stratum is 760 m thick at the DSI borehole;

- The Aptian, which consists of finely crystallized limestone and is 500 to 600 m thick;

- The Albian, which consists of a sandstone-dolomotic series interbedded with the fine sandstone or calcareous sandstone with sandy limestones of varying degrees of dolomitization and limestone-sandstone dolomites. It is 1,600 m thick at the Mbour borehole;

- The Cenomanian-Turonian, which begins with a thin sandstone stratum overlain by clays with a fairly high quartz content, the upper part consisting of fairly sandy lamellar black clays. It is about 450 m thick at the DSI borehole;

- The Lower Senonian, which consists of alternating layers of grey or black lamellar or plastic clays and calcareous sandstones, with a thickness of more than 1,500 m;

- The Campanian, which begins with black lamellar sandy clays and sandstones and continues with strata similar to those of the Maestrichtian; it is 150 m thick at the Mbour borehole;

- The Maestrichtian, which over most of the country consists of more or less coarse sand and argillaceous sands with bands of sandstone and sandstone-limestone; the facies becomes more argillaceous and deeper in the west (900 m thick at Diourbel and 1,800 m at Rufisque). The Maestrichtian outcrop of Ndiass horst complicates this picture to some extent;

- The Paleocene, which consists of more or less karstified marly limestones and zoogenic limestones and is 50 to 100 m thick;

- The Lower Eocene, or Ypresian, which consists mainly of whiteish to grey clays and marls, with increasingly marly limestone at the top. It varies in thickness from 100 to 400 m in the west of Senegal where it is best known;

The Middle Eocene or Lutetian, which consists of marly limestones with phosphated layers at the base and of limestones which become marle-limestone in places and marly at the top. It can be up to 200 m thick.

- The Upper Eocene, which consists of generally eroded clays which are encountered only in certain deep boreholes;

- The Oligocene-Niocene, which is found in the western part of Casamance (south-west of the country); the boreholes drilled in this area have penetrated limestones and marls of Oligocene age and argillaceous sands of Miocene age;

- The continental terminal, which outcrops extensively over almost all the country but disappears in the west. The commonest facies consist of argillaceous sands of various colours interbedded with clay or sandstone. The thickness is variable and reaches 150 m at the Lombi Sud borehole. The continental terminal consists mainly of Miocene formations of marine origin which underwent continental alteration; most of it is covered with a lateritic shell of varying thickness.

- The Quaternary,which consists of formations of varied origin: gravelly cuirasses covering the continental terminal, lacustral limestones and more or less gravelly deposits which form the bedrock of the fine and recent alluviums, and immobile red sands in the Niayes region and the lower valley of the Senegal.

Lastly, reference must be made to the marines silts and sands formed during marine transgressions, which occur as barren pockets and in the mangrove area; there are also the fluvial alluviums, more or less argillaceous sands covering the gravel along the Senegal below Bakel and along the Casamance and the Gambia.

III. GROUND WATER

Development and management of water resources

Senegal has a large number of government and other bodies responsible for ground water. They work in particular in the areas of water supply, health, rural development, urban development and housing, in trial development. capital investment, and scientific and technical research; there are also a number of international bodies (OMVS and OMVG), research organizations (ORSTOM and BRGM) and private organizations (CARITAS). These different bodies are concerned with the problem of water from various aspects; however, it is the Water Department which determines Senegal's water policy. Within the Ministry of Water and Sanitation, the Department of Water and Rural Development (DGHER) has specific responsibility for carrying out this policy through its technical offices:

- The Water Studies Office (DEH), which is responsible for surface and ground-water prospecting and research; it determines the programmes for exploitation and use of water resources;

- The Office of Urban and Rural Water Supplies (DHUR), which is to become the Office of Rural Water Supplies when the Ministry is reorganized. In conjunction with the Maintenance Office, it will be responsible for preparing the studies to be made by the DEH in agreement with the local people concerned. It has the particular task of monitoring project implementation and dealing with all the problems arising from the implementation of water-supply works in rural areas;

- The Office of Rural Development (DER), is responsible for implementation studies and the construction and control of hydro-agricultural installations; it is also responsible for the control of all the companies working in the area of agricultural water supplies;

- The Maintenance Office (DEM) which was established recently to take charge of the maintenance of water installations and equipment; there will be 600 installations with motorized pumping stations on the conclusion of the projects in operation;

- The Office of Urban Water Supplies and Sanitation, which deals with problems of water supply and sanitation in urban areas in conjunction with the national Corporation for Water Exploitation of Senegal (SONEES).

Background of hydrogeological research

The climatic conditions and the lack of fresh surface water over much of the country (between the Senegal and the Gambia) made it necessary to seek almost all the people's water supply in the subsoil; hence the impressive number of documents on the hydrogeology of Senegal produced since the creation of the Hydrological Mission of Senegal in 1917. This Mission studied the underground hydrology of Senegal, in particular in the regions between Dakar and Thiès, partly to solve the problem of a drinking-water supply for the towns of Dakar, Yang Yang and Diourbel, and partly to provide watering for livestock in the Louga region.

The depth of detail of the studies depended on the region. While the water resources of Cape Verde are fairly well known, this is not true of the other regions of the country which have been afflicted by drought over the past 10 to 15 years.

The most recent studies include:

- Hydrological study of the valley of the River Senegal", between Bakel and Rosso, with an inventory of water points, geophysical prospecting, study of aquifer-river relations, and hydrochemical and isotopic research, carried out under a UNDP project (1973) and implemented by the United Nations Food and Agriculture Organization (FAO);

- Study of the continental terminal of Senegal-Gambia", a comprehensive hydrogeological summary including the results of geophysical and hydrogeological prospecting and hydro-climatic observations and many chemical analyses, which establishes a mathematical model for management (1976);

- "Hydrogeological study of the Lutetian limestone between Bambe and Louga" (2nd phase); a hydrodynamic and hydrochemical interpretation of the aquifer and a mathematical model (1979) were established on the basis of the mass of data produced during the works and studies (levelling, geophysical prospecting, chemical analyses);

- "Study of water resources to supply Dakar, 1970-79", which included a large programme of test drilling, geophysics, isotopic geochemistry and modelling as part of a UNDP project executed by the World Health Organization (WHO);

- "Hydrogeological study of the Paleocene limestones of the Mbour region" which made a comprehensive summary of the data obtained during the CARITAS-Senegal drilling programme (140 boreholes) which ended in 1987;

- Hydrogeological study of Casamance" made between 1979 and 1987. This list shows that the whole of the country has been subjected to very intensive hydrogeological investigation.

The Water Ministry has a national hydrogeological service which was established in 1977 and is assisted by units responsible for studies, technical control of new boreholes and collection and storage of ground-water data. Within this service the Water Resources Inventory Office collects and standardizes all the data for computerization and automatic processing.

Aquifer systems

Distinctions must be made between surface, semi-deep and deep aquifers in terms of the conditions of the deposit.

Surface aquifers (shallow ground water)

These include:

- The Quaternary alluvial formations of the three big rivers (Senegal, Siné-Saloum, Casamance) which hold large quantities of water, especially along the Senegal;

- The Quaternary sandy formation of the north coast and Cape Verde, where the thickness and the hydraulic properties permit a drawoff in the order of 120,000 m^3/day;

- The "continental terminal" aquifer system, which is from 10 to 150 m thick and is found in the centre and east of Senegal. Its yields are small but sufficient for the many villages which tap it;

- The Lower and Middle Eocene system which contains aquifers of varying capacity, depending on the degree of karstification of the limestone strata. This aquifer outcrops and suboutcrops in the western area where it is exploited by village wells;

- The superficial alteration formations of eastern Senegal, which are believed to contain only small amounts of water, the wells being usually drilled in the alteration fringe of the Precambrian crystalline rock. However, given the modern drilling techniques used in neighbouring countries and the results of the 1979 campaign, good yields may be expected from the fractures in the non-altered substratum which will at any event be sufficient for village water supplies;

- The superficial Maestrichtian aquifer, which is located in the upper levels of the Maestrichtian formation and extends over the whole area of the Ndiass mountains and their northern extension as far as Lake Tanma. This aquifer is exploited by village wells and it provides the recharge for the aquifers around the Dome de Ndiass.

Semi-deep and deep aquifers

- The Paleocene aquifer has been identified mainly in the western region of the country where the strata are karstified and contain large quantities of water. It is unconfined in places, but mostly confined beneath Lower Eocene formations. This aquifer is exploited in the areas of Sébikotane, Pout and Mbour and yields about 60,000 m^3/day, which may be taken as the limit of its capacity.

- The deep Maestrichtian aquifer extends over almost the whole of the Senegal basin and constitutes an immense reservoir. The aquifer is confined and lies at depths of between 100 and 350 m. It is exploited by several hundred boreholes, some of which can yield up to 200 m^3/h and more to supply the population and livestock. This aquifer is bounded in the east by the basement rock and in the west by salt water; it rests on salt water trapped in the Cretaceous formations.

The hydrological and discharge properties of these aquifers are set out in the annex.

Chemical and physical properties of the water

Considerable data is available on the chemical composition of Senegal's ground water, and a general picture can be given of the water quality of the main aquifers.

In the area of the ancient basement rock the water usually has a low mineral content (dry residue below 0.5 g/1); it is mainly bicarbonated and slightly acid. Harder water is found in a few places in schistic formations, with facies ranging from calcium chloride to sodium chloride. Concentrations of nitrates above 20 mg/1 have been reported, as have concentrations of fluorine of up to 1 mg/1.

The central part of the Maestrichtian aquifer is heavily mineralized with dry residues between 2 and 10 g/1 and two lateral zones of softer water with a dry residue below 0.5 g/1. The most heavily mineralized water is of sodium chloride facies and relatively basic (up to pH 8), while the less heavily mineralized water (0.5 to 1 g/1) is of calcium bicarbonate facies and slightly acid. In the western part of the aquifer the fluorine content is fairly high - up to 5 mg/1/

The Paleocene aquifer, in the region where it shows its individual characteristics and is exploited, forms a narrow corridor a few dozen kilometres wide containing fresh water. This corridor is surrounded by water with an overall mineral content in excess of 2 g/1 of dry residue. The fresh water is mainly of the calcium and magnesium bicarbonate type. The mineralized water is of sodium chloride facies with a fairly high sulphate content.

The water of the Oligocene-Miocene semi-deep aquifer of Casamance is very soft and mainly calcium bicarbonate in the east; it is more saline towards the west, ranging from sodium bicarbonate to magnesium sulphate.

The chemical content of the water of the surface aquifers is also very varied.

The formations of the continental terminal in the east and south of the country contain fresh water with a dry residue below 0.2 g/1. In Casamance the water becomes increasingly mineralized close to the ocean and acquires a sodium chloride facies.

In the coastal dunes as well, with the exception of the areas of brackish lagoons, the water is fairly fresh although of the sodium chloride type. In the centre-west of the country between the rivers Senegal and Siné-Saloum the well water has a very varied salt content which can often be fairly high, especially when the wells reach the Eocene formations or the Ancient coastal formations. Nitrates are often reported in the surface aquifers, especially in the western part.

No information is available at present about the most toxic elements (arsenic, heavy metals), but it can be said that the problem of quality does exist with respect to Senegal's ground water owing to the heavy mineralization and the presence of certain harmful elements (iron, fluorine, nitrates) in sufficiently large amounts to be toxic. This problem of water quality arises to a greater or lesser extent with respect to all the country's aquifers.

IV. EXPLOITATION OF GROUND WATER

Bodies responsible for exploitation

Ground water is exploited in Senegal in both urban and rural areas. Two bodies are responsible for exploitation.

In the case of urban water supplies, the National Corporation for Water Exploitation of Senegal (SONEES) is responsible for the production and distribution of water in the Cape Verde region, including Dakar, and in 37 interior towns. Its status is to be changed to strengthen its operational capacity and relieve the State, which will still retain control of the choice of the main directions of development, of an increasingly heavy burden. The long-term goal of the current reform (1984-1985) is for SONEES to assume responsibility for all matters of urban water supply.

The Rural Water Department of the Water Ministry is responsible for preparing studies in collaboration with the rural people concerned under the auspices of the DEH, and to supervise the implementation and monitoring of the projects; the Maintenance Department is responsible for establishing a structure capable of coping with the present maintenance requirements of water installations throughout the country.

Drilling companies

Senegal has four water drilling companies;

- The National Drilling Corporation (SONAFOR), a joint company in which the Government holds 35 per cent of the shares and which is being comprehensively organized;

- INTRAFOR-COFOR AND SASIF, which are accredited private companies with efficient equipment and qualified staff;

- CARITAS, which is also a private company but specializes in the area of agricultural water supplies. Its drilling experience dates from 1978.

Water use

The per capita average daily consumption is 110 litres in the Cape Verde region and Dakar, seven litres in rural areas and 40 litres in the interior towns.

The water supply for settlements of more than 5,000 inhabitants is furnished by SONEES, about 80 per cent from ground water and 20 per cent from surface water. Every day the town of Dakar (Cape Verde region) consumes 130,000 m^3 of water from various sources (Dakar aquifers: 12.8 per cent; Thiaroye aquifers: 7.9 per cent; Sébikotane aquifers: 20.6 per cent; Pout-Sud aquifers: 10 per cent; Pout-Nord aquifers: 5.2 per cent; Kirène aquifers: 11 per cent; Kelle aquifers: 12.5 per cent; and Lake Guiers: 20 per cent). A total of 37 boreholes contribute to the water supply of the Cape Verde region, which has about 1.2 million inhabitants. The town of Saint-Louis is supplied from surface water in a daily amount of 5,000 m^3, which was to reach 12,900 m^3 in 1985. The other interior towns (about 30 of them) are supplied, like Dakar, mainly from ground water in a total amount of about 40,000 m^3/day (57 boreholes and about 10 surface-water installations).

The requirements of rural settlements (fewer than 5,000 inhabitants) are met exclusively from ground water, which is supplied free of charge. However, owing to the shortage of equipment and of water points, there is a serious supply deficit. Senegal has about 1,050 modern installations (wells, boreholes, tubewells) and 40,000 traditional wells which provide an average yield of 23,000 m^3/day for a population of about 3.2 million and 2,525,000 head of cattle, 2,825,000 sheep and goats, 345,000 pigs, 324,000 horses, 22,000 donkeys, 7,200 camels and 900,000 head of poultry.

Ground water is used for agricultural purposes mainly in two areas - the Atlantic coast and the groundnut basin.

- On the Atlantic coast (Grande Niayes and Petite Côte as far as Mbour), an area of market gardens and fruit farms, there are two types of operation: small plots usually hand-watered with water drawn from the surface aquifer (céanes, lined sumpholes); and modern farms run by companies and equipped with sprinkler and drip watering systems using water taken from boreholes in the deep aquifers. These two types of operation drew off about 45,000 m^3/day in 1980. Since then there has been a considerable increase in the drawoff. Under the CARITAS agricultural water supply programme many boreholes have been drilled in the Paleocene aquifer in the Petite Côte.

- In the groundnut basin, a semi-arid area, ground water is used mainly for market gardening in the dry season.

The total daily production of ground water is currently estimated at 500,000 m^3, a figure which may have to triple in 20 years to meet the requirements. This will entail the installation of an additional 1,000 productive boreholes.

Problems

The main problem is the wide gap between resources and requirements. The Cape Verde peninsula has a growth rate of five per cent and the region's aquifers are being overexploited for 100 km around.

This overexploitation may produce sea-water intrusion in all the aquifers in the western part of the country. Another effect is the drop in the water table throughout the country; this has already reached alarming proportions in the most heavily exploited aquifers (in the area of Ndiasshorst, for example).

Surface water from 250 km away (Lake Guiers) is therefore used to supply Dakar, and the amount was to have doubled before the end of 1985. Other possible solutions, i.e. the recycling of waste water to recharge the infrabasaltic aquifer and, if absolutely necessary, the desalination of sea water entail a financial effort which the State does not seem ready to make.

Water quality is the second largest problem of ground-water exploitation; many people consume brackish water, especially water with very high fluorine content, and this could have harmful effects in the long term.

Integrated exploitation of surface and ground water

As stated above, at present surface water is used for human supply, either exclusively or in conjunction with ground water. Moreover, several hydro-agricultural projects are using surface water in the areas where it is available.

As for the artificial recharge of aquifers from surface runoff, only the Pantior system is at present in operation. However, new installations are planned, especially for the Cape Verde peninsula and the Lake Tamna region. The studies have been made and the financing is being sought. There is also a programme for storage of runoff in the Siné-Saloum region.

V. CONCLUSION

The disastrous consequences of the recent drought (1981-1983) have prompted Senegal to adopt an emergency water supply programme to supply drinking water to rural areas to enable the people to increase their agricultural and livestock production and to slow down the rural exodus, which is a serious problem.

The terms on which water is supplied are different in urban and rural areas, mainly in that in the towns the water is sold to the consumers, while in the countryside it is free.

In the urban sector the 1977 tariff system distinguishes between private consumers, municipalities, market-garden enterprises, and other users; the price of a cubic metre of water ranges from CFAF 36.25 to CFAF 140.56.

The Water Code which came into force in March 1981 established public control over all water exploitation and made it subject to prior authorization. This measure is designed to facilitate an accurate census of water installations and an accurate assessment of the requirements, with a view to achieving a balance between resources and needs, which must be the main goal of any programme of ground-water research and exploitation.

VI. REFERENCES

There is considerable documentation available on ground water; for example, about a hundred publications and studies have been collected for 1961-1983 which deal with every conceivable aspect of the hydrogeology of Senegal.

Therefore, in addition to the studies cited in the text, reference will be made here only to the most recent studies and surveys which are available for consultation:

- Study of the Maestrichtian formation at Taiba (1980-1982);
- Hydrogeological study for the rehabilitation of 11 interior centres (1980-1981);
- Study of the Djifère sands (1980);
- Study of the southern and eastern edge of the Ferlo;
- Simulation of the Maestrichtian aquifer;
- Establishment of a village water supply programme in northern Senegal;
- Establishment of a village water-supply programme in the Sine-Saloum region;
- Establishment of a water-supply programme in the Thiès region.

ANNEX

AQUIFERS OF SENEGAL 1/

Surface aquifers	Geology	Lithology	Permeability	Specific yield ($m^3/h/s$)	Comments (water quality)
1. Ferlo	Continental terminal	Sands and sandstones with varying amounts of clay	fairly good	several $m^3/h/m$	fresh water
	Eocene-Lutetian	limestone) marly limestone) sandstone)	often poor		sometimes high mineral content
	Ypresian	marls) marly-limestone)			
2. Delta	Recent Quaternary	argillaceous alluviums, sands, laterites	poor	poor	very saline aquifer
3. Niaye	Eocene-Lutetian Early Quaternary	limestone red dunes	variable		fresh water
4. Louga-Coki	Eocene-Ypresian Eocene-Lutetian	marls limestone	low good	poor up to 60	mineralized water 0.5-1 g/l
5. Thies-Tivaouane	Conti. terminal Lutetian	sands, calcareous limestones	good		low mineral content
6. Khombole-Bambey-Diourbel	Lutetian Ypresian	limestone marls	good	up to 50 and 75 at Bambey	fairly mineralized aquifer, polluted pumps
7. Baol	Ypresian	marls marly limestone	low	very low	
8. Sine-Saloum delta and associated islands	Quaternary, continental terminal	argillaceous and saline alluviums	low	very low	often saline

1/ According to BRGM.

AQUIFERS OF SENEGAL (continued)

Surface aquifers	Geology	Lithology	Permeability	Specific yield ($m^3/h/s$)	Comments (water quality)
9. Nioro du Rip Kaffrine-Koungheul	Continental terminal	argillaceous sands and sandstones	good	up to 6	low mineral content
10. Caramana	Continental terminal	argillaceous sands laterite layers sands and alluviums	good (favorable region)	average	(under 0.2 g/1)
11. Eastern Senegal	Quaternary	alluviums (not extensive) laterite cover	good but formation thin, good in vacuolar areas	small resources	
	Crystalline rocks " "	granite schists quartz ultrabasic rocks	nil	very low	
12. Cape Verde	Quaternary	infrabasaltic sands of tip of peninsular	very good	15-20,000 m^3/day	low mineral content
	Quaternary	Thiaroye sands	very good	10-15,000 m^3/day	iron content 0.4 g/1
Ndiass Horst	Central part Maestrichtian	sand-stone	very good		
	Paleocene	karstic limestones of Sebikotane		30,000 m^3/day	low mineral content
		karstic limestones of Pout		20.000 m^3/day	
	Maestrichtian	sand-sandstone 200-250 m thick	average K: 1 to 5 x 10^{-4} m/s	up to 200 m^3/h/m (borehole)	fresh to slightly saline water

SIERRA LEONE

Area; 71,740 km^2

Population: 3.47 million (United Nations estimate, 1983)

I. BACKGROUND

On the basis of the interpretation of aerial photography, Sierra Leone has been divided into 44 natural units, grouped in four main regions, each subdivided into a number of subregions.

Natural regions

Two-thirds of the country are occupied by plains and plateaus deeply incised by the hydrographic network, from which rise a number of hills and mountain chains and massifs. The plains and plateaus correspond to areas of ancient erosion and their characteristics are similar to those of the higher parts of the country. A feature of this landscape is the large number of narrow valleys covered with alluvial and colluvial deposits which are transformed into swamps by flooding in the west season. The coastal area has swamps and escarpments flanked by marine terraces.

The coastal plain covers 15 per cent of the country. It can be divided into four subregions.

1. The marine terraces (about 50 per cent of the coastal plain) line between 2 and 40 m above sea level and consist of unconsolidated material incised in all directions by small watercourses. There are also some higher terraces bordering the mountains of the peninsula close to Freetown, the capital. They are about 70 m high and are overlain by a lateritic crust and eroded by the beds of swift-flowing rivers.

2. The estuarial swamps lie at the foot of the marine terraces. They were formed by the deposition of silt and clay along the main rivers. Areas of mangroves are common in this region.

3. The alluvial plains lie at an altitude of up to 15 m above sea level in the area of fresh water which follows the courses of the little Scarcies and the Sewa. Here the grassy plains can be very large in extent.

4. The coastal zones lies along the edge of mountain chains of various ages and it can be up to 20 km wide, in Turner's Peninsula for example. In places these coastal chains have deflected the courses of large rivers such as the Wanji and Sewa, which converge in their lower courses and flow as a single stream for 50 km along the coast.

The lower plains cover some 43 per cent of the country and form its largest geographical region. The plains rise steadily from west (40 m) to east (up to 200 m). They are divided into two main subregions.

1. The "undulating plains" in the west, gently sloping terrain covered with a thick compact layer of lateritic and pyrolytic gravel which has a high iron content. The swampy areas are less common and narrower than in

the plains in the east. In their central part these plains are formed of hard sediments of the Rokel group. The swampy areas spread out into depressions called "bolis".

The eastern plains lie on a complex of granitic rocks. The slopes are gentle and the swamps narrow; isolated hills are found throughout the region. In the north near the escarpment the terrain is more broken with many low rocky hills.

2. The "Bolilands" extend for 50 km at their widest point. These are depressions and low terraces along the Rokel and the Mabole resting on Rokel sediments. The extensive floods of the rainy season are due to the lack of development of the hydrographic network, which causes the rivers and streams to overflow.

The plateaus lie on granitic rocks at an altitude of 300 to 700 m and cover 22 per cent of the country in the north-east and south-east. The variations in relief and the heavy erosion lend variety to the landscape. There are two main subregions.

1. The high rolling plains in the east. They resemble the interior plains and are dotted with low isolated hills.

2. The plains and hills of the western and southern parts of the plateaus. Narrow alluvial plains are found here adjacent to the courses of the main rivers. Altered granite suboutcrops in the north; this area has many rocky outcrops. The southern hills are covered with a thick layer of ferrous and lateritic gravel.

The hills and mountains form two subregions:

1. The hills of basic and ultrabasic rocks resistant to erosion. These chains of hills are generally quite high. The high iron and alluminium content of the rocks has produced crusts of laterite and compact ferralitic gravel. Chains of hills formed of these gravels are common at altitudes of 470 to 800 m in the Freetown peninsula and along the escarpment. The isolated mountains or mountain chains are covered with laterites and rise above the plateau and the interior plains.

2. The hills of acid granitic rocks, which rise to 1,000 m along the perimeter of the plateau. In the escarpment area the hills are eroded by the water courses.

Mt. Bintumani rises to 1,945 m in the Loma mountains; it consists of "young" granites and is capped with dolorites.

Climate

The climate is generally of the wet tropical type with six separate seasons: three dry and three wet.

First dry season: end-November/mid-December, with temperatures of 21°C to 32°C, west winds and high humidity: storms sometimes occur at night and a dense wet mist in the morning.

Cool dry season: mid-December/mid-February, with hot days (32°C) and cool nights (below 7°C in the north). In the afternoon the air is very dry, sometimes with a relative humidity of only 10 per cent. An east wind, the harmattan, blows constantly from the late morning until the afternoon. This wind can be strong, especially in the north and north-east of the country. In the coastal areas the nights are never cool. The minimum temperature rarely falls below 18°C and the harmattan rarely blows hard.

Dry hot season: mid-February to early April in the south-east and to mid-May in the north and some coastal areas, until the first storms come. The days and nights are hot and increasingly humid. The winds blow from the west. The springs dry up.

Early squalls: April-July, characterized by sudden storms from the east with strong winds, heavy rain, thunder and lightning. The squalls occur in the early evening in the south-east and increasingly later in the centre and west. The days are not and the nights cool after the squalls have passed.

Main rainy season: July-September, cloudy, humid and fairly cool. There is frequent heavy rainfall but little thunder; the sky is usually overcast. The winds blow from the west. This is the season of food crops.

Late squalls: mid-September/mid-November, similar to the season of early squalls, with gradually diminishing rainfall.

The coast has the heaviest rainfall (3,000 mm or more). There is another zone of heavy rainfall in the centre of the country. In the north the rainfall declines to 200 m or even less. December-January and March-April tend to be the coolest and hottest months respectively, with annual temperatures ranging from 24°C to 28°C. The seasonal variations are less marked on the coast than in the interior. Temperature ranges of 5°C to 7°C are common along the coast, while in the interior they can reach 15°C in the harmattan season.

Evapotranspiration

There is little reliable data on evapotranspiration. FAO has calculated the potential evapotranspiration figures by computer on the basis of data for season, latitude, rainfall, number of hours of sunshine a day and wind strength. The figures vary with latitude from 1,300 to 1,400 mm a year. They are higher in the dry seasons than in the rainy season. The heavy evapotranspiration and negligible rainfall in the dry seasons have a serious effect on the moisture content of the soil, particularly in the north. The soil has a moisture deficit from mid-January to the end of April in the north and from February to mid-March in the south. Crops cannot be grown during these periods unless the land is irrigated, and crops grown on bushes suffer from lack of water.

Surface water

The size, flow and annual discharge of the main river basin 1/ are given in the following table.

Basin	Area (km^2)	Flow Min	(m^3/s) Max	Annual discharge (mm)	Station	Year
Little Scarcies	12,870	5.1	1,801	1,787	Mange	1971
Seli/Rokel	10,620	1.7	416	782	Bumbuna	1971
		2.2	784	1,112	Bumbuna	1972
Pampana/Jong	7,511	3.1	389	1,275	Magburaka	1972
Sewn	14,140	12.4	771	1,028	Jaiama	1972
		418	601	860	Jaiama	1973
Moa	9,220	34.5	978	511	Kenema	1971

The river basins are fairly small. However, owing to the heavy rainfall the flows are very large, as is the discharge coefficient, which represents 20 to 40 per cent of the total annual rainfall.

The main rivers continue to flow during the dry season and overflow their banks in the rainy season.

Most of the watercourses have a "young" profile in their upper courses where they erode the basement rock. Accordingly, with the exception of the Upper Mongo, it is only in their lower reaches that alluvial plains of any size are found, the most important being those of the Scarcies and the Sewa. However, the flood waters in the rainy season are an obstacle to agricultural development in these plains. A striking feature of the country's river network is the large number of swampy areas in the upstream valleys. They are the result of regional orogenic movemen ts which blocked the valleys and allowed alluvial and colluvial material to accumulate. The resulting narrow flat-bottomed valleys are drained only by temporary slow-flowing watercourses which usually become swamps for six months of the year. These watercourses tend to flow more quickly in their lower reaches in the better drained alluvial valleys.

Another important feature is the "Bolilands" area described above. This complex of seasonal swamps seems to have resulted from the formation by silt deposits in early times of a common delta of the Mabole, Rokel and Pampana, which flowed into the ocean north of their present estuaries during a period when the level of the sea was higher. The hydrographic network draining the Bolilands is very caried, and the extent and depth of the floods in this region are a major obstacle to agricultural development.

Geology

Introduction

Most of the country lies on series of ancient folded crystalline rocks of varied lithology dating from the Archean epoch, a subdivision of the Precambrian

1/ Except for the Little Scarcies, the data concern only part of the basin.

period, which occurred more than 2,100 million years ago. They are overlain in discontinuity by the Rokel and Saionia Scarp groups, which vary in age from Upper Precambrian to Upper Ordovician, and by the much later sediments of the Bullom group of Tertiary to Recent age. Intense igneous activity occurred in the Mesozoic period before the deposition of the sediments of the Bullom group, giving birth to the Freetown gabbros and the associated silts and minor dykes.

Archean rocks

The Archean formation can be divided into two main units in Sierra Leone. The Liberian complex of granites and green rocks which runs north-south and the Kasila group which runs north-west/south-east.

The first unit includes a series of rocks with high iron and magnesium content of metamorphized to amphibolitic facies (Sula group) resting on a basement of quartzitic granite.

The degree of metamorphism in the basement rock tends to increase towards the edges of the Sula group, producing local zones of granulite (Mano-Moa formation). Younger granites penetrated the series after the period of most intense deformation which occurred about 2.7 million years ago at the edges of the Sula group.

The Kasila group consists of a series of granulites flanked by amphibolites which developed in an area of intense deformation and came to form the south-west edge of the Archean basement rock. In the early stages of the formation of the Kasila group a section split off eastwards, producing the schists of the Marampa group.

Rokel and Saionia Scarp groups

A long time elapsed between the deformation of the Archean basement rock and the formation of the River Rokel trench very late in the Precambrian period. Fossil-free sand-clay sediments were deposited in this trench (Rokel group), forming a band of rock thirty kilometres wide and 225 km long running south-south-east from the frontier with Guinea.

Volcanic episodes occurred from time to time in this epoch, producing lava and ash (Kasewe Hills formation).

The Saionia Scarp group rests in discontinuity on the Rokel group and includes a non-fossiliferous sequence of horizontal strata of sedimentary sandstones and schists, which are usually attributed to the Upper Ordovician period.

Igneous intrusions

Two periods of igneous activity occurred when the Gondwana continental mass divided into several units in the Upper Mesozoic period. The first period was contemporary with the first big fractures and was marked by the intrusion of the Freetown gabbros in the form of superimposed strata and series of silts, dolorites and dykes running parallel to the Atlantic coast.

The second period of igneous activity produced the intrusion of dykes and pipes of kimberlite about 90 million years ago. These intrusions are found mainly in the east of the country, particularly at Tongo and Yengema where they produced the diamond-bearing alluvial deposits of Sierra Leone.

Bullom Group

The band of Tertiary sediments running along the coastal zone can be up to 40 km wide. These formations rest in discontinuity on the Kasila group and the Freetown gabbros and include almost horizontal strata of gravel, sand and clay of marine, fluvial and estuarial origin. Laterites are found in places at the heart of relatively unconsolidated sediments, thus forming resistant outcrops, for example along the Bullom coastline north of Freetown.

Sediments of rivers and streams

Quaternary sands and gravels contemporary with the most recent Bullom deposits are found in all the country's valleys. They are themselves covered with recent alluvial and colluvial deposits in the valleys and along the estuaries of the coastal rivers. The alluvial deposits tend to be silty or argillaceous, while the colluvial deposits are mainly sandy. Both types helped to produce the relatively fertile formations which play an important role in the present and potential use of land in Sierra Leone.

II. GROUND-WATER RESOURCES

Government bodies

Freetown was the first town to be equipped with a water distribution network. Other towns followed. The Ministry of Public Works (Water Supply Division) was initially responsible for the planning (studies), construction, management and maintenance of these networks. This responsibility was subsequently transferred to the Ministry of Energy and Power established in 1984 with a Water Supply Division (WSD), to take charge of water supply and sanitation for towns (except Freetown) and rural areas. The WSD's Rural Water Supply Unit (RWSU) was organized with the help of the United Nations Development Programme in June 1980 to take charge of the monitoring and co-ordination of all water supply programmes in rural areas. Most of these programmes have been implemented at the local level through several integrated rural development projects carried out by the Ministry of Agriculture and Forestry with the assistance of the World Bank, the European Economic Community, the Government of the Federal Republic of Germany and the African Development Bank. In addition to assisting the RWSU, UNDP is helping with the implementation of a pilot construction programme as part of the programme in operation in the Koinadugu area, where new methods and materials are being tested and the supervisory and technical staff are receiving appropriate training. UNICEF is taking part in water quality monitoring operations and in health education activities under all the rural water supply programmes. The Division of Water and Land Resources of the Ministry of Agriculture has carried out very limited hydrogeological studies, but no systematic study of any size has been undertaken.

The World Bank is assisting with the institutional reorganization of the WSD. Initially the efforts will focus on the urban sector and will seek to reduce to a minimum the financial support required from the State budget by applying sales tariffs which will at least cover the operating and maintenance costs of the installations. Parliament is expected to produce appropriate legislation to give either the Ministry of Energy and Power or a new water distribution corporation (water authority) the legal powers to enforce with the necessary vigour the regulations and tariffs applicable to drinking water supplies and sanitation.

Background of ground-water research

As Sierra Leone has an abundance of surface water, ground water has so far received little attention. In March-April 1980 the UNDP/FAO project on the study of land resources sent a mission to train in the field a team consisting of a civil engineer, a geologist and an associate expert. They were taught to operate and maintain the electrical-resistivity geophysical prospecting equipment. The methods included electrical sounding and air-borne prospecting; the measurements were interpreted with an eye to the identification and development of ground water.

Three geological zones, each with its specific hydrogeological problems, were chosen for this research: the Bullom series in the coastal plain, the Bolilands in the Rokel area and the granitic series which covers more than half the country. The study was carried out in the middle of the dry season, so that the team would encounter the most difficult conductivity conditions. The best period for prospecting is October to December after the rainy season when the soil is wet and is a good conductor. A total of 12 electrical soundings were made, including seven deep ones (AB = 800 or 1,000 m) and four shallow ones (AB = 140 or 400 m); five geological units were identified:

- A thin layer of lateritic crust (resistivity 1,200 - 1700 ohm/m);
- A conductive layer of clay and sand (20 - 30 ohm/m);
- A resistant layer of clay (80 - 400 ohm/m);
- A conductive deep layer of clay (25 - 30 ohm/m);
- The Precambrian basement rock at a depth of 70 to 150 m (900 ohm/m and higher).

The E5 sounding in the lateritic crust identified a thick layer (30 m) with a resistivity of 800 ohm/m, corresponding perhaps to a water-bearing sand stratum.

The lateritic crust may be 38 m thick at the E52 sounding.

The electrical soundings have furnished new information about the Bullom series, the Bolilands and the fracturation of the granite formations. The Bullom series consists mainly of clay but it contains a sand aquifer 30 to 40 m thick in its eastern part. The Precambrian basement rock is 250 m deep at Waterloo. Two soundings in the Bolilands penetrated thick clay strata in the first 30 to 40 m. The electrical soundings made in granite areas seem to indicate that the fractured and altered rock may be 30 to 40 m thick along the main faults. Boreholes drilled along these faults would probably yield several litres a second.

Aquifers

The geological and hydrogelogical knowledge of the country is very limited. But it is possible to give the following description of the aquifers on the basis of the available data.

In some places the lateritic crust is fairly thick and forms a ground-water aquifer which could perhaps be used, but only to supply small villages.

The alluvial deposits are usually very permeable. In the biggest valleys they can be sufficiently large to provide an adequate water supply for small towns or small irrigated areas. However, it remains to be established whether in Sierra Leone, where permanent surface water is abundant, ground water can compete economically with surface water in meeting the requirements for irrigation.

The Bullom series of Pleistocene age, especially the sand and clay strata, constitutes the country's main aquifer. Problems of sea-water invasion may arise near the coast and along the estuaries.

The Rokel series, which includes argillaceous sandstones and schists and covers a large area of the Bolilands, is probably water-bearing but to an as yet unknown extent. The metamorphic schists of Marampa might contain small aquifers in the fracture zones. The altered and fractured zones of the granite basement rock are water-bearing.

III. EXPLOITATION OF GROUND WATER

Government bodies

The Water Supply Division of the Ministry of Energy and Power is the Government body responsible for ground-water development. However, in 1983 it still had no hydrogeological or geophysical equipment. At the end of the 1960s it had a cable-tool drilling rig and a local drilling team which operated in only two towns. In 1968 two boreholes were drilled at Lungi with a depth of 22.5 m and a diameter of 150 m. In 1970 a third hole was drilled to a depth of 41 m by the Ministry of Public Works (at the milestone 91 miles from Freetown). Owing to the lack of spare parts the drilling rig was scrapped and the team broken up. There are no private drilling companies in the country.

The Ministry of Energy and Power is planning a hydrogeological study for the whole country and was seeking financing for it in 1983.

Urban water supplies

The towns listed below use ground water for their drinking water supplies. In most cases they take it from dug wells.

Two towns are supplied from boreholes equipped with submersible pumps.

Location	Maximum yield (m^3/h)	Water use	Type of Well	Comments
Lungi international airport	22.6	Domestic and commercial	Borehole	Insufficient yield
Milestone 91	4.0	Domestic	Borehole	Yield declines considerably in dry season
Bonthe	50	Domestic	Dug well	Yield varies slightly throughout the year
York	13.5	Domestic	Dug well	Insufficient yield
Newton	62.5	Domestic and agricultural	Dug well	No supply for two months a year in dry season
Songo	50	Domestic	Dug well	
Kamiendor	16.2	Domestic	Dug well	Water level declines considerably in dry season
Songo	50	Domestic	Dug well	
Kamiendor	16.2	Domestic	Dug well	Water level declines considerably in dry season
Gbinti	13.5	Domestic	Dug well	New station (1982)
Mambolo	50	Domestic	Ground-water spring	Supply sufficient for the population
Jimmi	50	Domestic	Ground-water spring	Supply sufficient for the population

Rural water supplies

The water supplied to the inhabitants of rural areas can be classified as follows:

1. Treated water supplied by the WSD through distribution networks. Seven towns receive these supplies: Makali, Batkanu and Mateboi in the north, Pendembu and Buedu in the east, and Serabu and Jimmi Bagbo in the south.

2. Untreated water drawn from dug wells by diesel pumps. Installation with storage tank, main pipes and standpipes. Most of these installations are managed by the WSD, the others by the people themselves.

3. Untreated spring water in the high plateaus. Installations piping the water to villages three to five kilometres downhill. Distribution by standpipes, most of which are maintained by the local people. For example:

Town	District	Region
Jaima	Kono	East
Gbawandu	Kono	East
Tombudu	Kono	East
Kayima	Kono	East
Panguma	Kenema	East
Royeima	Bombali	North
Yiffin	Koinadugu	North
Mabonto	Tonkolili	North
Bimbuma	Tonkolili	North

4. Water drawn by bucket from protected wells. Pulleys are sometimes installed to facilitate the drawing of the water; most of the well-construction programmes are carried out under integrated rural development projects implemented by the Ministry of Agriculture and Forestry with the assistance of the World Bank, the European Economic Community and the African Development Bank. The intensification of farming activities resulting from the implementation of these projects might cause the contamination of ground water by chemical fertilizers and pesticides; hence the "drinking water supply" element in these projects. The well programmes are operating in the following areas:

Area	Organization	Financing	Number of wells
Koinadugu	UNDP and KIADP	UNDP and EEC	220
Kenema and Kono	EIADP	World Bank	300
Bombali/Tonkolili	NIADP and International Plan	International Plan	700
Moyamba	CARE/MIRDP	ADB/CARE	225
Kambia	ILO/Danida	UNDP	14
Magbosi	MIADP	IFAD	300

A lined hand-dug well costs about US$ 4,000; this amount covers staff costs, depreciation of equipment and vehicles, spare parts, fuel and lubricants, and the cost of the equipment.

5. Water taken from unprotected wells, pools and watercourses. Most of the population is still supplied in this way.

Problems

Some of the water-supply systems described above have particular problems connected with the area and the type of system. In the case of dug wells, the recharge rate is very low, below 2 m^3/h for the most part. This is particularly true in areas of clay aquifers which have a very slow recharge rate. There may be villages close to the coast where the dug wells are affected by sea-water intrusion. In the case of the old gravity systems, the main problems are the deterioration of the piping, inadequate supplies and lack of maintenance.

IV. CONCLUSION

Generally speaking, Sierra Leone has sufficient water resources to meet its needs in the immediate future. However, it does not have enough drinking water. Only two per cent of the rural population, which makes up 75 per cent of the total population, has access to water of acceptable quality; the rest of the rural population takes its water from rudimentary wells, watercourses and pools and is thus exposed to water-borne diseases with their familiar harmful social and economic consequences. It must nevertheless be stressed that in Sierra Leone the exploitation of water in rural areas is planned as part of integrated rural development programmes which involve the local people closely in environmental training activities. A water quality control programme is incorporated in the construction programme and great importance is attached to sanitation; sanitation construction works are undertaken with the participation of the local people. Lastly, a maintenance programme is established.

It is envisaged that these water-supply activities in rural areas will be developed further over the next five years, in particular with respect to ground-water research. A proposal has been prepared for this purpose; it includes deep ground-water research and the drilling of boreholes, 100 of which will be equipped with hand pumps.

In the case of urban water supplies, the World Bank has proposed the establishment of a single distribution corporation for Freetown and 75 urban centres. The Guma Valley Water Company is to be gradually strengthened over three or four years and will eventually become the Sierra Leone Water Company (SLWC); the following measures will first have to be taken:

- Improvement of water supply management methods and of sanitation throughout the country except for Freetown;

- Study of ways of improving the qualifications of the staff, followed by their redeployment;

- Identification of means of covering costs and establishing realistic tariffs with a view to creating a viable financial situation and releasing the Government from the need to make grants.

If this transformation is to be accomplished, foreign technical assistance will be needed in the following areas: organization of services, vocational training of staff, preparation of legislation and regulations, repair of networks, and supply of essential equipment, including vehicles.

V. REFERENCES

Astier, J.L. — Report of Mission, Feb. 18 - March 5, 1978. UNDP/FAO Land Resources Survey Project in Sierra Leone.

Birchall, C.J., Bleeker, P. and Cusani-Visconti, C. — Land in Sierra Leone: A Reconnaissance Survey and Evaluation for Agriculture. Land Resources Survey Technical Report 1, UNDP/FAO, Freetown, 1979.

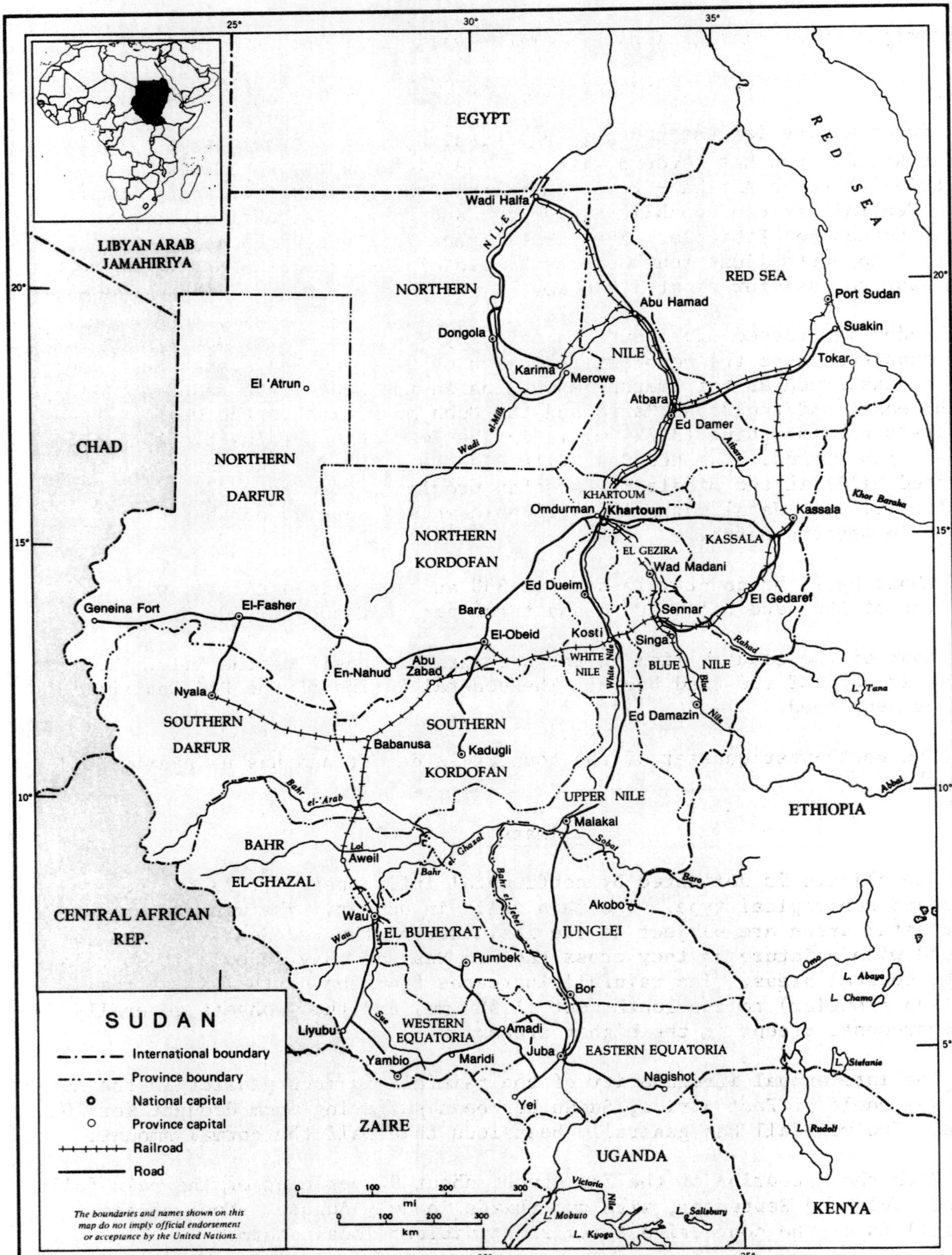

MAP NO. 1434 Rev.4 UNITED NATIONS
SEPTEMBER 1986

SUDAN

Area: 2,506.000 km^2

Population: 20.36 million (United Nations estimate, 1983)

I. BACKGROUND

Sudan is the largest country in Africa. It has a coastline of 700 km on the Red Sea and has borders with eight African countries: two Arab countries of North Africa - Egypt and Libya; three Central African countries - Chad, Central African Republic and Zaire; and three East African countries - Kenya, Uganda and Ethiopia. The country runs from south (close to the equator) to north (just south of the Tropic of Cancer) for about 2,000 km, and from east to west for about 1,300 km.

Sudan is crossed from south to north by the Nile Valley and it has vast flat expanses where the monotony is broken only by low hills and small mountain chains, such as the Imatong mountains in the south, the Red Sea Hills in the east, and the Jabal Marra and the Nuba mountains in the west. The country's highest point (3,224 m) is in the Imatong mountains close to the border with Uganda. The Red Sea Hills rise to 2,700 m. The Nuba mountains, isolated hills in the middle of Kordofan province, rise to 1,374 m in the Jabal Dair. The Jabal Marra is a volcanic massif running north-south for 200 km in western Dafur.

About half the country is between 300 and 500 m in altitude, with two per cent of the land below 300 m and three per cent above 1,500 m.

Most of the country lies in the hydrographic basin of the Nile. Western Darfur is part of the Chad basin. The coastal basins of the Red Sea should also be mentioned.

The north-west quarter of the country is desert and has no hydrographic network.

Climate

The climate is dominated by continental influences. It is of the tropical and subtropical type. The rain falls in summer. The winters are dry. The coastal areas are subject to the influence of the Red Sea; the winds are charged with moisture as they cross the Red Sea and they deposit it as rain in the coastal areas. The rainfall increases from the north (nil at the Egyptian frontier) to the south-west (1,500 mm) and the isohyets generally run east-west, except in the higher areas.

The interannual irregularity of the rainfall is much greater in the north. Like the whole of East Africa, Sudan has been suffering from drought for 10 years. The rainfall has generally been less than half the normal amount.

With the exception of the far south, about 90 per cent of the rain falls between July and September, with two-thirds in July-August. Most of the rainfall is of the convective type: short violent local storms.

Sudan is a hot tropical country. The highest temperatures are recorded in the central plain between Khartoum and Abu Hammad, with an annual average of 29°C. The highest temperature ever recorded in Sudan was 52°C at Wadi Halfa on 29 April 1903. The lowest (2°C) was also recorded at Wadi Halfa on 26 December 1917.

The temperature rises in January; the minimum temperatures are between 6°C and 20°C and the maximums between 24°C and 36°C. In summer (April) the central plain (Khartoum and Gezireh) have the highest temperatures; the maximum ranges from 34°C to 40°C and the minimum from 16°C to 24°C.

The ranges are wider in the dry north than in the wet tropical south. They are in the order of 18°C to 20°C in the northern and central plains and 12°C to 14°C in the south.

The part of the country which is drained towards the Chad basin contains the seasonal wadis - Wadi Kaya and Wadi Azum - which rise on the western slopes of the Jabal Marra. The eastern slopes of the Jabal Marra are drained by Wadis Bulbul, Kaya and Gendi towards the vast swamps of Bahr el Arab. The main coastal wadis on the Red Sea are Khor, Arbaat, Khor, Mog (near Port Sudan) and Khor Handoub.

There are two seasonal watercourses of great importance: the Khor El-Gash and the Khor Baraka in Kassala province, which rise in Eritra and have a very marked torrential regime. Their waters are muddy. Their courses terminate in deltas at the Red Sea.

The Nile is the country's only permanent watercourse; it is the dominant feature of the physical geography and from the human and economic standpoint. The main course of the Nile begins at Khartoum where the White Nile and the Blue Nile converge. Further north the Nile receives the Atbara then crosses the Nabri desert towards Egypt. The White Nile rises in the plateau of the lakes and it provides 16 per cent of the waters of the Nile. The Blue Nile drains the plateaus of Ethiopia and contributes 84 per cent of the water. The limits of the basins of the Blue Nile, the White Nile and the Atbara are not clearly defined.

II. GEOLOGY

Sudan has the following geological units and structures:

- Crystalline basement rock (Precambrian);
- Nawa formation and Paleozoic sandstones (Cambrian-Carboniferous);
- Nuri sandstone (Jurassic-Cretaceous);
- Basaltic outcrops (Tertiary);
- Coastal deposits (Late Tertiary);
- Umm Ruwaba formation (Pliocene-Pleistocene);
- Surface deposits (Late Pleistocene).

The country's geological history can be summarized as follows:

A long period of erosion which toninued up to the Upper Palezoic period removed most of the marine sedimentary cover which had been deposited in shallow water on the crystalline basement rock, with the exception of a few isolated pockets at Wawa (Kodofan), near the Chad frontier, near the Jabal Uweinat and in the north-west of the country.

In the Mesozoic (Jurassic-Cretaceous period plastic sediments of Nubian sandstone were deposited. These were eroded and have survived only in the depressed areas of the basement rock and its Paleozoic cover.

The tectonic movements of the Rift system (Middle-Upper Tertiary) led to the formation of vast structural basins such as those of Bara, Dinder and Baggara. A volcanic phase then produced the Jabals Marra, Meidob and Tasgabo and the basaltic flows of the Bayoda desert and the Gedarif region. This volcanic activity continued throughout the Upper Tertiary and Early Quaternary periods. In the Pliocene-Pleistocene period these basins received thick fluvial and lacustral deposits which constitute the formation known as Umm Ruwaba. The north of Sudan was subsequently subject to a dry climate. Strong winds caused heavy erosion of the Nubian sandstone and the most recent sediments, producing the vast expanses of sand and sand dunes which cover most of the northern part of the country and the regions of Kordofan and Darfur.

A description of the main formations now follows:

Crystalline basement rock

This consists mainly of orthogneiss and paragneiss and Precambrian schists which are divided into three main groups:

- Acid gneiss;

- Quartzites;

- Schists and grauwackes.

These rocks are heavily folded, faulted and injected with intrusive rocks, mainly granites and quartz veins.

Paleozoic sandstones and Nawa formation

Paleozoic sandstones outcrop mainly in the west of the country. Cambrian-Ordovician fluviatile sandstones are found in the upper basin of Wadi Hower which forms a continuation of the Ennedi plateau as far as the Chad frontier. Some of these formations are covered with sandstone sediments of the coastal and fluvial-deltaic type of Lower Siluran age.

Carboniferous sandstones rest in discontinuity on the crystalline basement rock on the Sudanese slopes of the Jabal Uweinat. They consist mainly of sandstone intercalations with intersected stratification which contain many remains of fossilized plants. The Nawa formation outcrops in very few places; it is found in the spoil of wells or boreholes in the area between latitudes 12°21' and 12°50' N on longitude 29°50' E. These rocks samples consist mainly of sandstone-arkose and brown or green fine or medium grained schists. The Nawa formation contains considerable amounts of mica, which distinguishes it from the Nubian sandstone.

Nubian sandstone

This formation covers 28 per cent of the country and outcrops mainly in Kordofan, Darfur and the Khartoum region. In the south it is overlain by thick unconsolidated sediments of the Umm Ruwaba formation, while in the rest of the country it outcrops in the plateaus or suboutcrops below a cover of surface formations of variable thickness. This formation is well stratified with layers of schist or conglomerate and with little or no slope. The elements are of average size and homogeneous. Its origin is disputed, but it was probably produced by deposits accumulated in fresh water. In the north the formation is divided into three groups:

1. Basic silicified sandstone and conglomerate;
2. Soft or hardened white, purple and variagated schist and sandstone;
3. Yellow or brown sandstone.

The Nubian sandstone varies greatly in thickness, attaining over 4,000 m in the deep structural trenches of the northern region.

It is generally thought to vary in age from Upper Jurassic to Lower Cenozoic.

Umm Ruwaba formation

This formation covers 20 per cent of the country, forming two big trenches in the centre and south; the main trench - Bara - covers a vast area in Kordofan, Darfur and the southern region; the other trench is found in the area of the Blue Nile and its tributaries, the Rahad and the Dinder.

These are unconsolidated series with little stratification consisting of sands, silts and clays. However, layers of pebbles can occur at the base of the series where it is in contact with the underlying basement rock or the altered surface of the Nubian sandstone. The sand or clay content is probably due to the Nubian sandstone, and the clay content to the basement rock. The Umm Ruwaba formation consists of lenticular sand and clay units with great lithological variation in the vertical-lateral direction and in the vertical direction. However, the following sequences have generally been identified, from top to bottom:

- Greenish clays with nodules;
- Sands with very coarse rounded grains;
- Greyish crumbly and soft clays.

The formation varies in thickness from 50 m (above the east-west uplift of the crystalline basement rock which divides the Bara trench into two sub-basins) to 1,400 m (along the main axis of the trench in the vicinity of the town of Bara and in the south). This great thickness and the composite and loose nature of the sediments indicate rapid sedimentation by short rivers in a continental or lacustral environment.

The age of the Umm Ruwaba formation is not known with any certainty; it is thought to date from the Pliocene to Pleistocene period. The sediments are younger than the peneplain attributed to the Middle Cenozoic period; moreover, they are covered by Pleistocene deposits up to the Recent black clays and Kordofan sands.

Surface deposits

These include the Nile alluviums, the fill deposits of the valleys, the Kordofan sands, the black clay plains and the coastal deposits of the Red Sea. All these deposits, which range in age from Pleistocene to Recent, cover most of the oldest formations with a variable but fairly thin layer.

The Nile deposits form the plains and the Ancient and Recent terraces of the river and of its tributaries to a depth of up to 60 m. They consist mainly of well-sorted silts and clays with occasional sandy strata. The nature of the alluvial fill accumulated by the temporary watercourses depends on the pattern of the watercourse. In the Piedmont areas these deposits consist of medium to coarse poorly-sorted sands with gravel and lenses of clay in places. The finer material is carried further downstream before being deposited. The clays and silts of the spreading plains are found only around minor arms along the river's course and in its delta; they can be up to 30-50 m thick.

The Kordofan sands consist mainly of eolian sand deposits in the form of vast layers of stabilized dunes covering most of the northern part of the country. These sands consist of fine or medium well-rounded grains of quartz, usually coloured light-brown or dark-red by the oxides. The thickness of the sand cover varies considerably, from a few centimetres to about a hundred metres. These dunes usually run north-south in line with the main direction of the wind. However, north-east transversal dunes of the Barkhane type are also quite common.

The black clays cover the plains of Gezireh, Rahad, Dinder and Butana, extending southwards to the Sudd region on the northern slopes of the Nuba mountains and to the vast plains north of the Nile-Congo watershed. This formation has a remarkably uniform clay content of 60 per cent, with few coarse elements but many limestone nodules and concretions. The black clays of southern Kordofan and Darfur usually rest in continuity on the unconsolidated sediments of the Umm Ruwaba formation, indicating that their sedimentation was the last phase of the deposition of this formation.

The Red Sea littoral formations consist of sequences of mixed marine and continental facies resting in discontinuity on Tertiary deposits and older formations. A highly developed network of wadis, including Khots Barka, Arbaat, Eit and Mog, drain the Red Sea Hills, carrying down large quantities of gravel, sand silt and clay and depositing them in the coastal area. These continental sediments are sometimes 50 to 75 m thick. The marine sediments along the coast consist mainly of complex deposits of coral reefs which form a layer 15 metres deep on top of the older sediments.

III. GROUND-WATER RESEARCH IN SUDAN

At only a little distance away from the Nile, ground-water aquifers provide the only permanent stocks of water in Sudan. They guarantee the supply and even the existence of almost 75 per cent of the population and the livestock, whose total annual needs are estimated at 500 million cubic metres. This figure can be adjusted to 375 million cubic metres, for during the three months of the rainy season the people have surface water available almost everywhere. But account must be taken of the needs of wild animals, supplementary irrigation and industry, as well as of the losses due to evapotranspiration, infiltration and wastage. Up to 1987 the bodies responsible for rural water supply produced about 90 million cubic metres a year, representing less than a quarter of the requirement.

Background

Sudan's first rural water supply service was established in 1946, when a soil conservation service was set up in the Ministry of Agriculture with responsibility for studying the erosion and drying-out of the soil and the water resources available to supply the population and livestock in rural areas; it was also required to recommend measures to solve the problems and in particular to propose new legislation and taxation measures; lastly, it had the task of establishing work programmes with a timetable and an estimate of the costs. Over the next 10 years it built about 300 hafirs (small surface-water reservoirs) with a total capacity of about nine million cubic metres in the rural areas where the rainfall was sufficient. Some of these reservoirs, built in haste, were filled in by the solid deposits carried down by the waters and rendered useless.

In 1956 a department of land use and rural development was established in the Ministry of Agriculture. It had three divisions: land use, development of surface water, and water drilling. Technical responsibility for the development of water resources and supervision of drilling operations was assigned to the geological service of the Ministry of Mineral Resources. This led to the construction of 100 tubewells, 200 hafirs and four small dams with an annual capacity of 11 million cubic metres.

The Rural Water and Development Corporation (RWDC) was established at the beginning of 1967 as a response in particular to the situation caused by the drought and the degradation of the soil and to satisfy the large requirements resulting from the implementation of the rural development programme. Its technical staff included hydrogeologists, drillers, civil engineers, hydrologists, engineers, mechanics, agronomists, soil experts, sociologists and economists, working in three departments responsible for ground water, surface water and rural development. Departments of pastures and forests were added later.

The establishment of the RWDC was followed by the Anti-Thirst Campaign which was the first step in the great advance made in the area of rural water supply. In the 20 years from 1967 to 1987 about 3,200 boreholes were successfully drilled and 350 hafirs and 10 small dams were built; these installations made it possible to store about 70 million cubic metres of water a year, representing 19 per cent of the total human and livestock requirements. During these 20 years the effort was concentrated on the exploitation of ground water, which is perhaps more difficult to develop than surface water but is of better quality and less vulnerable to pollution and, in the short term, to the effects of drought; furthermore, unlike surface water, it does not require large investments at the outset.

Prospecting and evaluation of ground-water resources

This kind of research expanded rapidly after the establishment of the RWDS. A particular effort was made in the Kordofan and Darfur regions, with financing provided by grants and loans. The actual research was entrusted to foreign study companies. The Ground-Water Research Section is responsible for the activities financed in this area by the Government.

A major project financed by the British Government was carried out in northern Darfur in 1968-1974 with the object of studying ground water; it included the following operations:

- Aerial photography and photo-geological interpretation;

- Evaluation of all available data on boreholes;

- Measurement of the water levels in the boreholes and observation of fluctuations in these levels and in the salinity of the water;

- Test pumping and interpretation;

- Geophysical prospecting by geo-electrical, gravimetric and seismic methods;

- Determination of the ground-water potential of the Sharggra basin, with a view to supplying El Fasher, and of the West Nyala basin to supply the town of Nyala.

In southern Darfur the ground-water studies dealt with the aquifers of the Nubian sandstone and Umm Ruwaba formations and with the main alluvial aquifers. These studies considered the possibility of small irrigation projects.

The second ground-water exploration programme in order of importance was carried out by a Czechoslovak company over an area of 80,000 km^2 in Kordofan from 1967 to 1976. It identified the limits of the ground-water basins and determined locations suitable for water boreholes in the areas of crystalline substratum of the Precambrian basement rock. The researchers used geophysical prospecting methods and then took core samples and made diagraphic studies. The holes were lined and equipped with filters. After test pumping the installations were handed over to the RWDC equipped with pumps, reservoirs and protective fencing. The geophysical operations consisted of magnetic and gravimetric measurements (6,100), electrical soundings (1,770) and seismic-refraction profiles (nine): 78 boreholes were installed with a total depth of 14,000 m and 30 were put into production. These studies brought about a considerable improvement in the knowledge of the ground-water basins and they identified several other basins such as Iyal Bakhit. Lastly they succeeded in establishing water points in areas considered barren.

Other work financed by bilateral aid programmes included the geophysical and hydrogeological research carried out between 1976 and 1985 by the Geological Service of the Netherlands at El Gedaref, El Jebelein, Kassala and Nyala. German missions have been studying ground water since 1958, for example at Kordofan and more recently at Khartoum (1979); in 1985 a scientific mission of the University of West Berlin began to study the Nubian sandstone aquifer in Sudan and Egypt.

The United Nations Development Programme has financed studies of the Jabal Marra (1958) and Kordofan (1967) and more recently a regional project on the main aquifer of north-eastern Africa (Nubian sandstone).

IV. MAIN AQUIFER SYSTEMS

River aquifers

The Nile silts and the fill deposits of the valleys are Sudan's largest surface aquifers. Most of them contain large quantities of water of good quality, particularly in the areas of crystalline basement rock, owing to their hydrological properties: high storage and transmissivity coefficients. The ground water drawn from these alluvial aquifers is an important resource used for domestic and irrigation purposes in several northern areas and at Kassala in Kordofan and in Darfur. The ground water of the alluvial formations is usually contained in unconfined aquifers at shallow depths (a few centimetres to 15 metres) below ground level. The ground-water horizon is highest during or shortly after the rainy season and it then gradually declines for the rest of the year. In some alluvial aquifers in exceptionally dry years the water level falls below the furface of the rock substratum. In contrast, when the static water level is above the top of the aquifer, pools and swamps are formed which can remain throughout the year. This is the case at Sudd in the southern part of the country.

In the alluvial fill the ground water flows in the same direction as the surface water, i.e. downstream.

During the high-water periods the flow is from the bed towards the banks, thus recharging the aquifer, and it is in the opposite direction during the dry season, with drainage from the banks towards the bed. The rate of flow within the aquifer depends on the surface gradient and the transmissivity of the alluvial aquifer. Accordingly, the rate of flow of the ground water is higher in the upper part of the aquifer than in the lower part; it is about 25 to 40 cm a day (Darfur, Kassala, Jabal Marra).

The results of the test pumping in the alluvial aquifers are given in table 1. The transmissivity coefficient ranges from 200 to 1,500 m^2/day. The storage coefficient is also high: 13 to 25 per cent, as is to be expected in an unconfined aquifer.

Table 1

Hydrogeological properties of a number of alluvial aquifers in Sudan

Location	Depth of well (m)	Static water level (m)	Specific yield (1/s/m)	Permeability K (m^2/d)	Transmissivity T (m^2/d)	Storage coefficient (%)
KASSALA						
K. El Gash	30	10		40	1,000	0.13
DARFUR						
W. Kutum	18	3.5	4.2	14	200	0.20
W. Azum	38	3	30	45	1,500	0.20
W. Aribo	20	1.5	30	50	1,000	0.25
W. Nyala	17	2	10	50	600	0.20
W. Bulbul	10.5	1.8	35	65	600	0.24
NORTH						
El Seleim Basin	93	6	40	10	500	0.15
Kerma Basin	94	5.7	65.8	12	10.50	0.15

Boreholes in the alluviums of the Blue Nile can deliver more than 500 m^3/h (the highest yields in the Sudan). The wells drilled at El Seleim and Kerma, where a fossil loop of the Nile overlies the Nubian sandstone, can deliver up to 500 m^3/h. The boreholes of Khor El Gash, Wadi Asum and Wadi Nyula can deliver 50 to 100 m^3/h. The ground water of the alluvial aquifers can be recharged by infiltration of flood water, rain water or surplus irrigation water. The discharge is mainly through evaporation and ground-water flow and, to a lesser extent, human activities, in particular irrigation. Recharge and discharge cause fluctuations in the water level which rises in the rainy season and declines in the dry season.

The safe annual yield of the El Gash alluvial aquifer is about 170 to 240 million cubic metres; at Wadi Kutum it is 20 million cubic metres, but the drawoff is below 50,000 m^3/year. Lastly, at Wadi Asum and Wadi Aribo, close to the town of Zalengei, it is 81 million cubic metres with a drawoff of about 50,000 m^3/year. The drawoff from the aquifers of Wadi Bulbul and Wadi Kaja in southern Darfur is about three million m^3/year. The ground water of the wadis is now used to irrigate small plots along their narrow beds. The ground water drawn from alluviums usually contains little mineral salt and is thus suitable for domestic and irrigation purposes. In the valley of Khor El Gash it has a dry residue of 180 to 270 mg/1, with a mainly calcium bicarbonate facies. Away from the bed the salinity increases to 2,000 mg/1 of carbonates, sulphates and sodium chloride. This high concentration is due mainly to the return of irrigation water.

The dry residue of the ground water of Wadi Azum is about 180 to 270 mg/1 of carbonates and calcium and sodium bicarbonates; it is below 150 mg/1 at Wadi Aribo, mainly calcium bicarbonate, and the situation is similar at Wadis Bulbul, Nyala and Kutum. The Bile alluviums in the El Seleim and Kerma basins sometimes contain water of better quality than that of the Nile itself, with a dry residue of 120 mg/1, which is perhaps due to the presence of layers of silt acting as a semi-permeable membrane trapping the salt of the irrigation water.

For the waters contained in the alluviums the sodium absorption rate is below 10, and they can therefore be classified as good or excellent for irrigation.

Umm Ruwaba aquifer

This aquifer is second in order of importance; it has lenticular water-bearing strata of sand and pebbles which are mostly in contact with each other. The areas composed of fine elements are unproductive or even barren.

The water of this aquifer is exploited mostly for domestic use and for livestock, occasionally for small-scale irrigation (Bara and El Kheiran). It covers an area of about 800.000 km^2, mostly south of the 12th parallel N. It may overlie altered formations of the crystalline basement rock (Bara basin) or altered Nubian sandstone or other older formations (Atshan, Baggara and Sudd basins).

Description of the aquifers

(i) Bara basin. This basin occupies a trench of 68,000 km^2 running north-west/south-east in central Kordofan and eastwards as far as the White Nile, which it meets south of Kosti. It is bounded in the north-west by the Nubian sandstone and on all its other sides by crystalline basement rock of the Sodiri-Mega anticline. The Umm Ruwaba formation rests in discontinuity on crystalline rock which is very unevenly altered. The basin is divided into two sub-basins by a north-east saddle produced by an uplift of the basement rock, but they remain in hydraulic contact with each other. The basement rock is uplifted in places to form isolated or aligned reliefs above the static water level; this produces barren or unproductive zones. The Bara basin is situated in the basin of the While Nile; it is recharged in the west and north-west and it has its outlets in the south-east, towards the White Nile.

(ii) This basin includes the southern part of the syncline located between the Bayoda horst and the Sodiri-Mega anticline; the northern part is occupied by Nubian sandstone. It is in the form of a triangular trench, the apex of which is the confluence of the Blue Nile and the Rahad, while the base terminates at the basement reliefs which run along the Ethiopian frontier. The Lithology is typical of Umm Ruwaba, except that it contains much more coarse material (sand and pebbles) than usual. Very little is known about the hydrogeology and it is hoped that the current petroleum research will provide a fuller picture.

(iii) Baggara basin. This basin covers about 150,000 km^2 in southern Kordofan and southern Darfur, south of the Nuba mountains and the saddle leading to the Umm Kaddada basin, i.e. south of the 13°30' parallel N. Its south-east limit is the Bahr El Arab. In the west it terminates at the high banks forming the central part of the Jub-El-Genina horst. This basin is part of the Bahr El Arab hydrographic basin. Almost all the seasonal streams south of the watershed are lost in desert deltas or continue their course in the Bahr El Arab. This is true of Wadis El Ghalla, Shalengo, Bardab, Bulbul, Ibra, Nyala and Kaja. Here have been identified three deep trenches (3,000 m), which are interconnected in a north-east/south-west direction, together with other shallower trenches containing oil resources, but the hydrogeology is not known.

(iv) This basin takes the form of a plain of 200,000 km^2 covered with fresh-water lakes and swamps at the confluence of the Bahr El Arab, the Bahr El Ghazel, the Bahr El Jebel and the Sobat. This basin is situated in the area where the Mega syncline covered with Umm Ruwaba sediments has just split into two. The outlet is towards the north through the White Nile. It is bounded in the east by the Ethiopian platea, in the west by the Juba-Genena horst and in the south by the plateau of the lakes.

The Umm Ruwaba aquifer which occupies the basin consists of fine sediments carried down by the Nile and its tributaries. It is thought that these sediments were deposited in the form of inland deltas made up of mixed strata of sand and clay.

The aquifer is unconfined at its edges with a static water level close to ground level (2 to 10 m deep). Around Bara it is exploited by hand-dug wells for domestic purposes and to irrigate small cultivated plots (vegetables and fruit trees).

The aquifer is saturated in the south and sometimes emerges in swamps and lakes. In the centre semi-artesian or artesian conditions can occur beneath the clay strata at depths of 200 to 400 m, with a static water level of 10 to 100 m above ground level. Temporary artesian conditions have been observed in some cases. The Umm Balagi borehole which is situated at a point where the ground-water flow is blocked by an uplift of the basement rock, has been artesian for more than 10 years.

The fluctuations in the piezometric surface of the Umm Ruwaba aquifer do not exceed 10 to 30 cm; this is attributed to the very small amounts of recharge and discharge.

Ground-water flow

In the Bara basin the ground water flows south-east across the basement saddle towards the town of Umm Ruwaba, and eastwards towards the While Nile. In the Akshan basin it flows in the same direction as the surface water, i.e. northwards. In the dry season in the Baggara basin it flows south-east towards the Bahr El Arab, and then eastwards towards the Sudd basin. During the rainy season and the high-water periods of the Bahr El Arab the habitual southward flow encounters the northward flow from the river and the two flows combine to run eastwards, i.e. towards the Sudd basin. The Sudd area thus acts as a drain for all the surface and ground-water flows of the basin. The Sudd area thus acts as a drain for all the surface and forming swamps. The northward ground-water flow is very slow owing to the poor permeability of the Umm Ruwaba formation and the shallow gradient.

The properties of the Umm Ruwaba aquifer are given in the following table:

Table 2

Location	TD(m)	SWL(m)	M(m)	Q(m^3/h)	SpC(1/s/m)	K(m/d)	T(m^2/d)
BARA BASIN							
Umm Shelikeit	233	86	91	5.5	0.55	1.2	105
Shawa	226	22	45.5	5.5	1.40	0.7	35
Hedeid	137	24	70+	15.2	0.73	2.9	75
Umm Ushra	233	11.4	200+	16.0	1.2	5.0	210
Umm Sheila	121	48	100+	5.5	0.5	1.9	50
BAGGARA BASIN							
Tiwal	220	25	200+	14.4	8	22	880
Umm Berida	153	90	70+	14.4	8	26	845
Hariza	155	100	52+	14.4	4.5	18	600
Shaffa	119	92	21	14.4	3.1	19	320
El Malam	225	70	150+	14.4	1	4	140
Serrir	268	67	200+	14.4	1	3.8	130

TDL Total depth of well
SWL: Static water level
M: Aquifer thickness
Q: Discharge of the well
SpC: Specific capacity
K: Permeability
T: Transmissivity

The transmissivity coefficient ranges from 30 to 200 m^3/day in the Bara basin. The lower values are found in areas of fine elements where some of the wells are of defective construction. The permeability coefficients range from a minimum of 0.7 m/day to a maximum of 5 m/day. In its centre the aquifer is artesian or semi-artesian, with a storage coefficient of 10^{-5} to 10^{-3}. The volume of water stored has been calculated on the basis of an average coefficient of 5×10^{-2}.

Since the Umm Ruwaba aquifer is in hydraulic contact with the underlying aquifers (Nubian sandstone and older formations), the transmissivity values of the wells penetrating the whole formation are higher than those of the boreholes which penetrate only the Umm Ruwaba formations. However, as the boreholes are not lined for the whole of their productive depth, the transmissivity and storage values are probably lower than the potential ones.

The storage coefficients obtained are not reliable. The average value of 5×10^{-2} mentioned above is perhaps a reasonable estimate.

Ground-water resources

It is thought that the Bara basin has a considerable recharge from direct infiltration of rainfall (in the El Kheiran area for example), and from surface runoff (for example, south of the town of Umm Ruwaba). The recharge is considerable in the El Baggara basin; it is provided by frequent temporary flows from the Nuba mountains, the slopes south of the watershed of the Nile and the Bahr El Arab, and the volcanic mass of the Jabal Marra.

Table 3 gives the figures for the ground-water resources of each of the basins of the Umm Ruwaba aquifer.

Table 3

Estimate of the ground-water potential of the Umm Ruwaba aquifer

Hydrogeological system	Ground-water stocks (km^3)	Annual Recharge (hm^3)	Annual drawoff (hm^3)
Bara	45	15	4
Atshan	23	70	20
Baggara	1,300	30	10
Sudd	110	340	20
TOTAL	1,478	455	54

This table shows that the aquifer is far from being exploited to its full capacity. In fact, the present level of drawoff is less than 30 per cent of the safe yield.

The water is usually of good quality; it has a low mineral content or is slightly saline. The dry residue is generally below 1,000 ppm, sometimes below 80 ppm. However, there are some saline pockets with concentrations of 6,000 ppm; the salts may have been deposited by evaporation in the lacustral areas prior to the deposition of the Umm Ruwaba formation itself. In the Bara basin the salinity is generally low (150 ppm) near the recharge areas such as El Kheiran; it increases gradually towards the basin's outlet. However, this phenomenon is reversed and low concentrations are found in the vicinity and to the south of the town of Umm Ruwaba owing to the recharge from Khot Abu Habb. Saline pockets (5,000 ppm) occur north of the saddle of the crystalline basement rock which divides the basin into two sub-basins. Other pockets of 6,000 ppm occur near the Jabal El Kon.

The average salinity of the aquifer in the Baggara basin is 340 ppm. The lowest salinity levels are found close to the recharge areas, i.e. in the north, west and south-west, close to the mouths of Wadi El Ghalla and the other wadis which drain the Nuba mountains.

Areas with salinity levels above 400 mg/1 are found in the deep trenches of the central part. They may be caused by the presence of strata containing evaporites. Owing to the stagnation of the ground water, the concentrations of salt in the Sudd basin are high or very high: 270 to 6,500 mg/1, with average values of 1,500 mg/1 which increase with depth and towards the north, i.e. towards the basin's outlet. The facies are mainly of the carbonate and sodium bicarbonate types when the salinity is low or moderate, and of the chloride and sodium sulphate types in the areas of high concentrations; in some cases there are heavy concentrations of nitrates.

Apart from the small areas of high salinity, the water of the Umm Ruwaba aquifer is usually fairly soft and is suitable for human and animal consumption. However, in some cases when the concentration of nitrates exceeds the admissible amount (35 mg/1), the water cannot be used and the wells and boreholes must be condemned. However, the water can usually be used for irrigation.

Nubian sandstone aquifer

The Nubian sandstone aquifer is the most reliable and the largest in Sudan and indeed in North-East Africa. The water-bearing strata consist of sandy beds with high storage and transmissivity coefficients. However, vast areas of Nubian sandstone outcrops are barren because they are too thin and the static water level of the aquifer is lower than the "wall" of the formation.

As a rule this aquifer is an under-used if not a virgin resource, for up till now it has been exploited in only a few places and only for domestic purposes.

This is a potential resource of considerable importance for the country's future.

The aquifer covers about a million square kilometres, mostly north of the 12th parallel N. In this area it includes a number of interconnected or isolated depressions. South of the 12th parallel N it is generally overlain by the Umm Ruwaba formations which fill the vast trench in the form of a Y covering southern Kordofan and Darfur and the Sudd region in southern Sudan.

Description of the aquifers

The Sahara basin (1) forms a rectangular trench two million square kilometres in area running north-north-west/south-south-east in northern Darfur. In the north the basin extends beyond the Libyan and Chad frontiers and may encompass the Kufrah oasis. Its lower limit is not known. It can be divided into three sub-basins: Wadis Hawa, SaniyaHaiyeh and Tima. The Umm Kaddada basin (2) covers about 100,000 km^2 in central and eastern Darfur and in Kordofan. The Nahoud basin (3) is a saucer-shaped area of 10,000 km^2 in central and western Kordofan filled with Nubian sandstone. It was produced by complex orogenic movements and erosion. In the south it has a subdivision (Nahoud-Saata-Gefauwi) of very considerable ground-water potential. In other parts the Nubian sandstone is barren.

The Iyal Bakhit depression is smaller in area and it is one of the regions of Sudan which suffers most from lack of water. The Nilo-Nubian basin (4) is the biggest basin, extending for 300,000 km^2 in central and northern Sudan. Its upstream part corresponds to the Nile-Bahr El Arab watershed which starts in eastern Darfur and crosses central Kordofan along the Kagmer mountains. In the north it extends beyond the Egyptian frontier. It is divided into several sub-basins: Gezireh, between the White Nile and the Blue Nile; Atbara, between the Atbara watercourse, the main course of the Nile and the Bayoda horst; and Wadi El Milk, a syncline of the downstream part of this wadi and of other sub-basins as well. The El Gedaref basin (5) is a circular area of 28,000 km^2 in central Kassala; it extends eastwards beyond the Ethiopian frontier. The Nubian sandstone forms horizontal strata crossed by thick sills of basic volcanic rock which are heavily faulted.

Conditions of deposit and flow of ground water

The thickness of the Nubian sandstone has been established from drilling data or estimated by geophysical prospecting. It can be either very thin or very thick - up to 4,000 m at the edges where the aquifer is mostly unconfined and the water level lies only about 10 m below ground level. As the thickness increases, the aquifer tends to become confined and the static water level can be as deep as 100 m. In some cases the ground water emerges as springs: at Ain Faro and Kologol, for example, where it is close to ground level and produces oases such as Bir Natron, Nekheila, Lequia, Liquia Omeran and Selima. A borehole drilled at Idd El Tein in the El Gedaref basin, where the Nubian aquifer is covered with basaltic sills, touched a pocket of carbonic gas under pressure, producing a gusher of gasified water which rose 50 m above ground level.

The seasonal fluctuations in ground-water levels in the Nubian sandstones are very small owing to the enormous volume of water stored. Near the recharge zones (upstream) where the free aquifer is in direct contact with the watercourses which feed it, these fluctuations are in the order of one metre. In the central parts of the basin where the aquifer is confined or semi-confined, the fluctuations do not exceed 30 cm. So far no permanent rise or fall has been observed in the static water levels of the various basins.

The flow of the ground water is in conformity with that of the surface water which recharges the aquifers: northwards in the Sahara and Nilo-Nubian basins. The gradient is steeper in the recharge areas, reaching 1/300 just north of Maidob and falling steadily northwards. The flow rate is about 1 to 12 m/year (Wadi Hower).

In the Umm Kaddada basin the flow is generally southwards, indicating that the main recharge is provided by the temporary wadis flowing from the basin's watershed. The flow gradient eases from 1/250 at the El Malha crater to 1/800 at Umm Kaddada, 100 km further south. In the Shagara and Sag El Naam sub-basins the water first flows southwards then veers east in the Saniya Karro strait. The flow rate is 1.5 m/year at Umm Bayada and 60 m/year in the Sag El Naam area.

In the Nahoud basin the water flows northwards and eastwards, demonstrating that the recharge comes from Wadi El Ghalla and Khor Abu Habl. The flow is slowed by the fine-grained sandstones in the aquifer. The north-south chain which crosses the Gedaref basin divides it into two with respect to both surface and ground water. The two sub-basins have different hydraulic gradients and diametrically opposite directions of flow.

The hydrogeological properties of the Nubian aquifer are given in table 4.

The aquifer's transmissivity coefficient ranges from 35 to 1,500 m^2/day, the permeability from 1 to 19 m^2/day, and the storage from 10^{-4} to 25 x 10^{-2}, indicating the transition from confined to unconfined aquifer. The well yield is between a few m^3/h and 400 m^3/h, so that the water can be used for irrigation.

Ground-water resources

The Nubian sandstones are recharged by direct infiltration of rainwater or temporary runoff from the wadis or even from the river system of the Nile. Permanent recharge is indicated by the fluctuation of the water levels close to the recharge areas and by the shape of the isopiestic curves. The low salt content of the aquifer's water also seems to indicate a constant fresh-water recharge. Furthermore, isotopic analysis indicates a mixture of ancient and recent water.

The aquifer can be exploited only in the upper levels of the formation, to a maximum depth of 200 m. The corresponding potential is given in table 5, estimated on the basis of an average coefficient of one per cent. The annual recharge has been calculated on the basis of the isopiestic curves, fluctuations in the water level and variations in the salinity of the water.

Table 4

Hydrogeological properties of the Nubian aquifer

Location	WD (m)	WL (m)	TA (m)	SY (m^3/h)	SY (l/s/m)	T (m^2/d)	K (m/d)	S
1. SAHARA								
W. Hower	37.5	6	300+	18.5	17	1500	17	0.20
Atron	236	3.3	200+	20.5	28	2500	5	
Sanya Hayey	96	53	100+	6.5				
Time	106	67	50	6				
2. UMM KADDADA								
Umm Bayada	94	25	70+	30	-	15	0.4	0.11
Al Nusub	106	74	50+	16.2	-	15	0.65	10^{-2}
Umm Kaddada	70	33	200+	13.7	0.16	20	0.86	10^{-4}
Shaqq Mamda	146	81	65+	13.7	0.12	15	0.43	10^{-4}
Shagara	263	44.5	164	24.6	6.2	150	3.1	10^{-3}
Sag El Naam	136	87	100+	22.0	20	560	19	0.14
Wad Banda	232	94	250+	5	0.23	300	2.2	10^{-3}
Sug El Camal	185	98	500	12	0.31	320	4.3	10^{-4}
Fula	241	98	2000+	12	3	200	1.5	10^{-4}
3. NAHOUD								
Howag	280	132	150+	10	1.6	220	0.84	10^{-3}
Saata	310	95.4	200+		1.3	250	1.1	10^{-3}
Cefauwi	310	102	250	5	0.5	400	1.4	10^{-3}
Maya	357	111	250	5	0.3	175	0.7	10^{-4}
Iyal Bakhit	188	111	75+	5	0.2	-	-	-
4. NILO-NUBIAN								
Khartoum	65	20	100+	85	4	250	4.3	10^{-2}
Korma	245	6.2	500+	280	9.5	1500	6.0	0.1
El Seleim	311	14.6	500+	150	3.7	600	7.1	0.1
W. El Milk	96	7.2	300+	17	2.5			0.15
5. GEDAREF	130	50	206	16	0.1	80	4	10^{-2}

WD: Well depth
WL: Water level
TA: Thickness of aquifer
WY: Well yield
SY: Specific yield
T: Transmissivity
K: Permeability
S: Storage

Table 5

Estimate of the ground-water potential of the aquifer's upper section (200m)

Hydrogeological system	Ground water in storage (hm^3)	Annual recharge (hm^3)	Annual drawoff (hm^3)
Sahara	400	27	3
Umm Kaddada	600	100	10
Nahoud	2	16	5
Nilo-Nubian	600	175	35
Gedaref	76	40	8
TOTAL	1,678	358	61

This table indicates the enormous volume of water stored and suggests that these resources are renewable to some extent, for the drawoff is much lower than the safe yield; accordingly, there is at present no risk of over-exploitation or exhaustion of the resources provided that they are used only at scattered points and this use is controlled when the water is for irrigation.

Chemical composition of the water

The chemical composition of the water of the Nubian aquifer in the various basins in Sudan is given in table 6.

Most of the ground water of the Nubian sandstones has a low mineral content. The lowest concentrations (below 1,000 ppm) are found near the re-charge areas, at Sanya Hayeh (50 ppm) for example, or along the permanent or temporary watercourses which periodically recharge the aquifer: the Sag El Naam (80 ppm) for example. The salinity increases gradually downstream to 300-400 ppm in the centre of the basin. Local pockets of high salinity are found in closed basins such as East Gedaref or in areas where the ground-water flow is slow, i.e. in the Jabal Hilla in eastern Darfur.

Near the recharge areas the main mineral salts found in the ground water of the Nubian sandstone aquifer are carbonates and calcium bicarbonate. The ground water near the recharge areas and the surface water have very similar chemistry. Further downstream the water tends to have a higher sodium car-bonate content.

The water of the Nubian sandstone is generally suitable for human and animal consumption. Its salinity is usually low and it contains no toxic elements; it is suitable for irrigation of all crops and for all soil types.

Table 6

Mineral content

Location	Dry residue (ppm)	Ph	Ca m.equiv.	Mg. m.equiv.	Na m.equiv.	CO_3 m. equiv.	HOC_3 m.equiv.	SO_4 m.equiv.	Cl m.equiv.	SAR and suitability for irrigation
1. SAHARA										
Air Faro	100	7.2	2.4	1.2	0.6	-	3.6	0.2	0.2	0.04 (C_1S_1)
Tima	134	7.4	1.4	0.5	0.7	-	1.8	0.3	0.1	0.72 (C_1S_1)
Sanya Hayey	54	7.1	0.5	0.1	0.3	-	0.7	0.4	0.1	0.55 (C_1S_1)
2. UMM KADDADA										
Umm Kaddada	314	7	1	0.5	1.9	-	0.9	0.7	1.2	2.2 (C_1S_1)
Sal El Naan	80	7.8	1.8	1.5	7.8	0.5	4.7	0.7	1.8	6.1 (C_1S_1)
Shagra	668	7.6	0.7	0.2	1.9	-	1.6	0.5	0.5	2.8 (C_2S_2)
Umm Bayada	530	8.5	0.4	3.7	3.5	0.9	4.9	0.4	1.4	0.23 (C_2S_1)
El Fula	500	8.2	1.2	0.8	7.0	-	4.8	2.1	1.6	6.34 (C_2S_1)
3. NAHOUD										
Nahoud	280	3.0	3.0	0.8	0.3	-	3.3	0.5	0.3	0.2 (C_1S_1)
Khmnas	160	8.0	2.1	0.5	0.2	-	1.9	0.3	0.5	0.2 (C_1S_1)
Umm Feis	220	8.1	2.6	1.1	0.3	-	2.9	0.5	0.6	0.8 (C_1S_1)
4. NILO-NUBIAN										
Khartoum	700	8	3.0	24	7.4	3	3.2	1.9	2.7	4.5 (C_2S_1)
Gezireh										
Seleim	200	8.4	1.5	0.8	0.7	1	1.8	0.5	0.7	0.5 (C_1S_1)
5. GEDAREF										
Eastern	2840	8.2	3.6	8.6	41.3	30	20.5	1.0	1.9	16.6
Western	940	8.8	0.3	3.5	5.6	4	3.4	0.6	1.4	4 (C_2S_1)

V. EXPLOITATION OF GROUND WATER

The drought which has afflicted Sudan almost without cease since 1967 has seriously damaged the country's natural resources and caused the desert to advance 100 km southwards. At present (1985-1986) the desert is advancing at one to five kilometres a year. Sand dunes have invaded most of the agricultural land in the northern savannah which receives little rainfall and they are advancing towards the southern part where farm production is high.

The future of these regions is a source of concern and steps must be taken to protect what can be protected and to rehabilitate part of what has been lost. The availability of water resources is the key to the success of any programme to improve the environment and restore the ecological balance.

Fortunately, Sudan is better placed than other countries of the Sahelian zone to combat desertification and develop its natural resources. It has abundant water resources: rainfall, surface water from the Nile and its tributaries, and ground water. The rainfall and surface runoff vary very greatly in step with the cycles of drought and rains which affect the country. Furthermore, this water is inefficiently used and much of it is lost through evaporation or drainage. Better techniques must therefore be introduced for the collection and use of surface water so as to avoid wastage.

Ground water is the most abundant resource and also the least vulnerable to weather conditions. It is available in aquifers of average or high yield over more than half of the country. The data given above show clearly that these aquifers are very little used and that their exploitation could be considerably increased without risk of exhaustion.

Domestic and irrigation uses

At present Sudan's ground water is exploited almost exclusively to meet human and livestock needs. About 70 per cent of the human and livestock populations are supplied with ground water from six-inch diameter lined boreholes equipped with diesel or turbine pumps. The water is pumped into a raised reservoir from which it flows through five small faucets to fill family tanks and three or four watering troughs with a capacity of one cubic metre.

Each of these "water yards" usually has one or several tubewells. Large settlements have standpipes and private outlets. It is difficult to estimate the costs owing to the fluctuations in the price of the imported equipment and materials and in exchange rates.

However, in rough terms it can be said that a water yard with a single tubewell costs about US$ 50,000, including US$ 22,000 for a 200 m hole of 6" diameter, lined and equipped with a filter, and US$ 11,000 for the pump and motor.

For two holes the cost is about US$ 80,000.

It is difficult to keep the installations in permanent operation, for their dispersal over such a vast area causes serious problems of maintenance and repair; furthermore, most of them are in places of difficult access. At present there are about 25 maintenance and repair stations and workshops in operation; they service 3,000 yards over an area of one million square kilometres - a difficult task. About 50 additional stations are to be built under a two-year programme (1985-1987).

A modest charge is made for the distribution of the water, in some cases insufficient to cover the cost of fuel. It must also be remembered that the water yards are usually subject to no controls, with the result that the system deteriorates and the quality of the services declines. Studies have therefore begun to determine the most appropriate management methods. These studies have produced the following recommendations:

1. Increased charges;

2. Increased participation by the people and community organization in the management and operation of the installations;

3. Establishment of a properly defined maintenance programme with periodic checks.

The fairly recent increase in the number of installations has encouraged the people to develop their livestock holdings. However, the concentration of herds around the waterpoints has led to overgrazing and sometimes to the total destruction of the vegetation cover for 20 km around. In these areas animals which would have died of thirst now die of hunger.

In future the establishment of new installations in these ecologically vulnerable areas must be planned as part of an integrated development strategy. It must satisfy the real needs and secure substantial improvements in the social and economic conditions. The following criteria should be used:

- New water points should be located at not less than 20 km from the nearest permanent water point;

- The prospective site should be studied to determine the impact of the increased water supply on other natural resources;

- Higher charges should be made for the water so as to hold consumption at a reasonable level.

Irrigation should be limited to areas where the aquifers are shallow and high-yielding. The following criteria should be used:

- Total depth of well (which determines the amount of investment needed);

- Depth to water (which determines the operating costs);

- Specific yield expressed in 1/s/m of drawdown (which affects the cost of both investment and operation);

- Chemical quality of the water (which must be suitable for the soils and the proposed crops.

In the light of these considerations, it can be said that Sudan does have areas which can be irrigated by ground water. Auxillary irrigation from ground water can be envisaged in other areas where the aquifers are fairly deep.

Exploitation of alluvial aquifers

Most of Sudan's alluvial aquifers meet the criteria set out above. In one way or another they have been exploited for irrigation but not to their maximum potential. The alluvial aquifer of the Gash basin is a typical example. Some 8,000 ha are now cultivated in the El Kassala area. The annual drawoff of water is 120 million cubic metres, while the safe yield is about 200 million, so that the irrigated area could be increased by 4,000 ha. Most of the wells are less than 35 m deep and the static water level is usually at 10 m below the surface and the specific yield in the order of 10 l/s/m. Accordingly, a well can usually irrigate eight hectares. The water is of excellent quality and suitable for many crops and many soil types.

The alluvial aquifer of the Wadi Kutum system in northern Darfur is exploited by many shallow large-diameter wells to irrigate small garden and orchard plots. The available capacity up to the safe yield is about 20 million cubic metres a year or 40 times the present drawoff. Thus, an additional 2,000 ha could be irrigated, with the ground water supplementing the rainwater. The specific yield is 4.2 l/s/m, and each well can therefore irrigate four hectares. The water quality is excellent.

The alluvial aquifers of Wadis Azum and Aribo west of the Jabal Marra could supply 80 million cubic metres a year, whereas at present only 100,000 m^3 is drawn off to irrigate small plots. The depth to be reached by the wells is less than 40 m and the depth to water is usually less than five metres. The specific yield is 30 l/s/m, and each well could irrigate 24 ha. The water is of excellent quality. The climate is fairly cool and suitable for Mediterranean crops.

The alluviums of the temporary watercourses of southern Darfur (Wadis Kaja, Bulbul, Ibra and Nyala) could supply 70 million cubic metres a year. At present these alluviums are partially exploited for irrigation and to supply the town of Nyala. This town consumes about 2,500 m^3/day, i.e. one million cubic metres a year. About 14 million cubic metres could be drawn off up to the end of the dry season. Accordingly, it is hardly possible to increase the drawoff without risk of exhausting the aquifer.

Hardly any of the alluviums of other wadis are exploited. However, they are fairly thin and the water levels are subject to great variations. Substantial exploitation cannot therefore be envisaged.

The aquifers of the El Selim and Karima basins are contained in fossil beds of the Nile, so that in principal they have constant recharge. This area used to be irrigated by a canal from the Nile. As the level of the Nile has declined over the past 15 years and the canal has not been maintained, the irrigated areas have reverted to desert. In recent times the farmers have increased irrigation from large-diameter wells or from tubes sunk in the soil and equiped with centrifugal diesel pumps. The wells are less than 50 m deep and the water level is at three to five metres, corresponding to the level of the Nile. The specific yield of a 50-meter borehole is 40 to 65 l/s/m. A single well can irrigate 30 to 50 ha. In many cases the water is of better quality than the Nile water and is therefore suitable for irrigation.

Exploitation of the Umm Ruwaba aquifer

About 450 million cubic metres a year could be drawn from this aquifer, but at present only about 40 million cubic metres are taken for domestic purposes and 15 million for irrigation. It supplies 800,000 m^3/year to 50 water yards; about 400,000 cubic metres are drawn from large-diameter wells scattered over the whole aquifer. The towns of Umm Ruwaba and Bara consume about 400,000 m^3/year.

In the El Kheira area small garden and orchard plots have been irrigated since time immemorial from large-diameter wells lined with bricks. Some of these wells have centrifugal diesel pumps. The wells are usually less than 50 m deep and the water surface is at 5 to 8 m. Each well can irrigate two to three hectares. The water quality is excellent and suitable for irrigation. It is surprising that the 300,000 m^3 of water flowing freely from two wells at Umm Balagi should be wasted, apart from a small amount used as drinking water. This amount could irrigate about 8 ha of market-garden crops and fruit trees.

The Umm Ruwaba aquifer in the Bara basin is thought capable of supplying three million cubic metres to the town of El Obeid by means of 12 boreholes; the investment is estimated at US$ 10 million, with annual operating costs of US$ 500,000.

In the El Atshan basin the Umm Ruwaba aquifer and the overlying Nile alluviums are intensively exploited to irrigate 400 ha under the Rahad project. The drawoff is 400 million cubic metres. The recharge is by drainage from the River Rahad. The ground water is of good quality.

The Baggara basin could supply 30 million cubic metres of water a year, as against the present drawoff of only 10 million cubic metres. The annual human and livestock requirements are about five million cubic metres. During the four-month rainy season these requirements are met from "rahads" and "turdas"bordering the bed of the watercourse. About two million cubic metres are provided in this way, with the balance of three million cubic metres drawn from hafirs are partially filled with silt and the total available capacity is no more than 750,000 cubic metres. There are 130 water yards supplying 2.3 million cubic metres of water. In theory there are sufficient water points. In fact a large number of them are not in service. A considerable effort must therefore be made to rehabilitate the water points and clear the hafirs.

The Baggara basin has a good irrigation potential. The most favourable areas are those where the ground water lies at the shallowest depths and where the specific yields are highest, for example in the southern parts of Umm Agaga (600,000 ha suitable for mechanized agriculture). The rainfall (600 mm) is sufficient for rain-fed crops; ground water could supply additional irrigation for certain crops. It lies at a depth of 25 to 30 m and the specific yield ranges from 200 to 700 m^3/d/m of drawdown. A single well can irrigate 60 ha and the water is of excellent quality.

In the Sudd basin the aquifer is hardly exploited at all; the little water which is drawn off is used for domestic purposes. Given the abundance of surface water, this is the only use which can reasonably be made of the ground water.

Exploitation of the Nubian sandstone aquifer

The available potential of the Nubian sandstone is about 360 million cubic metres a year, whereas the present drawoff is only 60 million cubic metres (17 per cent). This is a major resource available in northern Sudan to combat the devastating effects of the drought. It is located in an extremely arid area which includes northern Darfur and its extensions into Libya and Egypt. The potential of the three sub-basins (Wadis Hower, Saniya Haiyeh and Tima) is encouraging with respect both to domestic and to irrigation and industrial uses. For a long time now the three States concerned have been considering the construction of two major international roads, one linking Darfur to the Kufrah oasis in Libya, and the other following an ancient track (Darb El Arbain) linking northern Darfur and Kordofan to the "New Valley" of Egypt. Sudan has already taken the initiative of establishing five waterpoints along the proposed road to Libya; each has two boreholes equipped with pumps and a storage capacity of five cubic metres. The works at Wadi Hower on the road to Egypt are planned for the near future. The financing is to be provided by an Italian grant for the repair of water points in northern Darfur. These water points will encourage vehicular traffic between the three countries and at a later stage they will supply the water needed for the construction of the roads. They will also facilitate the exploitation of a number of isolated mineral deposits which have been identified and the expansion of mining research in these remote areas. They might also provide the driving force for the development of the ground-water and land resources of Wadi Hower and the Nekhila and Bir Natron oases, making it possible to grow high-value cash crops or forage and thus to combat the desertification of this region and reduce the overgrazing of central Sudan by increasing the export of animals to the markets of Libya and Egypt.

Ground water is found along Wadis Hower, Bir Natron and Nikheila and at the Laquia oasis. In Wadi Hower the wells are about 40 m deep and the water surface lies at about 6 m. The specific yield is about 20 l/s/m and each hole could irrigate about 10 ha of arable land along the banks of the wadis. As the area is inaccessible and no fuel is available, the ground water must be extracted by wind, solar or hand pumps to irrigate the small plots. The water is of excellent quality.

At Saniya Haiyeh the relatively mild climate makes it possible to grow corn and beans and other vegetables which are viable only in northern Sudan; this might contribute to the sedentarization of the Zaghawa, one of the tribes which have suffered most from the drought in 1984-1985. The depth to be reached by the wells is about 100 m; the water surface is at 50 m at most. The holes can produce 50 to 75 m^3/h and irrigate 8 to 10 ha. The water is of excellent quality. Similar conditions are found in the Tima sub-basin. The projects envisaged will facilitate the sedentarization of the Meidob tribe in northern Darfur. The ground-water potential of the Umm Kaddada basin is about 100 million cubic metres, i.e. ten times the present drawoff, which is used only for domestic purposes, including supplies for the towns of El Fasher and Umm Kaddada and the rural areas of central and eastern Darfur and western Kordofan.

The Nubian sandstone aquifer of some of the Umm Kaddada sub-basins (Umm Bayada, Saq El Naam and Shagara) offers good prospects for irrigation based on joint use of ground and surface water; most of the surface water is at present lost, whereas it could be stored in the clay soils for most of the growing season. This could be achieved by building earth banks to block the flows and facilitate infiltration in deeply ploughed soil; the ground water would be used for supplementary irrigation or crop diversification.

The efficiency of the operations and the certainty of having water when and where it is needed will depend mainly on the extent to which ground water can be used to supplement irrigation.

The surface and ground water at Umm Bayada can be used to increase the production of food crops and forage and to facilitate the sedentarization of the tribes seriously affected by the drought. If agriculture is shifted to the clay plains, the sandy areas will be able to regenerate their natural vegetation. The total depth of the wells is about 100 m and the water has to be piped for only 25 m. The yield of the wells is 30 m^3/h, which means that each well can irrigate four hectares. The water quality is excellent.

In the Sag El Naam area the aquifer is very productive, with a stock of 100×10^9 m^3 and an available potential of 20 to 60 million cubic metres. The holes are about 150 m deep and the water is reached at 80 to 90 m. The specific yield is about 20 l/s/m and each well can irrigate 20 to 30 ha. The water quality is excellent.

In the Sag El Naam area the aquifer underlies a clay plain of 50,000 ha of arable land which receives rainfall of 400 m. The plain has an intermittent watercourse (three months a year) - Wadi El Ku and its tributaries. Joint use of surface and ground water could meet a large part of Darfur's grain needs. It could also improve the living conditions of the people and slow down the rural exodus.

In the Shagara sub-basin the ground water of the Nubian sandstone is exploited to supply the town of El Fasher with 60,000 m^3/year drawn from seven boreholes; this is approximately the available yield. The town also takes water from the Golo reservoir on Wadi El Ku, west of El Fasher.

The available yield in the Nahoud basin is 16 million cubic metres a year, and only five million cubic metres are drawn off for domestic use, including drinking water. The most productive part is the Gefauwi-Saata trench. However, the depth of the wells (250 m), the depth to water and the moderate specific yields make it impossible to use the water for irrigation, especially as it will probably be needed to supply the towns of Nahoud and El Obeid.

The Nubian sandstones in the Nilo-Nubian basin contain large amounts of water. The available discharge is about 175 million cubic metres a year; the present drawoff is 35 million cubic metres, most of it used for irrigation.

The irrigation boreholes tapping the upper part of the Nubian sandstones in northern Gezireh and near Khartoum are 60 to 70 m deep and those which tap the lower part are 120 to 15 m deep. The yields are 60 and 140 m^3/h respectively in the best cases. The unit cost per cubic metre is about two United States cents - a little less for the lower aquifer, a little more for the upper one. Irrigation is thus more economical with deeper boreholes despite a higher initial investment. This is why the small farmers put their money together to install deep boreholes.

In the northern region irrigation now tends to be based increasingly on ground water. The reason for this is the decline in the level of the Nile and the ceaseless displacement of its course, developments which are accompanied by the advance of the desert and the invasion of sand dunes to cover the irrigated land. The area of arable land in this region declined from 400,000 ha in 1960 to 10,000 ha in 1978. This caused half the population to migrate to the big towns of Sudan or to other countries. Huge volumes of water are needed to combat this situation and it can come only from the Nile or the Nubian sandstones. It is estimated that the cultivation of about 100,000 ha by means of ground-water irrigation could produce an annual re-return of about 30 per cent on the investment. At El Seleim the boreholes would be about 120 m deep, the water lying between five and 14 m deep. The well discharge would be 200 to 500 m^3/h, so that each well could irrigate 20 to 50 ha.

VI. CONCLUSION

Sudan is a vast dry country of farmers and herdsmen and the existence of available ground water is a decisive factor in its economic and social development. Leaving aside the permanent watercourses, i.e. the arms of the Nile and its tributaries, there are 80 million ha of grazing and arable land accommodating 70 per cent of the country's population, or about 50 million people, and about 80 per cent of the livestock, requiring a total of 500 million cubic metres of water. The rain which falls for three months a year supplies 30 per cent of this requirement; the balance must be drawn from ground-water aquifers. To date about 3,000 pumping stations, 1,000 <u>hafirs</u> and 20 small dams have been built, supplying a total of about 100 million cubic metres a year - a little over a quarter of the requirement. In 1967 a decisive boost was given to the ground-water development programmes in three main hydrogeological units: the alluvial beds, the Umm Ruwaba formations and the Nubian sandstones; together these units cover almost half of the country and supply more than half the population with water. The alluvial aquifers, which have a good recharge from surface runoff and rainfall lie at shallow depths, are widely used for irrigation. The Umm Ruwaba aquifer covers 20 per cent of the country; it includes unconsolidated detritic elements, usually fairly fine, in a complex sedimentary sequence with great lateral and vertical variations. The aquifer is unconfined, semi-confined or confined depending on its location.

The Nubian sandstones are one of the world's largest hydrogeological units and cover more than a third of Sudan; they can be up to 4,000 m thick in the centre of the basin which they fill. The aquifer is unconfined, semi-artesian or artesian depending on the distance from the edge of the basin.

As a rule Sudan's ground water is chemically suitable for human and animal consumption and for irrigation. The exploitation is still low in comparison with the available potential. This is due both to the installation costs (US$ 50,000 per water yard), the operating costs and the problems of maintenance and repair, for the installations are widely scattered and usually difficult of access. A great effort has to be made to overcome these difficulties.

A water policy must therefore be established as part of an integrated development plan for natural resources for each region of the country. Some encouraging progress has already been made in this direction.

Ground-water irrigation offers promising prospects in some alluvial regions: the water of the El Gash alluviums irrigates 7,000 ha and could irrigate a further 4,000 ha. There are other opportunities in northern Darfur (Wadi Kutum and Wadi Hower) west of the Jabal Marra, where there is an irrigation potential of 4,000 ha, and in the north of the country in the area of the fossil beds of the Nile.

The Umm Ruwaba aquifer is exploited to only a limited extent for irrigation, notably at El Kheiran (Bara basin) and at El Atshan, along the Blue Nile. It is estimated that this aquifer could be exploited for the full or supplementary irrigation of 400,000 ha in the Umm Agaga plain and in southern Kordofan and Darfur.

The Nubian sandstones of the Sahara basin could supply some 27 million cubic metres a year for the construction of the international roads to Egypt and Libya and the creation of oases along these roads where the nomadic tribes stricken by the drought could settle.

In the Umm Kaddada basin the ground water could be used for the full or supplementary irrigation of 50,000 ha at Sag El Naam.

The Nubian sandstones are exploited to irrigate several areas an important resource in the campaign against desertification in the north of the country (El Seleim and El Khewi) with respect to the creation of both wooded and grain-growing areas.

VII. REFERENCES

Abdul Salam, Y. — The Groundwater Geology of the Gezira, Sudan. Thesis MS - University of Khartoum, Sudan, 1966.

Ali, H.O. and Whiteley, R.J. — Gravity Exploration of Groundwater in the Bara Basin, Sudan. Geoexploration, Vol. 19, pp. 127-141. Elsevier Scientific Publishing Co., Amsterdam, 1981.

Doxiadis, I. — Land and Water Use Survey in Kordofan Province. UNSF/FAO Final Report, 1967.

El Boushi, I.M. and Whiteman, A.J. — The Nubian Sandstone Formation, the Major Aquifer in Sudan Republic: paper for seminar on Community Water Supply. Faculty of English, University of Khartoum, 1967.

El Tom, M.A. — The Rains of the Sudan. Khartoum University Press, Khartoum, 100 pp. 1975.

FAO — Land and Water Resources of Jebel Marra. FAO/SF 48 SUP 17, Rome, 1968.

________ — Savanna Development - The Sudan - Comprehensive Technical Report - AGS. DP/SUD/71/525, Rome, pp. 91. 1974.

Gishler, C.E. — Present and Future Trends in Water Resources Development in the Arab States. UNESCO-ROSTAS: prepared for Conference on Application of Science and Technology to the Development of the Arab States (CASTARAB), 1976.

Grabham, G.W. — Water Supply in the Anglo-Egyptian Sudan. Geol. Survey Bulletin, No. 2, 1934.

Ground-Water Survey, Netherlands — Ground-Water Investigation, El Gedaref, Rural Water Corporation open files, Khartoum, 1979.

________ — Ground-Water Investigation, El Jebelein. Rural Water Corporation open file, Khartoum, 1979.

________ — Water Resources Assessment, Kassala Gash Basin, Rural Water Corporation open file, Khartoum, 1984.

Hunting, Geology and Geophysics and Sir M. MacDonald — Geophysical Survey Report. FAO Savanna Development Project. London, 1973.

________ — Water Survey and Development Project in Darfur Province. London, 1970.

Hunting Technical Service — Jabel Marra Investigations - Report on Phase 1. Sudan Ministry of Irrigation, 1959.

________ — A Hydrogeological Study of the Qoz Resettlement Area, Eastern Kordofan Province. London, 1970.

Hunting Tech. Service and Sir M. MacDonald — Southern Dafur Land Use Planning Development Plan, Annex 2 - Water and Engineering. London, 1974.

———— Water Supply Development of El Obeid, Kordofan Province. London,1974.

———— Sag El Naam. Evaluation and Pre-investment Study - Darfur Province. London, 1976.

———— Agricultural Development in the Jabel Marra Area. London, 1977.

———— Southern Kordofan, Annex II - Hydrology and Hydrogeology. London, 1981.

Ibrahim F.N. — Ecological Imbalance in the Republic of the Sudan - with Reference to Desertification in Dafur. Institute of Geosciences, University of Bayreuth, FRG, 1984.

Ireland, A.W. — The Climate of the Sudan. Reprint from Agriculture in the Sudan, pp. 62-83. Oxford University Press, 1948.

Ishag, A.H. — Potential Recharge to Ground Water Aquifer, Kordofan Province, Sudan. Thesis MS, University of Arizona, USA, 1964.

Iskander, W. — An Appraisal of Ground-Water Resources of Zalengei Area, Darfur Province. Thesis MS, University of Arizona, USA, 1967.

———— A New Plan for Ground-Water Investigation Programme in the Sudan. Sudan Eng. Society Journal, No. 18. 1970.

———— Water Problems in the Rural Areas of the Sudan. UNEP, National Council for Research, ALESCO Symposium on Human Environment and Development, Khartoum, 1972.

———— Agricultural Development of Saqq El Naam Area, Darfur Province, Sudan. Paper submitted to Arkaweit Conference, El Obeid, Sudan, 1973.

———— Agricultural Development of Zalengei Area, Western Dafur Province, Sudan. Paper submitted to Arkaweit Conference, El Obeid, Sudan, 1973.

———— Ground Water in the Sudan (in Arabic): Resources of Groundwater in the Arab Countries. ALSCO Science Monograph, No. 2, pp. 66-97, 1973.

———— Feasibility Study on the Transnational Management of the Major Artesian Aquifers in NE Africa. "The Nubian Sandstone Aquifer in the Sudan". Study submitted to the Egyptian Remote Sensing Centre, Cairo, Egypt, pp. 32, 1976.

_________ Appraisal of Ground-Water Resources of Jongoli Canal Area. Study submitted to the Egyptian Remote Sensing Centre, Cairo, Egypt, 1977.

Iskander, W. and Tohami, M.S. Agricultural Development of Umm Agagga Area - South Darfur and Kordofan. Study submitted to AGRAR Consulting Firm on Mechanized Farming. FRG, 1974.

Karkanis, B.G. Hydrochemical Facies of Groundwater in the Western Provinces of Sudan. Thesis MS, University of Arizona, USA, 1965.

Kheiralla, M.K. A Study of the Nubian Sandstone Formation of the Nile Valley between 14°N and 17°42'N - with Reference to Ground-Water Geology. Thesis MS, University of Khartoum, 1966.

Kleinsorge, H. and Zscheked, J.G. Geologic and Hydrogeologic Research in the Arid and Semi-Arid Zones of the Western Sudan, Part I. Hanover, FRG, 1958.

Kleinsorge, H. and Kreysing, K. Geologic and Hydrogeologic Research in the Arid and Semi-Arid Zones of the Western Sudan, Part II. Hanover, FRG, 1959.

Klitzsch, E. Northwestern Sudan and Bordering Areas: Geological Development since Cambrian Time: Berliner Geowissenschaftliche Abhandlungen, Reihe A/Band 50 - Research in Egypt and Sudan. West Berlin, FRG, 1984.

Mabrouk, B. Geologic and Hydrogeologic Studies of East Kordofan Area, Sudan. Thesis MS, University of Khartoum, Sudan, 1972.

Mansour, A.O. Sources of Water Supply as Related to Geology in Kordofan Province. Sudan Geol. Survey Bull. No. 8, 1961.

Mitwali, M.A. Interpretation of the Low Gravity Anomaly in N-E Kordofan, Western Sudan. Bull. di Geofisica Teorica at Applicata, Vol.XI, No. 4, pp 119-125, 1969.

Mohamed, Y.A., El Semmani, M.O. and Shaddad, M.Z. The Northern Kordofan Rural Water Supply Base Line Survey. Study prepared by the Institute of Environmental Studies, University of Khartoum, Sudan, 1982.

Rodis, H.G. and Iskander, W. Ground-water in the Nahoud Outlier of the Nubian Series, Kordofan Province, Sudan. US Geol. Survey Prof. Paper 475-B, 1983.

Rodis, H.G., Hassan, A. and Wahdan, L. Ground-Water Geology of Kordofan Province. Sudan Geol. Survey Bull. No. 14, 1964.

Saaid, E.M. Ground-Water Appraisal of the Gash River Basin at Kassala, Kassala Province, Sudan. Thesis MS, University of Arizona, 1968.

__________ Geological and Hydrogeological Studies of Khartoum Province and its Neighbourhood, Sudan. Thesis MS, Geol. Dept., University of Cairo, 1974.

__________ Hydrogeology of Khartoum Province and Northern Gezira Area. Geological and Mineral Resources Dept. Bull. No. 29, Khartoum, 1976.

Salama, R.B. Ground-Water Resources of Sudan. Rural Water Corporation open file, 1978.

Salama, R.B. and Salama, M.N. Geology and Hydrogeology of the Southern Provinces. Technical Paper. Rural Water Development Corporation open file, 1977.

Sandford, K.S. Sources of Water in the North-Western Sudan. Geogr. J., Vol. 135, 1935.
Geological Observations on the N-W Frontiers of the Anglo-Egyptian Sudan and the Adjoining Part of the Southern Libya Desert. Geol. Soc. of London Quarterly Journal, Vol. 19, pp. 323-381, 1935.

Satakopan, V. Water Balance in the Sudan. Sudan Meteorological Service, Memo 5. Khartoum, 1965.

Shawki, M.K. Rural Water and Development in the Sudan. Rural Water Development Corporation, Book No. 3, 1967.

STROJEXPORT Geophysical - Geological - Hydrogeological Research, Central Kordofan, First Stage. Final Report, Prague, Czechoslovakia, 1970.

__________ Ground-Water Research - SW Kordofan Province, Second Stage. Final Report, Prague, Czechoslovakia, 1972.

__________ Geophysical Investigation of Hydrogeological Structures - W Kordofan Province, Third Stage. Final Report, Prague Czechoslovakia, 1975.

__________ Geophysical Investigations of Ground-Water Structures - W Kordofan & E. Darfur, Fourth Stage. Final Report, Prague, Czechoslovakia, 1976.

Geophysical Investigations of Ground-Water Structures - Central & Northern Parts of the Upper Nile, Fifth Stage. Final Report, Prague, Czechoslovakia, 1977.

Sudan-FRG Ground-Water Resources in Khartoum Province, Vols. A, B. Bundesanstalt fur Bodensforschung, Hanover, 1979.

Suliman, Y. The Hydrogeology of Part of Eastern Sudan (Gedaref District). Sudan Geol. Survey Bull. No. 16, 1968.

Texas Instrument Inc. — Report on a Gravity-Magnetic Survey in Sudan. Sudan Geol. Survey open file, 1962.

Thorweihe, U. — New Aspects of Hydrogeology in Southern Egypt. Berliner Gewissenschaftliche Abhandlungen, Reihe A/Band 50 - Research in Egypt and Sudan. West Berlin, FRG, 1984.

Whiteman, A.J. — Nubian Group Origin and Status. The American Assoc. of Petrol. Geology Bull., Vol. 54, No. 3, 1970.

__________ — The Geology of the Sudan Republic. Oxford Press, London, 1971.

ANNEX

UNDP Country and Intercountry Projects on Ground-Water Resources

Project No.	Project title	Duration	Executing agency	Importance of ground water 1	2	3
RAB/82/013	Transnational Project on Major Regional Aquifer in NE Africa	5 years	UN	x		
SUD/85/U/03/96/01	Rural Water Supply Programme for Drought-Affected Areas - Kordofan	1 year	UN	x		
SUD/84/009/A/01/12	Establishment of Shelter Belts in the Northern Province	5 years	UN/FAO	x		
SUD/84/006	Establishment of a Water Point Maintenance Unit in the West Bank - Equatorial Region	3 years	UN	x		
RAB/84/023	Training of Technicians in Water Resources Management and Technology in Least Developed Arab Countries	?	?	x		

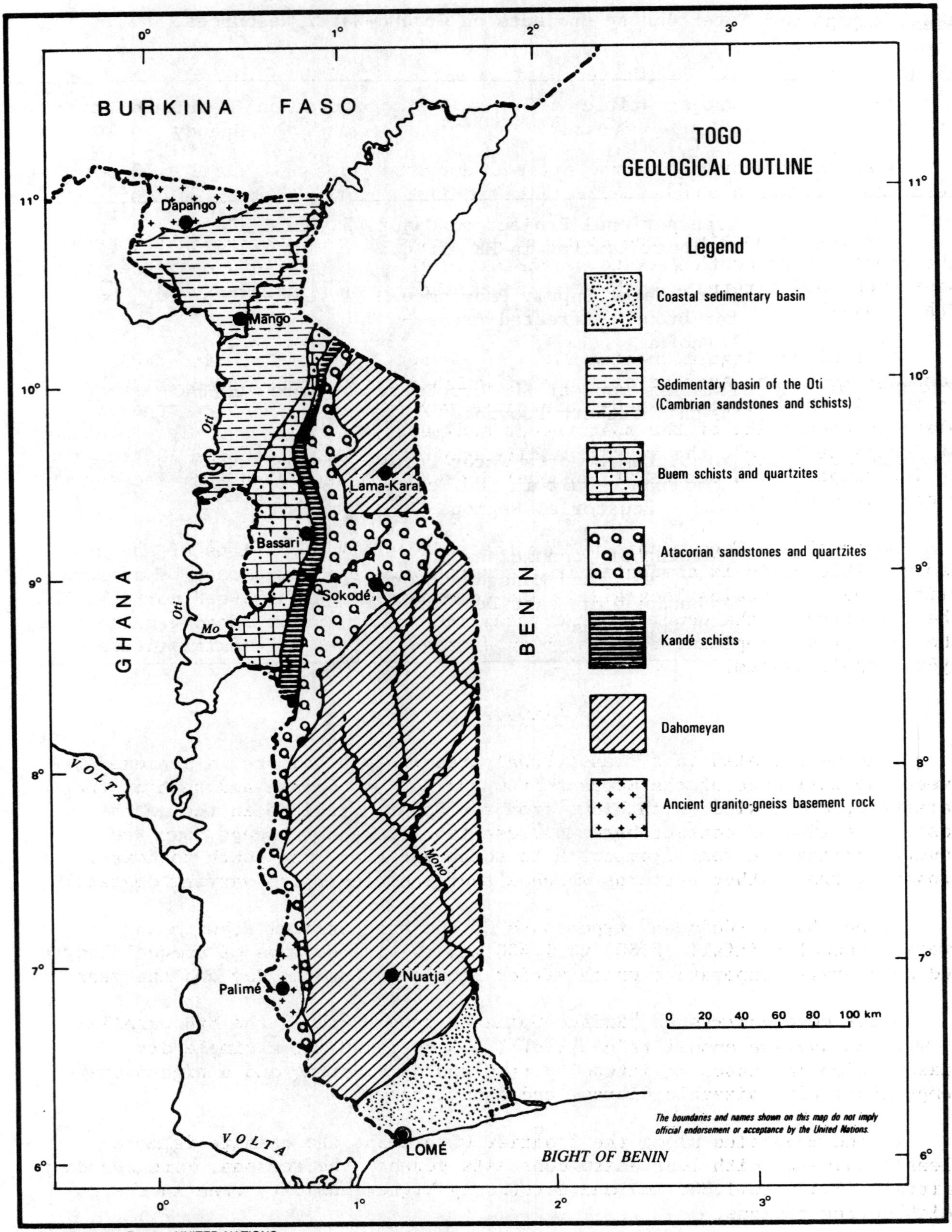

MAP NO. 2848.1 Rev.1 UNITED NATIONS
APRIL 1987

TOGO

Area: 56,000 km²

Population: 2.76 million (United Nations estimate, 1983)

I. BACKGROUND

Togo extends over a narrow strip of land 600 km deep on the Greenwich meridian. It has a southern coastline of 55 km on the Gulf of Guinea.

The Togo mountains, the southern extremity of the Atakora chain, cross the country from south-west to north-east. They are over 900 m high in the south-west and alternate with plateaus incised by deep gorges south of the 8th parallel N.

East of the Togo mountains lies the peneplain of the granito-gneiss basement rock, the average altitude of which declines from 450 m in the north to 50 m in the south; the peneplain gives way in the south to an area of plateaus consisting of the most recent sedimentary formations. This plateau of "terre de barre", the average altitude of which declines from 100 to 20 m, is interrupted by a series of lagoonal depressions separated from the sea by a coastal sandbar.

North of the Togo mountains lie the sedimentary formations of the Oti basin. This basin is dominated in the north-west and south-east by sandstone cliffs and it contains long glacis and some patches of cuirassed surface. In the far north of the country the crystalline formations of Dapaong again take the form of a peneplain, 300 m high on average, eroded by a fairly dense hydrographic system.

Climate

Togo is situated in a transitional area where a balance is achieved between the influence of the monsoon from the Gulf of Guinea and that of the harmattan, a dry wind which blows from the Saharan regions in the north-east. The line of contact between these two influences sweeps down the country during the year from north to south and then from south to north, producing two weather patterns which affect the country to varying degrees:

- The "Baoulo-Guinean" type south of the 8th parallel N, with an average annual rainfall of 800 to 1,400 mm, two dry seasons of unequal length and an average temperature which varies little during the day and the year;

- The more contrasted "Sudano-Guinean" type north of the 8th parallel N, with an average annual rainfall of 1,000 to 1,500 mm, a single dry season which increases in intensity from south to north, and a high average temperature with sizeable diurnal and annual ranges.

In the mountains along the frontier with Ghana the climate is more clearly Guinean, with less sharp contrasts between the seasons, more abundant rainfall (1,500 to 1,800 mm) and constantly higher humidity than in the neighbouring regions.

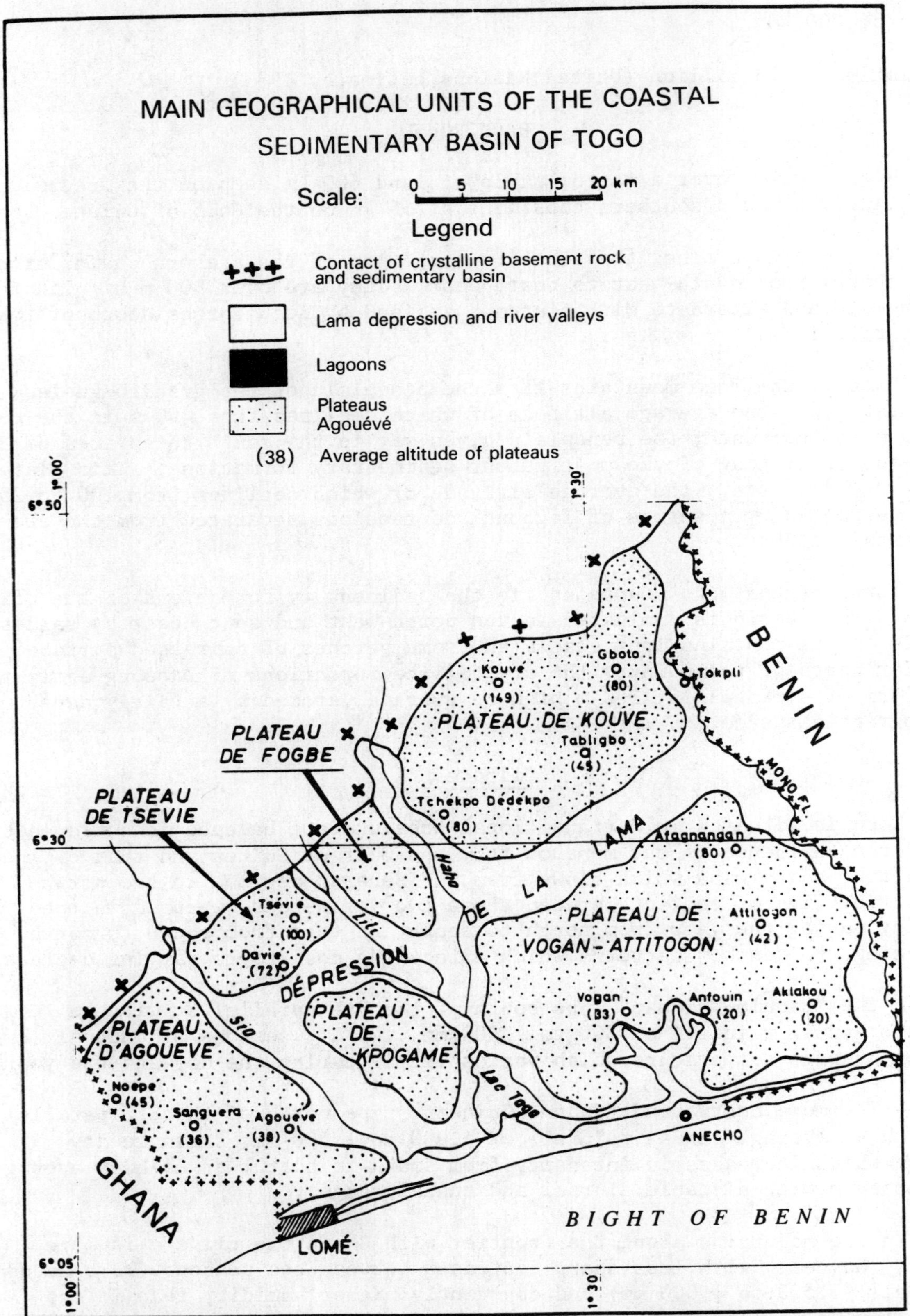

3436.10x

Source: Togo, Prospection des eaux souterraines dans la zone côtière Agency Terminal Report DP/UN/TOG-70-511/1 New York, 1975

The average interannual temperatures vary little from south to north (between 22°C and 28°C. The average monthly temperatures have two clear maximums in March-April and October-November in the tropical climate in the north; the secondary maximum in October-November declines considerably towards the south as one approaches the transitional equatorial climate of the coast. The annual thermal range also declines from north to south.

The evaporation measured in the Colorado pan ranges from under 1,000 mm/year in the mountains in the south-west to over 2,400 mm/year in the north-east.

The evaporation potential (Penman) ranges from 1,500 to 2,000 mm/year and from 80 to more than 200 mm/month. The positive balance of monthly evapotranspiration and rainfall is fairly high in the uplands but low everywhere else, particularly in the coastal zone.

Hydrological system

Togo is drained by four hydrographical units:

- The Volta basin and its main tributary the Oti, covering 26,700 km^2 (47.3 per cent of the country) north of the 9th parallel N and in the west of the country;

- The Mono basin, covering 21,300 km^2 (37.7 per cent of the country) in the central and eastern parts;

- A number of "coastal" rivers covering 8,000 km^2 (14.1 per cent of the country) in the southern region, the main ones being the Sio and the Haho;

- The basins of the Ouémé and the Couffo, covering 500 km^2 near the frontier with Benin.

The Mono basin is situated almost entirely in Togo, but the Oti basin extends into Benin, Burkina Faso and Ghana. The watershed of these two basins is formed by the ridge of the Togo mountains which determines the hydrological regime of the country's rivers.

The mean annual flows as ratios of the area of the recharge basins are as follows:

(millions of km^2)

Water course	Stations	1	2	5	10	20	50
Oti	5	-	-	-	-	3.0	5.0
Kara and Kéran	6	(17)	13.5	10.0	(8.0)	-	-
Mono-Anié-Amou	9	10.0	8.2	6.5	6.5	5	
Sio	3	5.3	4.9	-	-	-	-

The mean monthly flows expressed as percentages of the mean annual flow (coefficient of monthly volume flow) show a clear distinction between the seasonal distribution of the flows of the Oti and the Mono:

Month	June	July	August	September	October	November	Year
Oti (5 stations)	1.3	5.2	18.7	42.1	26.6	4.1	100.0
Mono (5 stations)	2.6	12.3	24.6	36.8	18.6	3.5	100.0

There are few permanent rivers. They are the Oti at Mango, the Kéran-Koumangou, the Mono below its confluence with the Amou, the Amou and the Sio.

Geology

Togo's geological formations can be classified in three main groups:

- The Precambrian granitogneiss baserock (almost three-quarters of the country's area) in the centre and the far north, with two main units - the Dahomeyan and the Voltaian;

- The primary sediments of the Oti basin in the north;

- The sedimentary coastal basin.

The "Dahomeyan" is the area (east of the Togo mountains) of gneiss and migmatities in formations of Birrimian age, variously affected by the tectonics and the metamorphism of pan-African orogenesis. The crystalline and crystallophyllitic rocks of this formation are very varied: acid and basic paragneiss and orthogneiss, migmatitic gneiss and migmatites, and granitoids.

The "Voltaian" includes:

- Non-metamorphic sedimentary formations: argillites with sandstone intercalations of the Oti formations, sandstones, quartzitic sandstones, quartzites and silexites;

- Metamorphic formations: chloritic and sericitic schists with intercalations of sandstone schists, silicified rocks and quartzites (Kande schists), with large amounts of micaschists and quartzites (Atacorian).

The sedimentary basin occupies the coastline over a 40 km strip. The Paleocene and Eocene formations consist mainly of limestones, interbedded with sands and clays. The Upper Cretaceous (Maestrichtian) formation consists mainly of sandy and argillaceous facies. These marine formations are covered in discontinuity by the continental terminal which includes azoic sand-clay deposits and varies in thickness from 20 to 60 m. Recent alluvial or dune formations are also present. The crystalline basement rock is broken up by a series of major faults running more or less parallel to the coast and it sinks rapidly towards the south-east (3,000 m in 50 km).

II. GROUND WATER

The Ministry of Planning has overall responsibility for the development of ground water. To varying degrees several other ministries (Water Resources, Transport, Health, Rural Development and Agricultural Engineering) are also involved in the study, evaluation, development, exploitation and conservation of these resources. The need to co-ordinate all these activities led to the establishment by Presidential Decree in 1970 of an inter-ministerial committee, the National Water Committee (CNE). The technical commission of this committee plays a consultative role with decision-making bodies. The Committee's secretariat is provided by the Department of Water and Energy.

The Department of Mines and Geology, assisted by the National Office for Mining Research (Ministry of Public Works, Mines, Energy and Water Resources) has so far (1984) played the main role in the study and evaluation of ground water. It served as the local host structure for the first hydrogeological research carried out in Togo, with finance from the United Nations Special Fund, in conjunction with mining research (references 1 and 2). The hydrogeological section created at that time was involved in the subsequent hydrogeological reconnaissance financed by UNDP (reference 3) and CIDA (reference 4). The capacity and resources of this section, mainly in drilling, were considerably strengthened during these works, which concentrated on the sedimentary coastal aquifers. As a logical extension, the latest hydrogeological reconnaissance financed by UNDP in the basement rock area (reference 10) was designed to diversify the section's activities and increase its efficiency.

All the equipment from that time and part of the personnel have been transferred to the Department of Water and Energy, which has particular responsibility for ground-water exploitation. On the conclusion of project TOG/75/008 the hydrogeological section of the BNRM still had a complete drilling unit (rotary down-the-hole hammer) and the geophysical prospecting equipment to continue the evaluation of the ground-water resources of the basement rock and handle the various techniques introduced during implementation of the project: down-the-hole hammer, test pumping, piezometric checks, photogeology, and geophysical prospecting (electrical, seismic and magnetic).

It is not yet possible to make a simple classification of the aquifers of various petrographic facies of the basement rock. The term "basement rock" includes, by virtue of the similarity of their hydrogeological behaviour, the crystalline and metamorphic formations of Precambrian and Lower Cambrian age and the primary formations of the Oti basin. The basement rock so defined outcrops over an area of 53,150 km^2, i.e. 94 per cent of the country. Little is yet known about it from the geological standpoint, but it encompasses very varied petrographic facies. The now exclusive use of boreholes to tap the water circulating in the deep fractures of the basement rock and the recent introduction of geophysical drilling methods have prompted a reconsideration of the knowledge of the ground water of the basement rock, which had formerly been deduced exclusively from well observations. The simultaneous use of electrical and seismic prospecting methods means that the boreholes can be accurately located. The seismic investigations sometimes extend to a depth of 40 m. However, these methods do not circumvent certain hydrogeological traps, such as the almost total blockage of the fracturation or its very poor development and extension, the frequency of which depends logically on the characteristics of the fracture system and thus on the petrographic facies. Depending on the local facies, which are sometimes of unknown size and difficult to determine, it is possible to obtain success rates of 75 to 95 per cent and average specific yields of 2 to 10 m^3/h, sometimes as high as 70 m^3/h.

The resources stored in the basement rock are small and the renewable resources extremely small. This is confirmed by a set of observations, in particular test pumping (transmissivity, assumed storage coefficient, residual drawdown), and piezometric checks. The only productive zones are those covered with a layer of alteration or alluvium sufficiently thick to recharge the underlying fissured zone.

About 300 boreholes (reconnaissance, piezometric, producing), very unevenly distributed, have been drilled in the sedimentary coastal basin (3,450 km^2). The recent work (reference 4-7 and 8) confirm the conclusions of the basic document (reference 3).

About a third of the area of the basin, i.e. 1,300 km^2 is thought to recharge the aquifers in an annual amount of 66 to 120 x 10^6 m^3. Despite the possibility of tapping the large stocks of the clay strata and the evidence of considerable drainage from deep recharging aquifers, it is accepted that in the areas in question the exploitation must not exceed the rate of recharge. The continental terminal aquifer has very large stocks (more than a billion cubic metres), part of which, in excess of the average rate of recharge, can probably be tapped as it is replenished during periods of high recharge. In some places at the coast the exploitation of the continental terminal aquifer must take very careful account of any increase - difficult to forecast - in the salinity of the water.

III. EXPLOITATION OF GROUND WATER

The Department of Water and Energy was originally responsible for determining priorities, planning urban and rural water-supply projects and controlling their implementation. Since 1979, in step with the intensification of the village water-supply programmes, its activities have been extended and its human resources bolstered and diversified by the recruitment of hydrogeologists, geophysicists, pump maintenance mechanics, etc. Furthermore, it has the drilling facilities inherited from the BNRM (two operational units: percussion and rotary); this has enabled it to implement under State supervision, with financing from EDF and BOAD, a number of drilling programmes in the coastal sediments at a rate of about 50 installations a year. It has also acquired responsibility for updating the records of the country's waterpoints (coded classification) and for the chemical and piezometric monitoring of the sedimentary aquifers. The acquisition of these new functions means that it now plays the role formerly played by the BNRM in research and evaluation.

The National Water Authority of Togo (RNET) is responsible for the maintenance, replacement, extension and improvement of urban water-supply installations. The rapid recent development of water-supply networks is causing the RNET difficult structural problems; it has to adapt its human, technical and financial resources to the requirements of the new situation. In conjunction with the identification of the technical and practical means of modernizing the Lomé water-supply system, a recent project spelled out the steps which need to be taken if the RNET is to perform its functions and, first and foremost, have the resources to maintain the installations and finance their extension.

Togo cannot bear the full cost of the drilling programmes and maintenance of the equipment (of varied age and origin) and it has to make occasional use of drilling rigs which are difficult to repair and work at a slow rate.

In 1979 half of the 18 towns equipped with a water-distribution network were supplied from ground water. Five of them - Lomé, Anécho, Tsévié, Vogan, and Tabligbo, exploit the aquifers of the coastal sediments. The town of Lomé (production of 28,000 m^3/s and consumption of 23,000 m^3/s) consumed two-thirds of the water billed by the RNET in 1979.

The sedimentary coastal basin is small in size but it accommodates almost a third of the country's population.

The drawoff from the aquifers of the continental terminal and Eocene-Paleocene formations of the western sector is about 30,000 m^3/s, most of it to supply Lomé (Caravelli), with the balance going to smaller stations (Sanguéra, Noépé, Sanyra, Amedenta, Loqbossito, Dekpo, Abobo, Lébé, Benin Breweries). The aquifer of the Maestrichtian sands is exploited by the stations of Ahépé, Kouvé, Gabotto, Tabligbo and Afangnagan (2,000 to 3000 m^3/day). The continental terminal aquifer is tapped by the stations of Davedé (Berie), Tchékpo and Mission-Tové in the northern sector (1,000 to 2,000 m^3/s and by the stations of Fyoata, Vogan, CTMB and Attitogon in the eastern sector (5,000 m^3/s.

The available discharge from the coastal sedimentary aquifers is estimated at 170,000 m^3/s, i.e. 62 x 10^6 m^3/year. In 2000 the drinking water requirement of the population of the coastal basin will be 43.5 x 10^6 m^3/year, including 4 x 10^6 m^3/year for the rural sector. This requirement alone therefore represents 70 per cent of all available ground water. It has been evaluated on the basis of fairly moderate unit consumption:

- For the rural areas, per capita consumption of 20 l/day;

- For the medium-sized towns (more than 5,000 inhabitants; 24 to 36 per cent with private piped supplies), consumption of 66 to 77 l/day for private subscribers;

- For Lomé (60 per cent piped supplies), per capita consumption of 100 l/day for private subscribers. For a total production of 90,000 m^3/day the requirement of the town of Lomé in 2000 will break down as follows: domestic - 54,000 m^3/day; industry - 11,000 m^3/day; public sector - 8,000 m^3/day.

Present estimates of the available resources of the basin indicate an average per capita consumption of 100 l/s for the expected population of the coastal region in 2000.

The annual drinking-water requirement of settlements in the basement rock zone will be about 11.5 million cubic metres in 2000. On the basis of a consumption of 50 litres per inhabitant per day, a borehole delivering 10 m^3/h can meet the requirement of a town of 5,000 inhabitants. In view of the properties of the ground water of the basement rock, a very careful individual study will have to be made for each settlement. Except in very special local cases, it is unlikely that the available maximum yields (up to 70 m^3/h) will be permanent; accordingly, it will very soon be necessary to turn to surface water to supply the settlements.

Several financing bodies (EDF, FAC, BOAD, USAID) are involved to varying degrees in the installation of a thousand water points to supply the rural population in the basement rock zone. But this programme will meet only part of the requirement of 5,000 to 6,000 water points by 1990 (per capital consumption - 20 litres a day; daily drawoff per installation - 6 m^3/s. excluding the coastal region).

IV. CONCLUSION

The development prospects of the coastal region argue for the immediate introduction of rigorous planning of the use of the ground-water resources of the sedimentary coastal basin, with priority given as in the past to supplying the urban and village population. The long-term programmes for hydro-agricultural development, which are big consumers of water (62 hm^3 is sufficient for no more than 5,000 ha), must be implemented with the greatest care. In these circumstances it is essential to continue and intensify the research into the geometrical and hydrodynamic properties of the various aquifers of the coastal basin. These data are essential to the proper evaluation of the actual available resources and to the determination of the most suitable methods of exploitation, bearing in mind in particular the risk of sea-water intrusion.

The initial evaluations of the water resources of the basement rock must be confirmed and if possible broken down in terms of geological formations and climate, in particular by means of improved piezometric monitoring of changes in water levels brought about by the recent exploitation. The village water-supply programmes must be compatible with the estimated resources: reduction of the density of water points and limitation of the drawoff rate. Priority must be given to consideration of the possibility of using surface water to supply settlements as part of the overall water-supply policy.

V. REFERENCES

ITALCONSULT — Etudes des eaux souterraines du Togo. United Nations Special Fund, 1967.

United Nations — Recherches minières et en eaux souterraines. United Nations Special Fund. Project TOG 4, 1971. Results, conclusions and recommendations.

United Nations — Prospection des eaux souterraines dans la zone côtière. United Nations Development Programme. Final report of project 71/005, 1975.

Gendron-Lefebvre — Etude de la factibilité d'approvisionnement en eau potable pour la ville de Lomé et neuf villages avoisinants. Canadian International Development Agency, 1977.

BURGEAP — Construction de puits sur financement FED. 1060/TO, 1978.

BRGM */ — Travaux d'hydraulique villageoise au Togo. West African Development Bank, 1981.

BRGM */ — Surveillance piezométrique et chimique des nappes du bassin sédimentaire côtier. Synthèse des connaissances hydrogéologiques. Fonds d'Aide et de Cooperation, 1982.

LCHF */ — Stratégie d'aménagement des eaux du Togo. United Nations Development Programme, 230/78, 1982.

__________ — ATLAS - "Ressources et besoins en eau", 1982.

__________ — Rapport final de synthèse de la confrontation des ressources et besoins en eau du Togo. 1982.

United Nations — Stratégie d'aménagement des eaux du Togo. United Nations Development Programme, TOG/75/00, 1984. Final report - Conclusions and recommendations.

__________ — Reconnaissances hydrogéologiques entre 7° et 9° de latitude nord. Rapport technique d'ensemble, 1984.

---

*/ Bureau de Recherches Géologiques et Minières (Geological and Mining Research Office).

**/ Laboratoire Centrale d'Hydraulique de France (Central Hydraulics Laboratory of France).

TUNISIA

Area: 163,000 km^2

Population: 6.89 million (United Nations estimate, 1983)

I. BACKGROUND

Tunisia has a long coastline of 1,500 km on the central Mediterranean with a varied shoreline, including a large peninsula in the north-east (Cap Bon), deep gulfs in the east (Gulfs of Hammamet and Gabès) and fairly large islands in the south-east (Kerkennah and Djerba islands).

The centre of the country is occupied by an east-west series of depressions called chotts (Chott El Rharsa, Chott Djerid, Chott Fedjedj). The northern part of the country is crossed diagonally (south-west/north-east) by the Tunisian Dorsale, a mountain chain of moderate altitude rising to 1,554 m in the Jabal Chambi, which forms the eastern end and the convergence of the Tell Atlas and the Saharan Atlas of Algeria. There are therefore two geographical units:

- The north-western part, with broken relief bordering a checkerboard of high plains;

- The south-eastern/part, with lowlands and valleys stretching to the coast (Sahel).

The southern part is dominated by the vast expanses of the Saharan platfordm, the higher eastern edge of which (Matmata mountains, 600 m) delimits two separate areas: a narrow coastal plain, the Gefara, in the east; and a plateau, the Dahar, which slopes gently westwards to disappear under the sands of the Great Eastern Erg.

Climate

Tunisia has a Mediterranean climate - semi-arid in the north, arid in the centre, and desert (Saharan) in the south.

The rainfall distribution is determined by the direction of the prevailing winter winds (north-west) and the disposition of the mountains; the rainfall is therefore heavy in the northern regions (over 400 mm/year and up to 1,500 mm/year in the far north-west), declining towards the centre (150 to 300 mm/year) and becoming very light (less than 150 mm/year) in the southern part; the rainfall is less than 50 mm/year in the far south, with some years of no rain.

Surface water

With the exception of the far north of the country, most of the watercourses rise in the north-west and flow eastwards; the north has the main permanent watercourses which reach the sea, while in the arid central region the only watercourses are of the torrential type and flow for only a few weeks or days a year. In the south surface flow is rare and meagre, limited to the piedmont area of the Matmata mountains. None of these temporary flows reaches the sea. The water infiltrates as it crosses the plains or runs off towards the depressions (sebkhas), which are usually found well inland from the sea, to which they are linked on very rare occasions of heavy floods.

The basin of Wadi Medjerda in the north is the largest in the country both in terms of area (23,700 km^2) and of recharge (one billion cubic metres a year). This recharge comes from flood waters, which often exceed 500 m^3/s and in March 1973 reached 3,500 m^3/s, delivering 943 hm^3 in six days (26 March to 1 April 1973).

Wadi Zeroud (9,100 km^2) is the main watercourse in the centre of the country and provides a typical example of the irregularity of the flow patterns in the arid areas, with 2.5 billion cubic metres in the space of two months (September and October 1969) and a flood crest of 17,000 m^3/s; the estimated mean annual flow is 95 million cubic metres.

In the south Wadi Oumezessar had a flow of 12 million cubic metres at the Koutine station (285 km^2) in March 1979, with a flood crest of 1,475 m^3/s; the catchment basin of this wadi is in an area which receives average rainfall of 200 mm/year and its annual module is three million cubic metres.

The salinity of the water is generally low in high-water periods (under 1 g/1) but it increases rapidly in the low-water period, especially in central Tunisia where it can be as high as 3 to 5 g/1.

The sediment discharge is very high in Tunisia. The violent rains, the soft terrain and the scanty vegetation cover all encourage the erosion of the slopes; Wadi Zeroud carries down an average of five million t/year 1/ of sediment, but the notion of average has little meaning as it was calculated during the flood crest of October 1964 (191 hm^3), when the 12.5 million tons of sediment was equivalent to the total discharge of the four previous years.

The sediment discharge is smaller in the Medjerda basin, which has denser vegetation cover; it has been calculated that an average annual flow of 716 hm^3 means a sediment discharge of 18 million tons.

The main morphological, climatic and hydrological factors described in the following table, composed from geological and hydrogeological data, were used as the basis for dividing the county into three main natural regions (north, centre and south), which can be used for the presentation of the data.

Surface water

Region	Average volume (hm^3/year)	Average volume (percentage of total)
Northern Tunisia	2,085	80
Central Tunisia	325	13
Southern Tunisia	170	7
Total	2,580	100

1/ t = metric ton.

II. GEOLOGY

Tunisia is essentially a country of sedimentary formations.

- The south is the area of the Saharan platform characterized by geological series developed regularly from the Triassic to the Quaternary period in the form of a large monotonous monoclinal structure sloping gently down towards the Sahara as a distinct unit. The coastal plain of Gefara is a subsidence zones.

- The north-east, a heavily tectonized area of the Tunisian Atlas which generally runs south-west/north-east, is characterized by anticlines and synclines aligned in series and by small fault zones. This phenomenon reaches its highest point towards the northern coast, where Triassic diapiric activity is apparent, together with overthrusting in the stratum of Numidian flysch (Oligocene). In the coastal Sahel, however, the terrain is less broken. In the plains there are two big sedimentary basins of Miocene and Pliocene-Quaternary age.

The north, a heavily folded region, with formations of poor permeability (Oligocene flysch, Senonian marls) has most of the country's hydrographic network and surface-water resources.

The area crossed by the Atlas chain, called North-West Tunisia, is dominated by Cretaceous series, especially Senonian, with a limestone sequence outcropping above a marly episode, and to a lesser extent Eocene, especially Lower Eocene Limestone, and a particular facies (reef limestones) of the Aptian epoch. The Aptian, Senonian and Lower Eocene limestones form the skeleton of most of the region's mountains, which are separated by the high plains.

In central Tunisia the Miocene formation has a facies of continental sandstone which forms an excellent aquifer; this sandstone is disposed in perched synclines or forms the base of the Quaternary plains or rests directly on the Senonian limestone through a gap in the sedimentation peculiar to the region; the folds are narrower and the plains extend over a larger area, also with a continental facies of the Miocene-Pliocene period.

The region of the coastal Sahel includes Miocene formations outcropping in a few low reliefs delimiting the Miocene sedimentary basin of Sfax in the south, in the Pliocene-Quaternary subsidence plain of Kairouan in the north, and in a few smaller folds bordering the sea.

In the south, at the western edge of the coastal subsidence plain of Gefara which consists of Senonian and Miocene limestone series, the geological outcrops include series of Triassic sandstone, dolomites and gypsum of no great hydrogeological interest. Next come the series of Jurassic marls, limestones and dolomites overlain by a Lower Cretaceous formation with a continental facies of sands, limestones and clays characteristic of southern Tunisia. This is the "continental intercalated", the region's main aquifer. The Upper Cretaceous formation outcrops in a limestone series of particularly interesting lithological nature and considerable extent. A particular detritic facies has been identified in the Lower Pontian formation in the Djerid and at the south-west edge of Chott Djerid; these two latter formations form the aquifer of the terminal complex.

III. GROUND WATER

Government bodies

All water-resources activities in Tunisia are at present (1985) the responsibility of the Ministry of Agriculture and in particular of its Department of Water Resources, whose functions include the inventory of surface and ground-water resources, the establishment of measuring and observation networks, and studies and research for the evaluation of the resources and planning and control of their exploitation and management.

The utilization of water resources is controlled by two other departments: for surface water, the Department of Water Research and Major Works; and for ground water, the Department of Agricultural Engineering; lastly, a national corporation - the National Corporation for Water Exploitation and Distribution (SONEDE) - is responsible for urban water supplies. Most of the water drilling activities are entrusted to the Water Drilling Authority, which is attached to the same Ministry.

Background

The first body with specific responsibility for water resources was established in 1930 in the Special Water Service of the Public Works Administration; it was called the Office for Inventory of Water Resources (BIRH) and was responsible for the inventory of the country's water resources and for ground-water studies; surface water did not receive the serious attention of the authorities until 1941.

After the country's independence in 1956 the intensification of hydro-agricultural development led to the transfer of the Special Water Service, which had become the Rural Water Supply and Development Service (HAR), to the Ministry of Agriculture, where it was to remain although undergoing some changes. In 1965 the BIRH was given responsibility for all water-resources activities, many of which had been entrusted up till then to study offices. The BIRH then developed rapidly with the recruitment of additional Tunisian hydrologists and hydrogeologists and the training of middle technical staff to constitute the Division of Water Resources, which was attached first to the Department of Water and Soil Resources and then from 1984 to the Department of Water Resources (DRE); this Department now (1985) has 10 hydrologists, three geophysicists, 23 hydrogeologists, 26 assistant engineers, 62 agents and technical assistants and 80 technicians.

Hydrogeological studies and research

This work is entrusted to hydrogeologists who have received specialized training in geology and hydrogeology and usually have a post-graduate degree; they are assisted by technicians trained in technical colleges or on special hydrology courses in Ministry of Agriculture schools.

Electrical prospecting is preceded by field reconnaissance, preparatory geological research and study of the inventory of the water points of the various hydrogeological structures; it is used systematically to determine and locate the most suitable formations for mechanical sounding; the hydrogeologists and technicians monitor all the reconnaissance sounds and, after electrical logging, they tap the wells, carry out test pumping and determine the hydrodynamic characteristics of the aquifers; samples of water are taken from the wells and boreholes to determine the geochemical properties of the aquifers; isotopic analyses are sometimes made.

All these data are analysed for the preparation of hydrogeological studies which evaluate the available resources and formulate proposals for well or borehole programmes; the use of mathematical models is becoming widespread as a means of checking the basic assumptions, simulating the flows and optimizing the exploitation of the available ground water.

An observation network is then established to monitor ground-water levels and exploitation; regular checks are made to ensure the proper management of the ground water.

Ground-water experimentation is also increasing, especially with respect to the artificial recharge of aquifers under different conditions, as part of the integrated management of surface and ground water.

Description of the main aquifers

The geology of the north and centre of the country is divided into sections with various separate small aquifers recharged by infiltration of rainwater and surface runoff. In the south there are very large aquifers which enclose virtually fossil horizons where the recharge is now very small.

Northern and central Tunisia

Here the aquifers are very varied but can be classified in three main groups: the alluvial plains, the limestone formations and the Miocene sandstone formations.

There are several alluvial plains in northern and central Tunisia; they are covered by detritic formations of Quaternary age which contain semi-deep to deep aquifers, often forming part of the same aquifer system. The thickness of these formations varies from 20 to 400 m. They include, for example, the Ghardimaou plain (northern Tunisia), the Kalaa Khasba plain (northern Tunisia) and the Kairouan plain (central Tunisia). These aquifers are recharged by direct infiltration of precipitation, but most of the recharge comes from infiltration of flood water from the wadis.

The ground water flows towards the outlets provided either by sebkhas, from which it evaporates, or by the beds of wadis, towards which the water can in certain conditions provide a constant minimum flow. The salinity is variable, but most of the water is of good quality (less than 1.5 g/l) in the north and the centre; the salinity is much higher in the coastal Sahel (3 to 5 g/l); it is often higher in the downstream part of the flows. The water is drawn off by a large number of large-diameter (3 to 5 m) surface wells with small yields (4 to 8 l/s on average) and by boreholes of moderate depth (100 to 300 m) with larger yields (20 to 50 l/s).

A series of limestone structures of limited size is scattered throughout northern and central Tunisia, depending on the geological and tectonic conditions; these formations are mostly of Lower Eocene and Senonian age, less frequently Aptian and Turonian. The ground water circulates in the tectonic fissures in the limestone or in actual karstic systems, emerging as large springs with varied regimes. The type and position of the formation determines the mode of deposit of the aquifers and the suitability of the springs for exploitation by borehole. As the north and centre of the country have quite high precipitation, this type of aquifer has a good recharge, with an infiltration coefficient of around 20 to 50 per cent and substantial discharges through the outlets. At their mean levels, the best springs have discharges of about 100 litres a second. Except in a few cases, the water of the limestones is of good quality (0.5 g/1) and is suitable for human consumption, especially as natural mineral water.

The traditional mode of exploitation is by tapping of the springs; when the structural position is suitable, the tapping can be improved by galleries (for example, Jabal Zaghouan) or even better by boreholes in the case of deep formations (sunken karsts).

In central Tunisia the Miocene formation consists of sandstone strata interbedded with clays; the sandstone consists of developed continental series and the clays are good aquifers.

The clays sometimes rest directly through a sedimentation gap on Senonian limestones which are themselves aquifers; the resulting indirect recharge helps to regularize the regime of some of the limestone aquifers; this is particularly true of the Sidi Merzoug-Sbiba structure. The Miocene formation often takes the form of synclinal structures occupying the plains beneath a Pliocene-Quaternary cover; it is also found in plateaus and perched synclines, such as the Oum Ali-Thelepte syncline, the Kasserine plateau and the Hadjeb-Djilma formation.

The formations are recharged by infiltration of surface runoff and precipitation, as well as by drainage from other underlying aquifers when the stratigraphic conditions are suitable, as in the case of the superposition of Miocene sandstones and Senonian limestones.

The ground-water flow terminates in springs or other outlets. Tectonic accidents place the less permeable formations in contact with the water-bearing sandstones, causing the ground water to emerge at the surface either as springs or as permanent flows in the beds of the wadis rising from hydraulic shelfs. Examples of this are the Kasserine plateau and Wadi Darb (400 1/2), the Sidi Merzouk-Sbiba and Wadi Sbiba formation (200 1/s) and the Hadjeb El Aioun formation and its springs.

The water of the Miocene sandstones is of good chemical quality (less than 1.5 g/1). The large permanent flows have prompted the users to built diversion dams to collect this fresh water and facilitate the passage of flood waters; the Sbiba and Wadi Darb dams are examples of this; but the commonest installation is the traditional borehole (200 to 500 m) which allows the exploitation to be more flexible in response to the needs and facilitates intensive drawoff when the hydrogeological and economic conditions so permit.

Southern Tunisia

Southern Tunisia has two big aquifers lying under the Dahar and extending into large areas of southern Algeria under the Great Eastern Erg and the Great Western Erg through vast and regular geological formations; they are:

- The aquifer of the terminal complex (TC);

- The aquifer of the intercalated continental (IC);

A third unit lies under the coastal plain of Gefara, between Gabes and Zarzis; this is the Gefara aquifer which receives most of its recharge from the IC aquifer.

Both the IC and the TC aquifers have very large stocks which must have accumulated during the rainy periods of the Quaternary epoch when this region had a much wetter climate than today and abundant fauna and flora.

<u>The aquifer of the terminal complex</u> (TC) is so called because of the geological series of which it consists; these series range in age from Senonian to Neogene and form a complex which is often interconnected.

This aquifer extends over 350,000 km^2, mainly in Algeria and in Tunisia, where it is represented by two units: the Nefzaoua (limestone Senonian aquifer) and the Djerid (Lower Pontian sand aquifer), with an average depth of 100 to 300 m in the Nefzaoua and 200 to 600 m in the Djerid.

Most of the flows of this big aquifer run south to north, converging on Chott Melrhir in Algeria and Chott Djerid in Tunisia which they recharge from bottom to top; these chotts have thick clay-sand sedimentation; they constitute evaporations systems and are considered to be the outlets of this aquifer.

The other outlets are provided by the large springs around the chotts, such as the Nefta (134 l/s, 1983), and Tozeur (165 l/s, 1983) springs, which are now declining considerably owing to the intensive exploitation of the aquifer by artesian borehole or pumping.

The salinity is varied - about 2.5 g/l in the Djerid and 1/5 g/l in the Nefzaoua, but with some higher values (4 g/l in the Kebili peninsula, 8 g/l at El Hamma in the Djerid). The water is increasingly exploited by artesian boreholes or boreholes equipped with pumps. In 1983 there were 142 boreholes in the Nefzaoua and 130 in the Djerid.

<u>The aquifer of the intercalated continental</u> (IC) consists of detritic series of continental origin deposited between the end of the Jurassic period and the marine transgression of the Cenomanian period; the aquifer extends over 600,000 km^2 in the northern Sahara, with a small area in Tunisia which includes the Tunisian Dahar and the Nefzeous and Djerid, in particular the Chott Fedjedj aquifer west of Gabes. Most of this aquifer is very deep, often more than 1,000 m, except in two zones at the edge of the outcrops; it is more than 2,000 m deep in the Djerid.

Part of the water flows towards south-west Algeria, but the main flow is towards the north-east to the outlet in Chott Fedjedj, which is blocked at its eastern end by the irregular formations of the South Atlas from which rise the El Hamma de Gabès springs; the ground-water flow then recharges the Gefara aquifer by direct contact. The salinity varies according to region; 2.5 to 3 g/1 in Chott Fedjedj, in the Nefzaoua and Djerid; 5/g/1 in the El Borma region in the far south-west of the country. Exploitation by large-diameter borehole is now on the increase; these boreholes often have large artesian discharges (300 1/s in Chott Fedjedj); the pressure is usually high (10 bars), as is the water temperature (70°C).

The Gefara aquifer is found in the coastal plain of Gefara between Gabès and Zarzis. It is contained in the Miocene sands and Senonian limestones which are directly superposed in the north and stepped in the south-east in a succession of faults which become deeper from west to east; its average depth is 100 to 300 m. Its main recharge comes from the IC aquifer through the El Hamma fault; it also has a local recharge from the Matmata mountains.

The water flows eastwards, passing under the sea, which it joins a few kilometres offshore. The main outlets are the Gabès springs, the yield of which is declining steadily (400 1/2 in 1970 to 150 1/2 in 1983) in step with the increase in the number of artesian boreholes and boreholes equipped with pumps (187 boreholes in 1983), which now supply most of the water used. The salinity is 3 to 4 g/1 in the Gabès region, increasing to 6 and 8 g/1 at Djerba and Zarzis.

The annual recharge of the TC and IC aquifers is small, but owing to their great size they still have very large stocks of water. Water can therefore be drawn from these stocks for the development of these regions, but at the price of depleting the springs, which entails increasing recourse to pumping and continued increases in the unit cost of the water.

Accordingly, the exploitation of these aquifers is subject to an economic constraint and to the technical constraint of the possible contamination of the aquifers by salt water from sea or chott.

The following work programme has therefore been established to determine the available amounts of water:

- Forecast horizon: 2000;

- Maximum pumping depth: 60 m in the aquifer of the terminal complex;

- Maintenance of the strong artesian conditions (large artesian yields from the IC boreholes).

The calculations made on the basis of mathematical models have produced the following results, taking 1970 as the base year:

Drawoff (m^3/s	1970		2000	
Region	TC	IC	TC	IC
Djerid	2.0	-	4.85	1.0
Nefzaoua	2.3	-	6.50	0.9
Chott Fedjedj	-	0.2	-	1.1
Gefara	4.2	-	4.7	-
Grand Sud	-	0.1	-	1.0

An evaluation of the country's ground-water resources in 1980 established the situation of the three main natural regions, with a distinction made between shallow ground water exploitable by well and deep aquifers exploited by borehole:

(millions of m^3/year)

Natural region	Shallow ground water	Deep ground water	Total
Northern Tunisia	287	151	438
Central Tunisia	162	224	386
Southern Tunisia	37	656	693
Total	486	1,031	1,517
	30/m	68/m	

The shallow aquifers contain about two-thirds of the total ground-water resources; the southern part of the country has about half the ground water, mainly in deep aquifers.

IV. EXPLOITATION OF GROUND WATER

Bodies responsible for ground-water exploitation

The urban water supplies of the whole country are the responsibility of the National Corporation for Water Exploitation and Distribution (SONEDE) of the Ministry of Agriculture, which carries out drilling operations, lays pipes, builds reservoirs, distributes the water and bills the subscribers.

At present SONEDE has 600,000 subscribers, i.e. a rate of cover of 68 per cent; it also supplies the industrial sector.

Two central departments of the Ministry of Agriculture are responsible for irrigation water:

- For ground water, the Department of Agricultural Engineering;

- For surface water, the Department of Studies and Major Hydraulic Works;

The water is exploited in accordance with the three master plans for the country's three big natural regions (north, centre and south). The master plans specify the uses of the water and the construction of the installations of the irrigated areas, the management of which is entrusted to regional irrigation offices, now numbering 10 for the whole country. In addition to the extensive hydro-agricultural development, which requires major resources and is therefore managed by the Ministry of Agriculture, the authorities also encourage intensive private farming; this involves the exploitation of shallow aquifers by a large number of wells (67,400 in 1983), including an increasing proportion equipped with motorized pumps.

Drilling companies

Up to 1982 most water drilling activities (about 100 boreholes a year) were carried out by a government service, the Water Drilling Authority (RSH) and the remainder by two private Tunisian companies; the DRE itself drilled some of the study and observation wells.

Since 1983 the Ministry of Agriculture has encouraged the creation of new private companies, while at the same time increasing the stock of equipment of the RSH to cope with the increasing drilling operations in the country. Six new Tunisian companies were established and their equipment added to the existing stock, which in 1984 consisted of:

Body	Number of rigs			
	Small rotary (200 m)	Medium rotary (200-700m)	Heavy rotary (700 to 2,000 m)	Total
RSH	12	15	3	30
Tunisian companies	7	8	5	20
Total	19	23	8	50

In 1984 the number of boreholes drilled and the metrage were as follows:

Body	Number of boreholes	Metrage
RSH	67	16,400
Tunisian companies (8)	47	15,000
Foreign companies	9	6,000
Total	123	37,400

To this must be added the number of piezometers installed for survey of the aquifers, monitoring the water levels or establishment of water points:

Body	Number of piezometers	Metrage
DRE	16	1,300
Regional administrations	6	450
Foreign companies	5	750
Total	27	2,500

As the tables above show, most of the activities are carried out by Tunisian companies or public bodies; however, owing to the urgency of certain work, occasional use is made of foreign drilling companies, especially for very deep boreholes.

The engineering staff of the RSH and the Tunisian companies is trained in the big schools or at the universities; a permanent training programme for technicians has also been established by the RSH in order to supply qualified personnel for the sector throughout the country.

Use of ground water

The evaluation of the country's water resources in 1980 established the following situation:

- Surface-water resources: 2,580 billion m^3/year;

- Ground-water resources: 1,520 billion m^3/year:

  Total 4,100 billion m^3/year

The 1980 balance sheet of the use of these resources was as follows:

Natural Region	Surface Water (hm^3/year) available	exploited	Ground Water (hm^3/year) available	exploited	Deep aquifers available	exploited
Northern Tunisia	2,085	1,366	287	227	151	56
Central Tunisia	325	130	162	137	224	133
Southern Tunisia	170	0	37	31	656	341
Total	2,580	1,500	486	395	1,031	530
				(80 %)		(53 %)

This balance sheet illustrates the already very high degree of exploitation of shallow aquifers in 1980 (80 per cent), with exploitation of only half the resources of the deep aquifers.

The exploitation of ground water had advanced considerably by 1983:

- Surface aquifers: 471 hm^3/year (96 per cent);

- Deep aquifers: 630 hm^3/year (63 per cent).

The water was distributed among the various users as follows:

- Drinking water (SONEDE): 140 hm^3 (1982) (including tourism)

- Irrigated agriculture

  - deep aquifers 468 hm^3 (1982)

  - shallow aquifers 471 hm^3 (1983)

- Industry (supplied largely by SONEDE).

The area irrigated was 171,400 ha in 1982, 70 per cent with ground water (shallow and deep aquifers).

The intensive exploitation of certain shallow aquifers at Cap Bon in the coastal Sahel and in central Tunisia by a rapidly increasing number of wells has caused sharp declines in the water levels and an increase in the cost price of pumped water, as well as increased salinity in the coastal areas (Cap Bon) and around the sebkhas.

Almost all the ground water is exploited, but this is not the case for surface water, some 60 per cent of which could be stored in reservoirs. More than a billion cubic metres of surface water a year is lost through evaporation or discharge to the sea, while only part of it recharges the ground-water aquifers. Accordingly, there must be an integrated policy for management of surface and ground water based on the damming of runoff water to ease the exploitation of the ground-water aquifers and increased recourse to artificial recharge of the aquifers. To this end, the following legislative and technological steps have been taken:

- A new Water Code was promulgated in 1975, giving the Administration the legal authority to establish protected or prohibited areas around sensitive zones, within which all exploitation of the aquifers is controlled;

- Recharge technologies have been tested and then applied over very large areas; for example, the shallow aquifer at Teboulba (coastal Sahel) has been fully recharged by injection of water into the wells from a reservoir; the shallow aquifer at Menzel Bou Zelfa (Cap Bon) which is being recharged through reservoirs and wells (1985). A start is about to be made on the large-scale recharge of the shallow and deep aquifers of the Kairouan plain with the waters of Wadi Zeroud by means of a dam which is being built on the wadi.

V. CONCLUSION

Can the figures offered in 1980 for water resources be increased to cope with the considerable growth in the demand for water entailed by the country's development? This is the question which is often put to the water resources services; the following answer can be given; the prospect of an increase in the amount of surface water available is very limited owing to the natural constraints; in contrast, there is considerable scope for development of ground water. This will entail:

- The continuation and intensification of research activities, especially soundings, to cover all the country's potential aquifers, including those at present considered to be of minor interest owing to their great depth and the high exploitation costs involved;

- The controlled development of the overexploitation of the surface and deep aquifers by means of increased use of systematic pumping in place of abstraction from springs or artesian boreholes;

The integrated management of surface and ground water, making best possible use of surplus runoff to ease the exploitation of ground water through the establishment of a national policy for construction of surface-water reservoirs and artificial recharge of the aquifers whenever technically possible.

These efforts to increase the resources must be accompanied by measures to preserve the quantity and quality of the water; the promulgation in 1975 of new water legislation (Water Code) has given the technical services of the Ministry of Agriculture the essential legal tool for attainment of this goal.

VI. REFERENCES

BIRH — Annuaire de l'exploitation des nappes profondes, 1983. Internal DRE report, Tunis, 1983.

Castany, G. — Carte géologique de la Tunisie au 1/500,000. Publication of the Mines, Industry and Energy Service, Tunis, 1953.

Division of Water Resources — Ressources en eau de Tunisie. Periodical bulletin published by the Division of Water Resources of the Ministry of Agriculture: No. 1 (1971); No. 2 (1972); No. 3 (1976); No. 4 (1978-79); No. 5 (1980-81); No. 6 (1982).

Fersi, M. — Estimation du ruissellement moyen annuel sur les bassins du sud-est, du sud-ouest et du Sahel-Sud. Internal DRE report, Tunis, 1979.

Kallel, R. — Bilan global des ressources en eau de surface de Tunisie. Internal DRE report, Tunis, 1981.

Ministry of Agriculture — Enquête: Périmètres irrigués, Campagne 1983-1984. Dept. of Planning, Statistics and Economic Analyses, Tunis.

Ministry of Agriculture — Code des Eaux. Tunis, 1978.

UNDP — Final report and technical reports of project TUN/69/528 - "Intensification of ground-water exploitation in northern and central Tunisia". United Nations.

UNDP — Updating of the study on the water resources of the Northern Sahara. Final report of project RAF/80/011, 1983.

UNESCO/Algeria/Tunisia — Study of the water ressources of the Northern Sahara (project REG-100). Report on the results of the project - conclusions and recommendations, 1972.

Zebidi, H. — Bilan des ressources en eau de la Tunisie - 1980. Internal DRE report, Tunis, 1980.

Zebidi, H. — Grands traits des ressources en eaux souterraines de la Tunisie. Internal DRE report, Tunis, 1980.

Zebidi, H. and Habib, A. — Les forages d'eau en Tunisie depuis l'indépendance - quelques statistiques. Internal DRE report, Tunis, 1984.

VII. HYDROGEOLOGICAL MAPS

1. Hydrogeological map of Tunisia - scale 1:500 000; southern part, 1964.

2. Maps of the water resources of Tunisia - scale 1:200 000:

   17 sheets published to date (1985)

   2 sheets in preparation

3. Hydrogeological map of Tunisia - scale 1:500 000. In preparation in accordance with the international key of hydrological maps.

Ground-water projects implemented under the auspices of the United Nations Development Programme

Country and project No.	Project	Implementation period	Executing agency	Importance of ground water phase */
ALGERIA				
ALG 66/509	Natural resources study: farm experimentation and demonstration in the Hodna region	1966–1975	FAO	b
BENIN				
BEN 66/503	Pilot ground-water development	1967–1970	FAO	b
BEN 78/004	Co-ordination of national rural water supply plan	1979–1984	United Nations **/	a
BEN 83/009 and BEN 85/004	Water resources master plan	1984–1987	United Nations	a
BURKINA FASO				
UPV 63/504	Mining and ground-water research	Concluded in 1968	United Nations	b
UPV 69/510	Hydrological and railway studies for mining development in north-west	1970–1971	United Nations	b

*/ a: preponderant or exclusive
b: substantial (30 to 50 per cent)
c: minor

**/ Since 1978: Department of Technical Co-operation for Development
previously: Department of Economic and Social Affairs (Office of Technical Co-operation)

Country and project No.	Project	Implementation period	Executing agency	Importance of ground water phase
BKF 73/013	Emergency and medium-term water programme	1974-1976	United Nations	a
BKF 74/022	Hydrogeological studies in the valleys of the Bougouriba and Poni	1975-1979	United Nations	a
BKF 80/001	Strengthening of water service	1981-1985	United Nations	a
BKF 83/003	Assistance for water drilling programme	1983-1986	United Nations	a
BKF 86/001	Support for "water" activities of the Five-Year Plan	1986-1990	United Nations	a
CAMEROON				
CMR 71/056	Pilot ground-water research and exploitation	1972-1975	United Nations	a
CMR 73/013	Ground-water soundings in the north	1975-1978	United Nations	a
CAPE VERDE				
CVI 75/001 and CVI 76/002	Ground-water research	from 1975	United Nations	a
CVI 79/001 and CVI 82/004	Strengthening of management and operation of ground-water services	to 1985	United Nations	a
CENTRAL AFRICAN REPUBLIC				
CAF 84/006	Ground-water exploration	1984-1986	United Nations	a
CAF 86/004	Technical support for village water-supply programmes	1986-1989	United Nations	a
CHAD				
CHD 71/510	Rural water supply	1972-1977	United Nations	a

Country and project No.	Project	Implementation period	Executing agency	Importance of ground water phase
CHD 72/004	Ouaddai regional and rural development - Phase I	1974-1975	United Nations	b
CHD 75/020	Ouaddai regional and rural development - Phase II	1975-1977	United Nations	b
CHD 76/009	Assistance to SERARHY	1977-1978	United Nations	a
CHD 83/016	Upgrading and repair of pastoral and village boreholes	1985 - 1986	United Nations	a
CHD 85/004	Ground-water exploitation in rural areas	1986-1990	United Nations	a
<u>COTE D'IVOIRE</u>				
IVC 70/518	Abidjan water-supply and sanitation	1970-1974	WHO	c
<u>EGYPT</u>				
UAR 49	Preinvestment studies in the north-west coastal region	1965-1968	FAO	c
EGY 71/561	Pilot project for ground-water use in "New Valley"	1972-1978	FAO	a
EGY 73/024	Master plan for water-resources development - Phase I	1977-1982	World Bank	b
EGY 81/031	- Phase II	1982-1986	World Bank	b
<u>GAMBIA</u>				
GAM 74/077 and GAM 82/008 and GAM 82/T01	Rural water supply	1976-1987	United Nations	c
GAM/CILSS/204	Water points	1978-1980	United Nations	a

Country and project No.	Project	Implementation period	Executing agency	Importance of ground water phase
GHANA				
GHA 71/251	Rural water supply	1971-1981	WHO	b
GUINEA BISSAU				
GBS 75/034	Ground-water development	1975-1978	United Nations	b
GBS 77/002 and GBS 82/007	Rural water supply	1978-1985	United Nations	a
MALI				
MLI 67/507	Strengthening of government ground-water research and exploitation services	1968-1974	United Nations	a
MLI 73/002	Emergency and medium-term water programme	1974-1976	United Nations	a
MLI 74/001 and MLI 76/004 and MLI 82/005	Rural water supply	1975-1985	United Nations	a
MLI 82/017	Preliminary studies for the transfer of population affected by construction of Manantali dam	1983-1984	United Nations	b
MLI 84/005	Ground-water planning, exploitation and management	1985-1989	United Nations	a
MLI 80/005	Creation of oases on the route du sel	1983-1989		
MLI 84/027	Ground-water exploitation in Azaouad-Sud	1985-1987	United Nations	a
MAURITANIA				
MAU 67/502	Strengthening of Ground-Water Service	1968-1977	United Nations	a
MAU 73/005	Emergency and medium-term water programme	1974-1976	United Nations	a

Country and project No.	Project	Implementation period	Executing agency	Importance of ground water phase
MAU 77/002	Water planning and use	1977-1981	United Nations	a
MAU 84/001	Well repair and maintenance	1984-1986	United Nations	a
MOROCCO				
MOR 70/022	Hydro-agricultural development in Souss valley	1971-1976	FAO	b
MOR 70/003	Strengthening of the Division of Sedimentary Basins	1983-1985	United Nations	a
MOR 70/002	Deep water exploration - Haouz pilot phase	1971-1985	Government	a
MOR 82/009	Supreme Water Council	1983-1987	UNDP	b
NIGER				
NER 74/507	Development of livestock production and water resources in eastern Niger	1964-1968	FAO	a
NER 70/508	Dallol-Maouri development studies - Phase I	1967-1970	FAO	b
NER 70/516	Pilot farming development in Dallol-Maouri	1970-1976	FAO	b
NER 72/006	Use of Davey drilling rig	1970-1975	FAO	a
NER 74/013	Assistance to Logistical Service of Office of Waters and Subsoil	1975-1977	United Nations	c
NER 73/007	Emergency and medium-term water programme	1974-1977	United Nations	a
NER 79/006 and NER 83/002	Ground-water exploitation in rural areas Phases I and II	1984-1986	United Nations	a

Country and project No.	Project	Implementation period	Executing agency	Importance of ground water phase
NIGERIA				
NIR 83/004	Water resources development in rural areas of Benoue State	1985-1987	United Nations	a
SENEGAL				
SEN 66/609	Establishment of water supply and sanitation master plan for Dakar and vicinity	1969-1972	WHO	b
SEN 73/003	Emergency and medium-term water programme	1973-1979	United Nations	a
SEN 72/004	Water resources study and technical assistance with sanitation	1973-1978	WHO	b
SEN 77/011	Water supply plan for Dakar and vicinity	1978-1980	WHO	b
SEN 81/003	Ground-water resources management and protection in Senegal	1981-1987	UNDP	a
SIERRA LEONE				
SIL 79/009	Rural water supply	1982-1986	United Nations	a
SUDAN				
SUD 61/504	Studies on land and water use in Kordofan province	concluded in 1966	FAO	b
SUD 71/525	Development of savannah zone	1971-1974	FAO	b
SUD 67/530	Hydro-agricultural studies in Jebel Marra region	1967-1974	FAO	c
SUD 68/001	Water drilling	1968-1975	United Nations	a
SUD 72/003	Strengthening of government water resources planning and exploitation services	1974-1975	United Nations	b
SUD 84/006	Organization of a water points maintenance service	1984-1987	United Nations	a

Country and project No.	Project	Implementation period	Executing agency	Importance of ground water phase
TOGO				
TOG 62/504	Mining and ground-water studies	1969-1971	United Nations	b
TOG 71/511	Ground-water research in coastal region	1970-1975	United Nations	a
TOG 75/008	Water development strategy	1978-1984	United Nations	b
TUNISIA				
TUN 69/528	Intensification of ground-water exploitation in northern and central Tunisia	1969-1974	United Nations	b
RAF 65/071	Study of water resources of Chad basin	1968-1970	UNESCO	a
RAF 67/100	Study of water resources in northern Sahara	1968-1971	UNESCO	a
RAF 65/061	Hydro-agricultural studies in Senegal basin	1966-1974	FAO	b
RAF 71/251	(Algeria, Mali, Niger)			
	Study of route of trans-Saharan road	1975-1977	United Nations	c
RAF 74/309	Co-ordination of emergency programmes for ground-water exploitation in Sahel	1975-1978	United Nations	a
RAF 74/067	Water drilling in Chad basin	1974-1981	FAO	a
RAF 74/313	Use of Davey drilling rig	1977-1978	FAO	a
RAF 80/011	Water resources in the countries of North Africa	1980-1986	UNDP	a
RAF 82/013	International project (Egypt-Sudan on the major aquifer of East Africa (Nubian sandstone)	1984-1986	United Nations	a

WATER RESOURCES: LIST OF UNITED NATIONS PUBLICATIONS

Report of the United Nations Interregional Seminar on Flood Damage Prevention Measures and Management, Tbilisi, Union of Soviet Socialist Republics, 1969. Publication symbol: ST/TAO/SER.C/144.

Integrated River Basin Development. Revised edition. 80 pp., 2 maps. Report of a panel of experts. English (French, Spanish out of print). Sales No. E.70.II.A.4.

Solar Distillation as a Means of Meeting Small-scale Water Demands. 86 pp., figures, tables. English (French out of stock). Sales No. E.70.II.B.1.

Triennial Report on Water Resources Development, 1968-1970. 202 pp., English only. Sales No. E.71.II.A.15.

Abstraction and Use of Water: A Comparison of Legal Regimes. 254 pp., French, Spanish (English out of print). Sales No. 72.II.A.10.

Proceedings of the Interregional Seminar on Water Resources Administration, New Delhi, 1973. Publication symbol: DP/UN/INT-70-371.

National Systems of Water Administration. 183 pp., figures, annexes. English only. Sales No. E.74.II.A.10.

Ground Water Storage and Artificial Recharge. Natural Resources/Water Series, No. 2. 270 pp., figures, tables. English, French, Spanish. Sales No. E.74.II.A.11.

Management of International Water Resources: Institutional and Legal Aspects. Natural Resources/Water Series, No. 1. 271 pp., figures, map, tables, annexes. Report of a panel of experts. English, French, Spanish. Sales No. E.75.II.A.2.

The Demand for Water: Procedures and Methodologies for Projecting Water Demands in the Context of Regional and National Planning. Natural Resources/Water Series, No. 3. 240 pp., tables, figures, annexes. English, Spanish (French out of print). Sales No. E.76.II.A.1.

Ground Water in the Western Hemisphere. Natural Resources/Water Series, No. 4. 337 pp., tables, figures, maps. English, French, Spanish. Sales No. E.76.II.A.5.

Guidelines for Flood Loss Prevention and Management in Developing Countries. Natural Resources/Water series, No.5. 183 pp., tables, figures, annexes. English only. Sales No. E.76.II.A.7.

Report of the United Nations Water Conference, Mar del Plata, 14-25 March 1977. 181 pp., annexes. English, French (Spanish out of print). Sales No. E.77.II.A.12.

A Review of the United Nations Ground-water Exploration and Development Programme in the Developing Countries, 1962-1977. Natural Resources/Water Series, No. 7. 84 pp. English, French, Spanish. Sales No. E.79.II.A.4.

Efficiency and Distributional Equity in the Use and Treatment of Water: Guidelines for Pricing and Regulations. Natural Resources/Water Series, No. 8. 175 pp., tables, figures, annexes. English (French, Spanish out of stock). Sales No. E.80.II.A.11.

Water Resources Planning: Experiences in a National and Regional Context. Report of a United Nations workshop convened in co-operation with the Government of Italy. English. Document symbol: TCD/SEM.80/1.

Rural Water Supply. Report of a United Nations interregional seminar convened in co-operation with the Government of Sweden, Uppsala, Sweden, 6-17 October 1980. English. Document symbol: TCD/SEM.81/1.

Ground Water in the Eastern Mediterranean and Western Asia. Natural Resources/ Water Series, No. 9. 230 pp., tables, figures, maps. Sales No. E.82.II.A.8.

Experiences in the Development and Management of International River and Lake Basins. Natural Resources/Water Series, No. 10. 424 pp., Proceedings of the United Nations Interregional Meeting of International River Organizations, Dakar, Senegal, 5-14 May 1981. Sales No. E.82.II.A.17.

Flood Damage Prevention and Control in China. Natural Resources/Water Series, No. 11. Report of a study and workshop in the People's Republic of China, 16-31 October 1980. Sales No. 82.II.A.13.

Ground Water in the Pacific Region. Natural Resources/Water Series No. 12, 289 pp., tables, figures, maps. Sales No. E.83.II.A.12.

Technical Co-operation among Developing Countries in Groundwater Resources Development: Report of a United Nations International Colloquium convened in co-operation with the Government of Yugoslavia in Zagreb, 23-28 May 1983, 123 pp., tables, annexes. English only. Document symbol: TCD/SEM.83/1.

Treaties Concerning the Utilization of International Water Courses for Other Purposes than Navigation. Natural Resources/Water Series No. 13. 98 pp., English, French. Sales No. E/F.84.II.A.7.

The Use of Non-Conventional Water Resources in Developing Countries. Natural Resources/Water Series No. 14. Sales No. E.84.II.A.14.

Ground Water in Continental Asia. Natural Resources/Water Series No. 15. Sales No. E.86.II.A.2.

Water Resources Legislation and Administration in Selected Caribbean Countries. Natural Resources/Water Series No. 16. Sales No. E.86.II.H.2.

Institutional Issues in the Management of International River Basins: Financial and Contractual Considerations. Natural Resources/Water Series No. 17 to be issued as a sales publication.

Application of Computer Technology for Water Resources Development and Management in Developing Countries. Document symbol: TCD/WATER/1. 95 pp.